Advances in discrete mathematics and applications

Ramanujan Mathematical Society Lectures Notes Series

Vol. 1: Number theory

Vol. 2: The Riemann zeta function and related themes

Vol. 3: Formal language aspects of natural computing

Vol. 4: Commutative algebra and combinatorics

Vol. 5: Convexity in discrete structures

Vol. 6: Number theory and discrete geometry

Vol. 7: Discrete mathematics

Vol. 8: Lectures on operator theory

Vol. 9: Essays on geometric group theory

Vol. 10: Techmüller theory and moduli problems

Vol. 11: Perspectives in geometry and topology

Vol. 12: The Iwasawa theory of totally real fields

Vol. 13: Advances in discrete mathematics and applications

Ramanujan Mathematical Society

Lecture Notes Series

Volume 13

Advances in discrete mathematics and applications

Proceedings of the International Conference on Discrete Mathematics, held at the University of Mysore, India, June 2008

Volume editors

B. D. Acharya
G. O. H. Katona
Juarez Nešetřil

International Press
www.intlpress.com

Ramanujan Mathematical Society
Lecture Notes Series, Volume 13
Advances in discrete mathematics and applications

Volume editors:
B.D. Acharya
G.O.H. Katona
Juarez Nešetřil

Published in 2011 by International Press, Somerville, Massachusetts, U.S.A. under license from the Ramanujan Mathematical Society.

This work is also published in the Republic of India, exclusively by the Ramanujan Mathematical Society.

ISBN: 978-1-57146-220-6

Printed in the United States of America.

15 14 13 12 11 1 2 3 4 5 6 7 8 9

Preface

Though human cognition appears to comprehend the continuum, human action by and large takes place in discrete epochs. Discrete mathematics is that part of mathematics which deals with systematic treatment and understanding of discrete structures and processes encountered in our day-to-day life, which are often inherently quite complex in nature but seemingly understandable. Illustration of this character of discrete mathematics is abound with excitements, occasionally even to a common man. Further, discrete mathematics cuts across wide range of disciplines of human understanding not only in areas of pure mathematics but also in a variety of application areas, so much so that no other single area of mathematics has perhaps this special character of 'transversality'.

The holding of the International Conference on Discrete Mathematics during June 6–10, 2008 (ICDM-2008) was hence aimed to bring together a sizeable number of active researchers engaged in foraying into the frontiers of the subject. To our delight, there were more than 250 participants, including more than a dozen experts from abroad (see the list of their names, with addresses, appended to this volume), each of whose presentations contributed substantially to raise the academic atmosphere filled with the spirit of innovation and interaction that kept the peers as well as the sessions lively throughout.

The last session on 10 June 2008 was observed as "FIFTH GRAPH THEORY DAY", with special activities such as expression of appreciation to the Invited Speakers and honoring Prof. E. Sampathkumar, the Founding President of the Academy of Discrete Mathematics (ADMA), India, on the occasion of his 72nd birth day. Also, it was decided and announced that a volume containing the best part of the technical proceedings of the ICDM-2008 would be brought out under the Ramanujan Mathematical Society's Lecture Notes Series in Mathematics (RMS-LNSM) with B.D. Acharya (SRC-IIIDMS, University of Mysore, Mysore, India), G.O.H. Katona (Hungarian Academy of Sciences, Budapest, Hungary) and J. Nešetřil (Charles University, Praha, Czech Republic) as Volume Editors.

Accordingly, today, we are happy to present this volume as a result of cooperation received from all the participants, Invited Speakers and, most of all, the patient and the perseverant referees (see the list appended to the volume) – we place on record our deep appreciation and thanks to each one of them.

Volume Editors
May, 2008

Members of the Advisory Board

Contents

Proc. ICDM 2008, RMS-Lecture Notes Series No. 13, 2010, pp. 1–18.

Domination in Hypergraphs II. New Directions

B. D. Acharya

Srinivasa Ramanujan Center for Intensification of Interaction in Interdisciplinary Discrete Mathematical Sciences (SRC-IIIDMS), University of Mysore, Mysore 560 005, India
Address for Correspondence: No. 22, 10th Cross, 5th Main, New Thippasandra Post, Malleshpalya, Bangalore 560 075, India
e-mail: devadas.acharya@gmail.com

Abstract. In this up-to-date creative review, we report recent advancements made on the notion of domination in hypergraphs that are not necessarily finite. In the sequel, some insight into the problem of determining finite hypergraphs H whose stability numbers $\alpha(H)$ and strong stability (or, 'independence') numbers $\beta(H)$ are equal is provided. Further, we establish few new results on finite hypergraphs concerning their independent domination number $\gamma_i(H)$, the disjoint domination number $\gamma\gamma(H)$, the disjoint independent domination number $\gamma^i(H)$, the inverse domination number $\gamma^{-1}(H)$ and the domchromatic number $d_\chi(H)$. Several new directions for further research are hence indicated by means of raising open problems and conjectures.

Keywords. Bipartite hypergraph, hypertree, constellation, domination, stability, independence, domchromatic number, domatic number, indomatic number.

2000 Subject Classification: 05C69 Hypergraph Theory

1. Preamble

For standard terminology and notation in hypergraph theory, we refer the reader to C. Berge [11]. A hypergraph $H = (X, \mathcal{E})$ as treated in [11] satisfies

$$\bigcup_{E\in\mathcal{E}} E = X$$

whereas we may in general have

$$Y = X - \left(\bigcup_{E\in\mathcal{E}} E\right) \neq \emptyset,$$

when the elements of Y are called *isolates* of H. However, unless mentioned otherwise, we shall consider simple hypergraphs without isolates as in [11] and possibly infinite. New terms and notations will be defined as and when found necessary for further discussion, as this approach is believed to aid an easy reading of the paper, not having to look for them back and forth in the text.

A set D of vertices in a hypergraph $H = (X, \mathcal{E})$ is a *dominating set* if for every $x \in X - D$ there exists $y \in D$ such that x and y are *adjacent*, that is, there exists

$E \in \mathcal{E}$ such that $x, y \in E$ [3] (*also see* [4], [5], [7]); it is said to be *minimal* if it does not contain any proper subset which is dominating and then it is *minimum* whenever its cardinality $\gamma(H)$ is the least possible for a dominating set of H. The set D is said to be *stable* if it does not contain any edge E of H with $|E| > 1$ and *strongly stable* if no two distinct vertices in D are adjacent in H (see [11]); throughout this paper, we will use the term '*independent*' instead of 'strongly stable' to be in consonance with the terminology widely used in graph theory (e.g., see [16] and [17]). Obviously, every nonempty subset of X is a dominating set if and only if H contains X as an edge. Therefore, for an interesting study of the phenomenon of domination in hypergraphs it is enough to consider hypergraphs H in which every edge is a nonempty proper subset of their vertex sets $X(H)$, which we will indeed assume here onwards. Hence, there can be interesting fundamental questions such as: What should be the structure of a hypergraph H in which every nonempty proper subset of $X(H)$ is a dominating set?

The following two results have been reported in [9], and will be found useful to derive new results; the first of these extends two similar results known for finite graphs (see [11], Proposition 2 and Theorem 1 on p. 309).

Theorem 1.1. *Let $H = (X, \mathcal{E})$ be any hypergraph. Then every maximal independent set of H is a minimal dominating set in H and, conversely, every independent dominating set of H is a maximal independent set in H.*

Lemma 1.2. *Let $H = (X, \mathcal{E})$ be any hypergraph without isolates and D be any dominating set of H. If D is minimal then, $X - D$ is a dominating set in H.*

Given any hypergraph $H = (X, \mathcal{E})$, as in [7] denote by $\mathcal{S}_\alpha(H)$ the set of all stable sets in H and by $\mathcal{S}_\beta(H)$ the set of all strongly stable (or, independent) sets in H. If H is finite, the *stability number* (respectively, *strong stability number* or *independence number*) of H, denoted $\alpha(H)$ (respectively, $\beta(H)$), is the maximum cardinality of a stable (strongly stable) set in H. It is noted in [7] that

$$\alpha(H) \geq \beta(H) \tag{1}$$

for any hypergraph H as also that the equality in (1) holds when H is a *graph* (see [16] for standard terminology and notation), since $\mathcal{S}_\alpha(H) = \mathcal{S}_\beta(H)$.

The main aim of this paper is to provide a creative review of the concepts, problems and results arising out of the notion of domination in hypergraphs.

2. Hypergraphs with $\alpha(H) = \beta(H)$

In general, the following problem (Problem 1.1 in [7]) was raised.

Problem 1. Characterize finite hypergraphs H for which equality holds in (1).

Existence of hypergraphs, that are not graphs, satisfying equality in (1) is guaranteed by the following.

Proposition 2.1. *For every integer $n \geq 2$, there exists a finite connected hypergraph H with $\alpha(H) = n = \beta(H)$.*

Proof. Consider the complete bipartite graph $K_{n+1,n}$, with its bipartition $\{V_1, V_2\}$ such that $|V_1| = n + 1$ and $|V_2| = n$. Define the hypergraph $H = (X, \mathcal{E})$, where $X = V := V_1 \cup V_2$ and $\mathcal{E} = E \cup \{V_1\}$ where E is the set of edges of $K_{n+1,n}$. Then, it is easy to check that any one-vertex-deleted subset of V_1 is a maximal stable set of cardinality n in H and V_2 is its unique maximum independent set. Now, it is not difficult to verify that $\alpha(H) = n = \beta(H)$. □

Notice that the hypergraphs constructed in the proof of Proposition 2.1 are *linear*, in the sense that in these hypergraphs any two distinct edges have at most one vertex in common (see [11]); it would be perhaps helpful to solve Problem 1 for this restricted class of hypergraphs. For, even in this class, there are hypergraphs which do not satisfy equality in (1) as, for instance, the hypergraph $H = (X, \mathcal{E})$, where

$$X = \{a, b, c, d, e, f\},\ \mathcal{E} = \{\{a, b, c\}, \{c, d, e\}, \{a, e, f\}, \{b, d, f\}\},$$

has $D = \{a, b, f\}$ as a maximum stable set but H has no maximal independent set of cardinality equal to 3.

An edge in a hypergraph H is a *hyperedge* if it contains at least three vertices; otherwise it is *ordinary*; H is a *pure hypergraph* if each edge of H is a hyperedge. The following result shows nonexistence of pure hypergraphs H that satisfy equality in (1).

Theorem 2.2 ([14]). *For every finite pure hypergraph* $H, \alpha(H) > \beta(H)$.

In a bid to characterize finite hypergraphs H that satisfy (1), K.A. Germina [14] pointed out that a hypergraph H satisfies (1) if and only if $\alpha(H) = \beta(H_{(2)})$ where $H_{(2)}$ denotes the standard 2-*section* graph of H (cf.: [11], p. 390, 396). However, this is not a 'good' structural criterion for a finite hypergraph in general that satisfies (1).

In fact, there exist finite pure hypergraphs H for which the difference $\alpha(H) - \beta(H)$ could be arbitrarily large. For instance, K.A. Germina [15] pointed out the following example. Let $H = (X, \mathcal{E})$ where $X = \{x_1, x_2, \ldots, x_k, x_{k+1}, x_{k+2}, \ldots, x_n\}$ and let $\mathcal{E}$ consist of the edges defined by the sets

$$E_j = \{x_j\} \cup \{x_{k+1}, x_{k+2}, \ldots, x_n\},\ 1 \leq j \leq k.$$

Then, it is easy to verify that

$$\alpha(H) = |\{x_1, x_2, \ldots, x_k, x_{k+1}, x_{k+2}, \ldots, x_{n-1}\}| = k + n - 1 - k = n - 1$$

whereas $\beta(H) = |\{x_1, x_2, \ldots, x_k\}| = k$. Therefore, keeping k fixed, we can enhance $\alpha(H) - \beta(H) = n - 1 - k$ arbitrarily, where $n - k$ is the number of vertices in the intersection $\cap_{i=1}^{k} E_i$.

We can even have such a hypergraph $H = (X, \mathcal{E})$ with $\cap_{i=1}^{k} E_i = \emptyset$. Towards this end, for any integer $n \geq 3$, let

$$X = \{x_1, x_2, \ldots, x_n, x_{n+1}, x_{n+2}, \ldots, x_{2n}\}$$

and $\mathcal{E} = \{E_1, E_2, \ldots, E_n\}$ where

$$E_i = \{x_i, x_{n+1}, x_{n+2}, \ldots, x_{2n}\},\ 1 \leq i \leq n.$$

Then,

$$S_i = \{x_i\} \cup \{x_{n+1}, x_{n+2}, \ldots, x_{n+i-1}, x_{n+i+1}, \ldots, x_{2n}\}$$

is a stable set of H with $|S_i| = n = |\{x_1, x_2, \ldots, x_n\}| = \beta(H)$. Now, consider the hypergraph $H' = (X, \mathcal{E}')$ where $\mathcal{E}' = \{S_1, S_2, \ldots, S_n\}$. Then, one can verify that

- $x_{n+i} \notin S_i$, for each i, $1 \leq i \leq n$,
- $x_{n+i} \in \cap_{j \neq i} S_j$, for each i, $1 \leq i \leq n$ and
- $\cap_{j=1}^{n} S_j = \emptyset$.

Clearly, $B = \{x_1, x_2, \ldots, x_n\}$ is a *β-set* (i.e., an independent set with $\beta(H)$ vertices) of the n-uniform hypergraph H' and

$$A_i = B \cup S_i = \{x_1, x_2, \ldots, x_n, x_{n+1}, x_{n+2}, \ldots, x_{n+i-1}, x_{n+i+1}, \ldots, x_{2n}\}$$

is an α-set of H'. Thus, $\alpha(H') = 2n - 1$ whence we have $\alpha(H') - \beta(H') = n - 1$.

Let $\mathcal{E} = \{E_1, E_2, \ldots, E_m\}$. A maximal collection $\{E_j\}_{j \in J}$, $J \subseteq \{1, 2, \ldots, m\}$ such that $\cap_{j \in J} E_j \neq \emptyset$ is called a *constellation* of H (see [1]); further, the set $\cap_{j \in J} E_j$ is called the *core* of the constellation.

Definition 2.3. *The edge-degree $e_H(x)$ of a vertex x in a hypergraph H is the number of edges containing x. A constellation $\{E_j\}_{j \in J}$ in a hypergraph H is edge-degree compatible (or, in short, EDC) if $|\{E_j\}_{j \in J}| = |J|$ and H itself is an EDC hypergraph if each of its constellations is EDC.*

We shall now introduce an important class of finite EDC hypergraphs each of which is constructed from an arbitrarily given finite connected graph of order $p \geq 3$ and the construction yields infinitely many varieties of finite pure hypergraphs H for which $\alpha(H) \geq 2\beta(H)$, raising more interesting questions.

A vertex in a hypergraph $H = (X, \mathcal{E})$ is called a *pendant vertex* if it belongs to exactly one edge of H. An edge of H is called a *leaf* if it contains a pendant vertex. In particular, if $G = (V, E)$ is any connected graph, a vertex x in G is *internal* if its degree $d(x) > 1$, otherwise it is called a *pendant vertex* of G. An edge uv in G is *internal* if both u and v are internal, otherwise, it is a *pendant edge* of G. Now, given a connected graph G of a finite order $p \geq 3$, we construct a finite pure hypergraph $H(G)$ as follows: Replace each internal vertex w in G by a set X_w with $|X_w| = d(w)$, replace each internal edge uv of G by the set $E_{uv} = X_u \cup X_v \cup \{uv\}$ where uv is a pendant vertex in E_{uv}, replace every pendant edge ab in G with $d(a) = 1$ by the set $E_a = \{a\} \cup X_b$; let $\mathcal{E}(H(G)) = \{E_a : d(a) = 1\} \cup \{E_{uv} : uv \in E\ d(u) \neq 1, d(v) \neq 1\}$ and $X(H(G))$ be the union of the members of $\mathcal{E}(H(G))$. The statements in the following proposition may be easily verified.

Proposition 2.4. *Given any finite connected (p, q)-graph $G = (V, E)$ the associated finite pure hypergraph $H(G)$ has the following properties:*

(i) *The cores of constellations of $H(G)$ are all pairwise disjoint.*
(ii) *Every edge e of $H(G)$ has exactly one pendant vertex e' and every nonpendant vertex of $H(G)$ is in some constellation of $H(G)$.*

(iii) *There is a one-to-one correspondence* φ *between the set of internal vertices of* G *and the set of constellations in* $H(G)$*; that is,* $d_G(u) > 1 \Rightarrow \varphi : u \mapsto \cap_{X_u \subset e \in \mathcal{E}(H(G))} e$.

(iv) *For each edge* e *in* $H(G)$ *there exists a unique vertex* $\xi(e') = u'$ *such that* $u' \in \varphi(u)$ *so that* $\xi : e' \mapsto u'$ *yields a matching* $\mathcal{M}_\xi = \{\{e', u'\} : e \in \mathcal{E}(H(G))\}$ *of adjacent vertices in* $H(G)$*; that is,*

$$i \neq j \Rightarrow \{e'_i, u'_i\} \cap \{e'_j, u'_j\} = \emptyset.$$

(v) $\sum_{d_G(u)>1} |\varphi(u) - \{u'\}| = q'$*, where* q' *is the number of internal edges* uv *in* G.

(vi) *Let* $H_\xi(G)$ *denote the hypergraph defined by the ordered pair*

$$H_\xi(G) = (X(H(G)) - Y_\xi, \mathcal{E}_\xi),$$

where

$$\begin{aligned} Y_\xi &= \bigcup_{d_G(u)>1} (X_u - \{u'\}) \\ &= \bigcup_{e \in \mathcal{E}(H(G))} \left(\bigcap_{X_u \subset e} e \right) - \{u' : \xi^{-1}(u') = e', e \in \mathcal{E}(H(G))\} \end{aligned}$$

and

$$\mathcal{E}_\xi = \{\{e'\} \cup \{u' : u' \in e\} : e \in \mathcal{E}(H(G))\}.$$

Then, $\alpha(H_\xi(G)) \geq 2\beta(H_\xi(G))$.

In general, the property of a set of vertices in a hypergraph being a stable set is a *hereditary property*, in the sense that every subset of a stable set is again a stable set. Further, every maximal stable set in a hypergraph is a transversal of the set of edges of the hypergraph. Therefore, given any maximal stable set A in a finite hypergraph H, it follows that by excluding all but one vertex from the intersection of each edge of the hypergraph with A, one by one, one can reduce A to an independent (i.e., a strongly stable) set of H. Thus, one sees the validity of the following important observation.

Proposition 2.5. *For any finite hypergraph* H *and any integer* k, $\beta(H) \leq k \leq \alpha(H)$, *there exists a stable set of cardinality* k *in* H.

3. New domination parameters

We begin with the following new result on domination in infinite hypergraphs.

Theorem 3.1. *Let* $H = (X, \mathcal{E})$ *be a hypergraph without isolates and in which every infinite descending chain of dominating sets has a nonempty intersection. Then, every dominating set in* H *contains a minimal one.*

Proof. Suppose that the theorem is false. Then, there exists a hypergraph H having a dominating set D_1 that contains no minimal one. Since D_1 does not contain any minimal dominating set, it in particular is not a minimal dominating set of H, whence there exists a dominating set $D_2 \subset D_1$. Next, by the same argument, since D_2 is not a minimal dominating set of H there exists a dominating set $D_3 \subset D_2$. Continuing in this manner, we get an infinite descending chain $D_1 \supset D_2 \supset D_3 \supset \cdots$ of dominating sets of H whose intersection is empty. This is contrary to the hypothesis. Thus, the theorem is established by contraposition. □

A hypergraph H is *locally finite* if the *vertex-degree* $d_H(u) = |N_H(u)|$ of every vertex u in H is finite, where $N_H(u) = \{v \in X : v \text{ is adjacent to } u\}$ is the *vertex-neighborhood* of u; we will omit henceforth the suffix 'H' in these notations when the hypergraph H is clear from the context.

Corollary 3.2 (Compare [19], Theorem 13.1.1, p. 206). *In a locally finite hypergraph $H = (X, \mathcal{E})$ without isolates, every dominating set contains a minimal one.*

The following result is well known when H is a finite graph [17].

Corollary 3.3. *In any finite hypergraph $H = (X, \mathcal{E})$ without isolates, every dominating set contains a minimal one.*

We note in passing here that, in general, it is an open problem to characterize isolate-free infinite hypergraphs (infinite graphs in particular) in which every dominating set contains a minimal one.

Since, according to Theorem 1.1, any finite hypergraph is trivially locally finite as also since any maximal independent set is a minimal dominating set, Lemma 1.2 and Corollary 3.2 together imply the following result.

Corollary 3.4. *In any locally finite hypergraph without isolates there are at least two disjoint minimal dominating sets.*

Note that if one allows isolates in a hypergraph H then Corollary 3.4 would not hold as every dominating set of H would then contain all the isolates.

Given a hypergraph $H = (X, \mathcal{E})$, let $\mathcal{D}(H)$ denote the set of dominating sets of H and $\mathcal{D}^m(H)$ denote the set of all minimal dominating sets of H. One interesting extremal problem that may be pointed out at this stage is the following.

Problem 2. Characterize isolate-free hypergraphs that have at least two disjoint minimal dominating sets.

If, in Problem 2, we want to relax the requirement of disjointness of the minimal dominating sets and require only 'distinctness' amongst them then we have the following interesting result.

Proposition 3.5. *For any isolate-free hypergraph $H = (X, \mathcal{E})$, $|\mathcal{D}^m(H)| \geq \chi(H)$, where $\chi(H)$ denotes the chromatic number of H.*

Proof. Let $\{X_1, X_2, \ldots, X_k\}$ be any chromatic partition of $X(H)$, i.e., $k = \chi(H)$ and X_i is an independent set of H for each i, $1 \leq i \leq k$. Let D_i be the maximal independent set containing X_i for each i, $1 \leq i \leq k$. Since any chromatic partition is 'complete', in the sense that there is an adjacent pair of vertices belonging to the union

$$X_i \cup X_j \text{ for all distinct } i, j \in \{1, 2, \ldots, k\},$$

$D_1, D_2, \ldots, D_k$ are all distinct. Hence, the proof follows by applying Theorem 1.1 to each D_i. □

Proposition 3.5 suggests the problem of determining hypergraphs H having exactly $\chi(H)$ distinct (respectively, disjoint) minimal dominating sets.

Further, for any finite hypergraph H, we denote by $\mathcal{D}^0(H)$ the set of all minimum dominating sets of H (or, in short, 'γ-sets') of H; the maximum cardinality of a minimal dominating set of H is called the *upper domination number*, denoted $\Gamma(H)$. Lemma 1.2, Corollary 3.2 and Corollary 3.4 motivate the following notions, generalizing their analogues in the theory of domination in graphs [18].

Definition 3.6. *Given any finite isolate-free hypergraph $H = (X, \mathcal{E})$ and $D \in \mathcal{D}^0(H)$, an inverse dominating set is any minimal dominating set contained in the subhypergraph $\langle X - D \rangle$ induced by the set $X - D$ in H; the set of inverse dominating sets with respect to D is denoted $\mathcal{D}_H^{-1}(D)$. Then, $\gamma^{-1}(H) = \min\{|S| : S \in \mathcal{D}_H^{-1}(D),\ D \in \mathcal{D}^0(H)\}$ is the inverse domination number of H. Similarly, $\Gamma^{-1}(H) = \max\{|S| : S \in \mathcal{D}_H^{-1}(D),\ D \in \mathcal{D}^0(H)\}$ is the upper inverse domination number of H. Further,*

$$\gamma\gamma(H) = \min\{|S_1| + |S_2| : S_1, S_2 \in \mathcal{D}^m(H),\ \text{and } S_1 \cap S_2 = \emptyset\}$$

is the disjoint domination number of H and

$$\gamma^i(H) = \min\{|S_1| + |S_2| : S_1, S_2 \in \mathcal{D}^m(H), S_2 \in \mathcal{I}(H) \text{ and } S_1 \cap S_2 = \emptyset\},$$

where $\mathcal{I}(H)$ denotes the set of maximal independent sets in H, is the disjoint independent domination number of H; a γ^i-pair is an ordered pair (S_1, S_2) such that $S_1 \in \mathcal{D}^m(H)$, $S_2 \in \mathcal{I}(H)$ and $|S_1 \cup S_2| = \gamma^i(H)$. The least cardinality of an independent dominating set in H is the independent domination number of H, denoted $\gamma_i(H)$.

The following set of inequalities is immediate from the foregoing results and Definition 3.6:

$$2\gamma(H) \leq \gamma\gamma(H) \leq \gamma(H) + \gamma^{-1}(H) \leq n, \tag{2}$$

for any hypergraph H of a finite order n. It would be of theoretical interest to characterize finite hypergraphs that satisfy equality in any one of the inequalities in (2). [S. M. Hedetniemi *et al.* [18] report that G. S. Domke *et al.* [13] have characterized finite graphs H for which $\gamma(H) + \gamma^{-1}(H) = n$.]

The following is a generalization of a result due to S. M. Hedetniemi *et al.* [18] and may be established mimicking the proof of their result for graphs.

Proposition 3.7. *For any finite hypergraph H without isolates, if $\beta(H) = \Gamma(H)$ then $\gamma^{-1}(H) \leq \beta(H)$.*

Theorem 3.8. *For any finite hypergraph $H = (X, \mathcal{E})$ without isolates,*

$$\gamma\gamma(H) \leq \gamma^{i}(H) \leq \gamma_i(H) + \beta(H). \tag{3}$$

Proof. Let I be a maximal independent set with $|I| = \gamma_i(H)$ and let S be a maximal independent set in $\langle X - I \rangle$. Lemma 1.2 and Theorem 1.1 applied to the induced sub-hypergraph $\langle X - I \rangle$ then implies that S is a minimal dominating set in $\langle X - I \rangle$. Note that S may not be a dominating set of H and, hence, the set D of vertices of I such that no vertex of S is adjacent to any vertex of D may be nonempty; also, $D \cup S$ is a maximal independent set of H. Since I is an independent dominating set of H (and hence a minimal dominating set of H, by Theorem 1.1), every vertex in D has at least one adjacent vertex in $(X - I) - S$. For each vertex in D, choose one such vertex in $(X - I) - S$ and let R be the set of so chosen vertices. Then, the set $R \cup S \in \mathcal{D}(H)$, with $|R \cup S| = |R| + |S| \leq |D| + |S| = |D \cup S| \leq \beta(H)$. Thus,

$$\gamma\gamma(H) \leq \gamma^{i}(H) \leq |I| + |R \cup S| \leq \gamma_i(H) + \beta(H),$$

and the proof is complete. □

The following conjecture restricted to the class of finite graphs without isolates is called the *Kulli-Sigarkanti Conjecture* (cf.: S. M. Hedetniemi *et al.* [18]).

Conjecture 3.9. For any finite hypergraph H without isolates, $\gamma^{-1}(H) \leq \beta(H)$.

The next result gives some classes of extremal hypergraphs that satisfy Conjecture 3.9.

Proposition 3.10. *For any finite hypergraph H without isolates and with $\gamma\gamma(H) = \gamma(H) + \gamma^{-1}(H)$, if $\gamma(H) = \gamma_i(H)$ then $\gamma^{-1}(H) \leq \beta(H)$.*

Proof. Invoking (3) and the hypotheses, we get $\gamma_i(H) + \gamma^{-1}(H) = \gamma(H) + \gamma^{-1}(H) = \gamma\gamma(H) \leq \gamma_i(H) + \beta(H) \Rightarrow \gamma^{-1}(H) \leq \beta(H)$. □

In fact, there exist vast variety of finite hypergraphs for which $\gamma(H) = \gamma_i(H)$ as demonstrated by an earlier result stated below. It requires the notion of (*strong*) t-*claw* in a hypergraph H as defined in [7]: It is a collection of t edges $\{E_1, E_2, \ldots, E_t\}$ such that $\cap_{i=1}^{t} E_i \neq \emptyset$ and there exist vertices $x_i \in E_i - \cap_{j=1}^{t} E_j$ such that $\{x_1, x_2, \ldots, x_t\}$ is a (strongly) stable set in H. We have then the following generalization of a well known 'Allan-Laskar theorem' for finite graphs (see [17]).

Theorem 3.11 ([7]). *If a finite hypergraph does not contain any strong 3-claw then $\gamma(H) = \gamma_i(H)$.*

If a finite hypergraph H satisfies $\gamma(H) = \gamma_i(H)$ then it is not necessary that it satisfies $\gamma\gamma(H) = \gamma(H) + \gamma^{-1}(H)$. For instance, take the hypergraph $H = (X, \mathcal{E})$ with

$$X = \{a, b, b_1, b_2, \ldots, b_r, c, d, d_1, d_2, \ldots, d_s, e\}$$

and

$$\mathcal{E} = \{\{a, b\}, \{b, b_1, b_2, \ldots, b_r\}, \{b, c\}, \{c, d\}, \{d, d_1, d_2, \ldots, d_s\}, \{d, e\}\}.$$

Then, $D = \{b, d\}$ is a γ-set of H, which is independent too. Thus, $\gamma(H) = \gamma_i(H) = 2$. However, $\gamma\gamma(H) \le |\{a, c, e\}| + |\{b, d\}| = 5$ whereas

$$\gamma(H) + \gamma^{-1}(H) = |\{b, d\}| + |\{a, b_i, c, d_j, e\}| = 2 + 5 = 7.$$

Nevertheless, we have $\gamma^{-1}(H) = 5 = \beta(H)$. Thus, we cannot relax any of the two conditions in the hypothesis of Proposition 3.10. Hence, determination of finite hypergraphs H for which $\gamma^{-1}(H) = \beta(H)$ might shed more light on resolving Conjecture 3.9.

Next natural question is whether every isolate-free hypergraph has two disjoint independent dominating sets. The answer to this question is in the negative according to the following known result for graphs, where $G = H^+$ is obtained from the given graph H by augmenting a new vertex v' for each vertex v in H and making the new adjacency vv'.

Theorem 3.12 ([18]). *If $G = H^+$ for some finite graph H, then G has two disjoint independent dominating sets if and only if H is bipartite.*

What is the hypergraph extension of Theorem 3.12? In essence, by having some reasonable extension of the notion of a bipartite graph to hypergraphs, can we have an extension of Theorem 3.12? We shall explore this possibility by making the following definition, motivated from the well known characterization of a bipartite graph.

Definition 3.13. *A hypergraph H is bipartite if every cycle in H is of even length.*

In [12], Chapter 5 (p. 155), C. Berge refers to these hypergraphs as 'hypergraphs without odd cycles', bringing to light a number of interesting properties of these hypergraphs. Here, we shall derive yet some new fundamental facts about the structure of bipartite hypergraphs.

Lemma 3.14. *Let $H = (X, \mathcal{E})$ be a hypergraph without cycles of any odd lengths. Then, for any three distinct edges E_1, E_2 and E_3 in H, $|E_1 \cap E_2 \cap E_3| \le 2$.*

Proof. If for some three distinct edges E_1, E_2 and E_3 in H, we have $|E_1 \cap E_2 \cap E_3| \ge 3$, then there exist distinct $x, y, z \in E_1 \cap E_2 \cap E_3$, whence we obtain the 3-cycle $(x, E_1, y, E_2, z, E_3, x)$ in H, a contradiction to the hypothesis. □

Lemma 3.15. *Let $H = (X, \mathcal{E})$ be a hypergraph without odd cycles. Then, for any cycle Z in H no edge of H contains three vertices of Z.*

Proof. Let $Z = (x_1, E_1, x_2, E_2, \ldots, x_{2k}, E_{2k}, x_1)$ and suppose there exists an edge E in H that contains three distinct vertices x_p, x_q, x_r of Z, where, without loss of generality, we may assume $1 \le p < q < r \le 2k$. Then, it is not difficult to verify that one of the three cycles

$$Z_1 = (x_1, E_1, x_2, E_2, \ldots, x_p, E, x_r, E_r, \ldots, x_{2k}, E_{2k}, x_1),$$

$$Z_2 = (x_p, E_p, x_{p+1}, E_{p+1}, \ldots, x_q, E, x_p)$$

and

$$Z_3 = (x_q, E_q, x_{q+1}, E_{q+1}, \ldots, x_r, E, x_q)$$

is of odd length, a contradiction to the hypothesis. □

Proposition 3.16. *Let $H = (X, \mathcal{E})$ be an isolate-free bipartite hypergraph having a cycle. Then, there exist two disjoint maximal independent sets in H.*

Proof. Suppose the proposition is false. Then, there exists an isolate-free bipartite hypergraph H having a cycle but having no pair of disjoint maximal independent sets. There exists, however, a maximal independent set X_1 in H. By our assumption then $X - X_1$ does not contain any maximal independent set of H. Since X_1 is a maximal independent set of H, by virtue of Theorem 1.1 it is a minimal dominating set and hence $X - X_1$ is a dominating set according to Lemma 1.2. Therefore, in accordance with Theorem 1.1, $X - X_1$ cannot be independent and hence cannot itself be a maximal independent set of H too. If there is no maximal independent set in H other than X_1, then it follows that every two vertices in $X - X_1$ must be adjacent. Therefore, if y_1, y_2, y_3 are any three distinct vertices in $X - X_1$ and the edges F_1, F_2, F_3 in H are such that $y_1, y_2 \in F_1$, $y_2, y_3 \in F_2$ and $y_1, y_3 \in F_3$ then the edges F_i cannot all be distinct, for otherwise $(y_1, F_1, y_2, F_2, y_3, F_3, y_1)$ would be an odd cycle in H, a contradiction to the hypothesis. Hence, suppose, $F_1 = F_2$. Then, $y_1, y_2, y_3 \in F_2 \cap F_3$, a contradiction to Lemma 3.14. The same contradiction is deduced if either $F_1 = F_3$ or $F_2 = F_3$. Hence, we see that $F_1 = F_2 = F_3 =: F$, say. Thus, we conclude that every three vertices in $X - X_1$ are in an edge of H. Now, by hypothesis, H has a cycle, say Z. Then, Z is a cycle of an even length, say $2k$ for some positive integer k. Let $Z = (x_1, E_1, x_2, E_2, \ldots, x_{2k}, E_{2k}, x_1)$. Since no edge of H has more than one vertex in an independent set of H, it follows that $X - X_1$ has all but one vertex of each edge E_i, $1 \leq i \leq 2k$. Thus, for any three consecutive vertices of Z, say x_i, x_{i+1} and x_{i+2}, indices reduced modulo $2k$, we have two cases to consider, *viz.*, (i) $x_i, x_{i+1}, x_{i+2} \in X - X_1$, and (ii) $x_i \in X_1$ and $x_{i+1}, x_{i+2} \in X - X_1$ (or, equivalently, $x_i, x_{i+1} \in X - X_1$ and $x_{i+2} \in X_1$).

Case (i). In this case, as shown above, there exists an edge F such that $x_i, x_{i+1}, x_{i+2} \in F$ and hence, by Lemma 3.15, this leads to the existence of an odd cycle in H, contradicting the hypothesis.

Case (ii). Without loss of generality, let $x_i \in X_1$ and $x_{i+1}, x_{i+2} \in X - X_1$. Then, $x_{i+3} \notin X - X_1$ for otherwise, as in Case (i), it would lead to a contradiction. So, $x_{i+3} \in X_1$. If there are no more vertices in H then $Z = (x_i, E_i, x_{i+1}, E_{i+1}, x_{i+2}, E_{i+2}, x_{i+3}, E_{i+3}, x_i)$, which implies $i = 1$, $k = 2$ and every edge of Z is an ordinary edge (whence $Z = C_4$), whereby x_i, x_{i+3} being in E_{i+3} are adjacent, contrary to the independence of X_1. Therefore, H has vertices other than x_i, x_{i+1}, x_{i+2} and x_{i+3}. In particular, therefore, Z must have at least three vertices in $X - X_1$. But, then, H has an edge containing three vertices of Z, leading to the existence of an odd cycle in H as shown in the proof of Lemma 3.15, contrary to the hypothesis.

The above argument implies that H has at least one more maximal independent set, say X_2. However, by our initial assumption, X_1 must intersect every other maximal independent set in H; in particular, therefore, $X_1 \cap X_2 \neq \emptyset$. Then, since X_1 and X_2 are minimal dominating sets of H (by Theorem 1.1), it follows by Lemma 1.2 that every vertex of $X_1 - X_2$ is adjacent to every vertex of $X_2 - X_1$. Further, since both X_1 and X_2 are independent no vertex of $X_1 \cap X_2$ can belong to any of the edges that contains a vertex of the symmetric difference $X_1 \oplus X_2 = (X_1 - X_2) \cup (X_2 - X_1)$. Also, by our very initial assumption, it follows that both $X_1 - X_2$ and $X_2 - X_1$ must be proper subsets of $X - X_2$ and $X - X_1$ respectively. Therefore, every hyperedge E intersecting both $X_1 - X_2$ and $X_2 - X_1$ must satisfy

$$|E \cap X_1| = |E \cap X_2| = 1 \quad \text{and} \quad E - \{a, b\} \subset X - (X_1 \cup X_2)$$

where $E \cap X_1 = \{a\}$ and $E \cap X_2 = \{b\}$. This implies, if Z has any two of its vertices x_i, x_j in $X_1 \oplus X_2$ then any hyperedge in H containing both x_i and x_j cannot contain another vertex of Z. Now, let $c \in X_1 \cap X_2$. Since X_1 and X_2 are both minimal dominating sets of H there must exist vertices $z_1, z_2 \in X - (X_1 \cup X_2)$ such that c is adjacent to both z_1 and z_2. Let $F_1, F_2 \in \mathcal{E}$ be such that $c, z_1 \in F_1$ and $c, z_2 \in F_2$. Also, since $X - X_1$ and $X - X_2$ are dominating sets of H there exist edges F_3 and F_4 such that $x_i, z_1 \in F_3$ and $x_j, z_2 \in F_4$. Let F_5 be an edge that contains both x_i and x_j (such an edge does exist as argued above). Clearly, F_3, F_4, F_5 are all distinct due to the fact that X_1 and X_2 are maximal independent sets in H. Now, if $z_1 = z_2$ then $(z_1, F_3, x_i, F_5, x_j, F_4, z_1)$ is a 3-cycle in H. On the other hand, if $z_1 \neq z_2$ then F_1 and F_2 are also distinct, and different from F_3, F_4 and F_5, whence $(c, F_1, z_1, F_3, x_i, F_5, x_j, F_4, z_2, F_2, c)$ is a 5-cycle in H. Thus, in either possible case, we have a contradiction to the hypothesis that every cycle in H is even in length and hence, by contraposition, it follows that $X_1 \cap X_2 = \emptyset$ and the proposition is established. □

The following result is a stronger version of Corollary 3.3 for any isolate-free bipartite hypergraph having a cycle.

Corollary 3.17. *Every isolate-free bipartite hypergraph having at least one cycle has at least two disjoint independent dominating sets.*

Remark 3.18. The converse of Proposition 3.16 is falsified by the following counterexample. Let $H = (X, \mathcal{E})$ where $X = \{1, 2, 3, 4, 5, 6, 7\}$ and $\mathcal{E} = \{E_1 = \{1, 4, 5\}, E_2 = \{1, 3, 6\}, E_3 = \{2, 4, 7\}, E_4 = \{2, 3, 5\}\}$. Then, $X_1 = \{1, 2\}$ and $X_2 = \{3, 4\}$ are two disjoint maximal independent sets in H. However, H is not bipartite because it has the 3-cycle $(5, E_4, 3, E_2, 1, E_1, 5)$. Thus, there exist non-bipartite hypergraphs having disjoint maximal independent sets.

Remark 3.19. Also, there are connected hypergraphs without cycles, viz., *hypertrees* (clearly, they are bipartite), not containing a pair of disjoint independent dominating sets: For instance, take the hypertree $H = (X, \mathcal{E})$ where $X = \{1, 2, 3, 4, 1', 2', 3', 4'\}$ and

$$\mathcal{E} = \{\{1, 2, 3, 4\}, \{1, 1'\}, \{2, 2'\}, \{3, 3'\}, \{4, 4'\}.$$

Then, $A_0 = \{1', 2', 3', 4'\}$, $A_1 = \{1, 2', 3', 4'\}$, $A_2 = \{1', 2, 3', 4'\}$, $A_3 = \{1', 2', 3, 4'\}$ and $A_4 = \{1', 2', 3', 4\}$ are the only independent dominating sets of H and it is easily seen that no two of them are disjoint.

So, we pose the following more general open problem.

Problem 3. Characterize hypergraphs having two disjoint maximal independent sets. In particular, determine such hypertrees.

From the foregoing analyses on bipartite hypergraphs, the point we shall hence proceed with is the fact that there are infinite classes of hypergraphs that possess pairs of disjoint maximal independent sets. Since maximal independent sets form a subclass of minimal dominating sets of a hypergraph, we give the following more general notion; when H is restricted to the class of finite graphs, the notion reduces to its graph theoretic analogue given by S. M. Hedetniemi *et al.* [18].

Definition 3.20. *Given any finite hypergraph H, its disjoint independent domination number $ii(H)$ is the minimum cardinality of the union of two disjoint independent dominating sets, if they exist, in H; such a pair $\{S_1, S_2\}$ is called an ii-pair if $|S_1 \cup S_2| = ii(H)$.*

The following is the hypergraph extension of a similar result due to S. M. Hedetniemi *et al.* [18].

Theorem 3.21. *For any finite hypergraph H having a pair of disjoint independent dominating sets*

$$\gamma\gamma(H) \leq \gamma^i(H) \leq ii(H) \tag{4}$$

Comparing the inequalities (3) (cf. Theorem 3.8) and (4) above, it is natural to think whether $\gamma_i(H) + \beta(H)$ could be an upper bound for $ii(H)$ too. This has been shown to be false when H is restricted to the class of graphs in general; however, in the case of trees T it has been shown that $\gamma\gamma(T) = ii(T)$ which yields $ii(T) \leq \gamma_i(T) + \beta(T)$ using (3) (see [18]). It would thus be interesting to determine finite hypergraphs H in general for which

$$\gamma\gamma(H) = ii(H). \tag{5}$$

In particular, is it true in the case of finite hypergraphs with *cyclomatic number* $\mu(H)$ (see [12]) equal to zero? We do not know yet! However, in the case of graphs, finite trees are the only class of graphs with cyclomatic number (or, the 'cycle-rank' as in [16]) equal to zero. It is possible to show that finite hypertrees H satisfy (5). Towards this end, first it is to be noted that hypertrees form a proper subclass of linear hypergraphs since existence of two distinct edges E_1 and E_2 with $|E_1 \cap E_2| \geq 2$ entails existence of two distinct vertices $x, y \in E_1 \cap E_2$ which would yield the 2-cycle (x, E_1, y, E_2, x). Thus, acyclicity of a hypergraph is inherited by its subhypergraphs. Secondly, one can show easily that a finite connected hypergraph H is a hypertree if and only if removal of any of its edges, say E, has exactly $|E|$ connected components; here, *removal* of E from H means to obtain the subhypergraph $H - E = (X, \mathcal{E} - \{E\})$; in fact, this new characterization of hypertrees is seen from its proof below to hold even when the hypertrees are not necessarily finite.

Lemma 3.22. *A connected hypergraph $H = (X, \mathcal{E})$, in which every edge is finite, is a hypertree if and only if for any arbitrary edge E of H, the subhypergraph $H - E = (X, \mathcal{E} - \{E\})$ of H has exactly $|E|$ connected components.*

Proof.
Necessity: Let $H = (X, \mathcal{E})$ be a hypertree and E be an arbitrary edge of H. Since every other edge of H has at most one vertex in common with E and H is acyclic, it

follows that for each $x \in E$ there is exactly one connected component H_x containing x in $H - E$. Thus, $H_x \mapsto x,\ x \in E$ is a bijection.

Sufficiency: Suppose the condition in the statement of the theorem holds and, yet, H contains a cycle $(x_1, E_1, x_2, E_2, \ldots, x_k, E_k, x_1)$. Choose any edge of this cycle, say E_k. Then,

$$(x_1, E_1, x_2, E_2, \ldots, E_{k-1}, x_k)$$

is still a path joining x_1 and x_k in $H - E_k$ whence the two distinct vertices $x_1, x_k \in E_k$ belong to the same component of $H - E_k$. Thus, the number of components in $H - E_k$ is one less than $|E_k|$, a contradiction to our assumption.

This completes the proof. □

Apart from the problems and conjectures proposed above, the following specific questions on the new 'pair domination' parameters are worth investigating for advancement of the theory of domination in hypergraphs.

Problem 4. Let $H = (X, \mathcal{E})$ be any finite hypergraph.

(i) Determine H such that $\gamma\gamma(H) = \gamma^i(H)$.
(ii) Determine H such that $\gamma\gamma(H) = ii(H)$.
(iii) Determine H such that $\gamma_i(H) + \beta(H) \in \{\gamma\gamma(H), \gamma^i(H), ii(H)\}$.
(iv) *Complement* of H is defined as the hypergraph $\bar{H} = (X, \bar{\mathcal{E}})$ where $\bar{\mathcal{E}}$ denotes the set of all nonempty subsets $\bar{E} = X - E,\ E \in \mathcal{E}$. Find attainable lower and upper bounds for $\gamma\gamma(H) + \gamma\gamma(\bar{H})$. [Even when H is a graph, a complete solution to this problem is unknown as at the time of posing this problem; see [18].]
(v) Determine H having a γ-set D such that $X - D$ contains a γ-set of H; in particular, to start with, explore if every hypertree has this property.

4. Domchromatic number of a hypergraph

As in [11], given a finite hypergraph $H = (X, \mathcal{E})$, its *rank* $r(H)$ is the maximum number of vertices in any edge of H. Next, given any positive integer $r \leq h = r(H)$, a set $A \subseteq X$ is defined to be a *clique of rank* r if either $|A| < r$ or $|A| \geq r$ and each subset of A with cardinality r is contained in at least one edge of H. Then, again, each subset of a clique of rank r is also a clique of rank r. Let $\omega_r(H)$ denote the maximum cardinality of a clique of rank r in H, called the *clique number of rank* r of H. Since $r \leq h$, and since an edge E_i with $|E_i| = h$ is a clique of rank r we must have

$$\omega_r(H) \geq |E_i| = r(H).$$

If the vertex set $X =: X(H)$ of H is itself a clique of rank r, then H is said to be r-*complete*. A clique of rank 2 is called simply a *clique* in H.

Let us now introduce a notion that is antithetic to that of a clique of rank r in a finite hypergraph. Towards this end, let $H = (X, \mathcal{E})$ be any finite hypergraph; its *antirank* $ar(H)$ is defined as the minimum number of vertices in any edge of H. Hence, given any positive integer $c > t = ar(H)$, a set $A \subseteq X$ is a *c-stable set* of

H if either $|A| < c$ or $|A| \geq c$ and no subset of A with cardinality c is contained in an edge of H. Clearly, then, every subset of a c-stable set is c-stable. Further, any nonempty set of vertices having cardinality less than or equal to $ar(H)$ is c-stable for any integer $c > ar(H)$. In particular, if $ar(H) \leq 2$ then a 2-*stable* set is nothing but an independent set of H.

Let $H = (X, \mathcal{E})$ be any hypergraph and c be any positive integer. By a (b, c)-*coloring* of H we mean a partition of X into b c-stable sets $S_1, S_2, \ldots, S_b$, each called a *color class*. A hypergraph is (b, c)-*colorable* if it admits a (b, c)-coloring. The c-*chromatic number* $\chi_c(H)$ of H is the least positive integer b such that H is (b, c)-colorable; $\chi_c(H)$ is defined to be zero whenever H is not (b, c)-colorable for any value of b. If, in particular, $ar(H) \leq 2$ then any $(b, 2)$-coloring of H is called simply a b-*coloring* of H and $\chi_2(H)$ is then denoted simply as $\chi(H)$ and is called the usual *chromatic number* of H.

The following result brings out a natural connection between the notion of c-chromatic number and the clique number of rank c of a hypergraph.

Proposition 4.1. *If a hypergraph H is (b, c)-colorable with $b = \chi_c(H)$, then $c \geq 2$ and*

$$\chi_c(H) \geq \frac{1}{c-1}\omega_c(H). \tag{6}$$

Proof. Since H is (b, c)-colorable, there exists (b, c)-coloring of H, say $S_1, S_2, \ldots, S_b$. Since $1 \leq ar(H) < c$, we have $c \geq 2$. Next, let Q be any clique of rank c in H such that $|Q| = \omega_c(H)$. Then, we claim $|Q \cap S_i| \leq c - 1$. Suppose this is false. Then, $Q \cap S_i$ must contain a set A of cardinality c, which being a clique of rank c must be contained in some edge E of H. Also, A is contained in S_i. Hence, by definition, it cannot be contained in any edge of H, a contradiction to the conclusion above that $A \subseteq E$. Thus, our claim holds, whence we get

$$\omega_c(H) = |Q| \leq b(c-1) = \chi_c(H)(c-1),$$

and the result follows. □

Next, we shall establish another fundamental property of colorings that has a link with the notion of domination in hypergraphs.

Proposition 4.2. *Every finite hypergraph $H = (X, \mathcal{E})$ possesses a χ-coloring in which at least one color class is a dominating set of H.*

Proof. Let $\chi(H) = k$ and $P = \{X_1, X_2, \ldots, X_k\}$ be a k-coloring of H. Suppose none of the color classes in P is a dominating set of H. In particular, then, X_1 is not a dominating set of H. Therefore, there exists $y \in X - X_1$ such that y is not adjacent to any vertex in X_1, whence $X_1 \cup \{y\}$ is independent. Now, let $y \in X_j$ for some $j, 2 \leq j \leq k$. If $X_j - \{y\} = \emptyset$ then $X_j = \{y\}$ whence $X_1' = X_1 \cup X_j = X_1 \cup \{y\}$ which is independent, whence $P' = \{X_1', X_2, \ldots, X_{j-1}, X_{j+1}, X_{j+2}, \ldots, X_k\}$ is a $(k-1)$-coloring of H, contrary to the minimality of k. Therefore, $|X_j| \geq 2$. We can thus transfer any vertex $y \in X - X_1$ that is not adjacent to any vertex of X_1 to X_1 so as to 'swell' X_1 to a maximal independent set X_1' so that the resulting partition

$P' = \{X'_1, X'_2, \ldots, X'_k\}$ is a k-coloring in which X'_1 is a maximal independent set of H and hence a (minimal) dominating set of H by virtue of Theorem 1.1, establishing the proposition. □

Proposition 4.2, which generalizes an earlier result in graph theory by H. B. Walikar, *et al.* [21], prompts the following new notion, generalizing its counterpart in graph theory studied recently by S. Arumugam *et al.* [10].

Definition 4.3. *Domchromatic number $d_\chi(H)$ of a finite hypergraph H with chromatic number $\chi(H) = k$ is the maximum number of dominating sets in a k-coloring of H.*

Clearly, for any finite hypergraph H,

$$d_\chi(H) \leq \chi(H). \tag{7}$$

We call H *domchromatic* if equality holds in (8) and a k-coloring of H in which every color class is a dominating set of H is called a *domchromatic partition* (or, equivalently, a 'domchromatic coloring') of H.

Definition 4.4. *A hypergraph $H = (X, \mathcal{E})$ with $\chi(H) = k$ is uniquely colorable if every k-coloring of H yields the same chromatic partition $\{X_1, X_2, \ldots, X_k\}$ of H.*

The following result is now straightforward observation from the proof of Proposition 4.1.

Proposition 4.5. *Every uniquely colorable finite hypergraph is a domchromatic hypergraph.*

One can easily construct hypergraphs as counterexamples to the question of validity of the converse of Proposition 4.5. Thus, characterizing domchromatic hypergraphs is a challenging open problem.

Let $H = (X, \mathcal{E})$ be any finite hypergraph with $\chi(H) = k$ and let $P = \{X_1, X_2, \ldots, X_k\}$ be a chromatic partition of H. Without loss of generality, we may assume that $\{X_1, X_2, \ldots, X_t\}$ are the dominating sets of H where $t = d_\chi(H)$. Since X_i is an independent dominating set of H, by Theorem 1.1, it is maximal independent and hence a minimal dominating set of H, $1 \leq i \leq t$. Thus, $|X_i| \leq \Gamma(H)$, $\forall i$, $1 \leq i \leq t$, where $\Gamma(H)$ is the upper domination number of H [5]. Now,

$$|X| = \sum_{i=1}^{k} |X_i| = \sum_{i=1}^{t} |X_i| + \sum_{i=t+1}^{k} |X_i| \leq t\Gamma(H) + (k-t)\beta(H).$$

This yields

$$\left\lceil \frac{|X| - d_\chi(H)(\Gamma(H) - \beta(H))}{\beta(H)} \right\rceil \leq \chi(H), \tag{8}$$

or

$$\left\lceil \frac{|X| - \chi(H)\beta(H)}{\Gamma(H) - \beta(H)} \right\rceil \leq d_\chi(H). \tag{9}$$

On the other hand,

$$|X| = \sum_{i=1}^{k} |X_i| = \sum_{i=1}^{t} |X_i| + \sum_{i=t+1}^{k} |X_i| \geq t\gamma_i(H) + k - t,$$

which implies

$$d_\chi(H) \leq \left\lfloor \frac{|X| - \chi(H)}{\gamma_i(H) - 1} \right\rfloor, \tag{10}$$

or

$$\chi(H) \leq |X| - d_\chi(H)(\gamma_i(H) - 1). \tag{11}$$

Thus, combining (9) and (12), we get

$$\left\lceil \frac{|X| - d_\chi(H)(\Gamma(H) - \beta(H))}{\beta(H)} \right\rceil \leq \chi(H) \leq |X| - d_\chi(H)(\gamma_i(H) - 1). \tag{12}$$

In particular, when H is a domchromatic hypergraph (i.e., when $d_\chi(H) = \chi(H)$) (13) yields

$$\left\lceil \frac{|X|}{\Gamma(H)} \right\rceil \leq \chi(H) \leq \left\lfloor \frac{|X|}{\gamma_i(H)} \right\rfloor. \tag{13}$$

Also, combining (10) and (11) we get

$$\left\lceil \frac{|X| - \chi(H)\beta(H)}{\Gamma(H) - \beta(H)} \right\rceil \leq d_\chi(H) \leq \left\lfloor \frac{|X| - \chi(H)}{\gamma_i(H) - 1} \right\rfloor. \tag{14}$$

Definition 4.6. *Given a finite hypergraph $H = (X, \mathcal{E})$, its domatic (respectively, indomatic) number, denoted $\partial(H)$ (respectively, $\partial_i(H)$) is the maximum number of pairwise disjoint dominating (respectively, independent dominating) sets that may be found in H. Further, H is dominable (indominable) if $X(H)$ can be partitioned into dominating (independent dominating) sets of H. A partial partition of $X(H)$ into $\partial(H)$ ($\partial_i(H)$) dominating (independent dominating) sets is called a domatic (indomatic) partition of H.*

The notions of dominable and indominable graphs were introduced in [20] and it has been shown that *indominable graphs do not have 'forbidden subgraph' characterization* (see [2]). It would be a challenging problem to extend these results to hypergraphs.

5. Conclusions and scope

Besides pointing out the fact that there could be rich harvest of fundamental notions and results in the theory of domination in hypergraphs, almost all the developments reported above are for finite connected simple hypergraphs without isolated vertices; relaxing any of these conditions is likely to lead to further understanding, revealing more about the nature of dominating sets in a hypergraph. Even in the case of finite hypergraphs, one needs to set out to establish deeper connections amongst various concepts and structure theorems in fundamentally different directions such as the ones arising from the following yet another new question suggested by Lemma 1.2: What is the maximum order of a dominating set D in a finite isolate-free hypergraph $H = (X, \mathcal{E})$ such that $X - D$ is a dominating set? Here, note that we do not require D to be minimal. Let us call such a dominating set D in H a *complementarily dominating* (or, simply, *c-dominating*) *set* and let $\Gamma^c(H)$ denote the maximum order of a c-dominating set in H. It is not difficult to see that

$$\Gamma(H) \le \Gamma^c(H) \le \frac{|X|}{2}.$$

Determining the finite hypergraphs attaining the bounds in these inequalities could be interesting; this notion appears to open up entirely new directions of research in the theory of domination in hypergraphs.

Precisely which hypergraphs represent the class of maximal families of pairwise intersecting minimal (minimum) dominating sets of a hypergraph? A similar question for graphs has been attempted in [8]. Some leads towards answering this question appear to be contained in [1].

6. Postscript

For a study of a different variation of the notion of domination in finite hypergraphs, see [6].

Acknowledgement

I am deeply indebted to the referees whose detailed and incisive comments enabled me to reset, prune and prepare the present revised draft of the paper.

References

[1] B. D. Acharya, Representing hypergraphs by families of sets with given properties, *Indian J. Math.*, **51(3)** (2009) 523–536.
[2] B. D. Acharya and H. B. Walikar, Indominable graphs cannot be characterized by a family of forbidden subgraphs, *Nat. Acad. Sci. Letters*, **4(1)** (1981) 23–25.
[3] B. D. Acharya, Interrelations among the notions of independence, domination and full sets in a hypergraph, *Nat. Acad. Sci. Letters*, **13(11)** (1990) 421–422.
[4] B. D. Acharya, Full sets in hypergraphs, Sankhyā, **54** (1992) 1–6.
[5] B. D. Acharya, Stable set covers, chromaticity and kernels of a hypergraph, *Bull. Allahabad Math. Soc.*, **17** (2002) 1–15.
[6] B. D. Acharya and Purnima Gupta, Weak edge-degree domination in hypergraphs, *Czech. Math. J.*, **56(131)** (2006) 99–108.
[7] B. D. Acharya, Domination in hypergraphs, In: Domination in Discrete Structures (Eds: B. D. Acharya, S. Arumugam and P. J. Slater), *AKCE Internat. J. Graphs and Combinatorics*, **4(2)** (2007) 117–126.

[8] B. D. Acharya, V. Swaminathan and A. Wilson Bhaskar, Intersection graph of the minimal dominating sets of a graph, In: Domination in Discrete Structures (Eds: B. D. Acharya, S. Arumugam and P. J. Slater), *AKCE Internat. J. Graphs and Combin.*, **4(2)** (2007) 153–159.

[9] B. D. Acharya, Domination in hypergraphs: Corrigendum, *AKCE Internat. J. Graphs and Combinatorics*, **5(1)** (2008) 99–101.

[10] S. Arumugam, I. S. Hamid and A. Muthukumatchi, Independent domination and graph colorings, In: Proceedings of International Conference on Discrete Mathematics (Eds.: R. Balakrishnan and C. E. Veni Madhavan), 195–203. Lecture Notes Series in Mathematics, Vol. **7**, Ramanujan Mathematical Society (India) 2008.

[11] C. Berge, Graphs and Hypergraphs, North-Holland, Amsterdam (1973).

[12] C. Berge, Hypergraphs, North-Holland, Amsterdam (1989).

[13] G. S. Domke, J. E. Dunbar and L. R. Markus, The inverse domination number of a graph, *Ars Combin.*, **72** (2004) 149–160.

[14] K. A. Germina, Personal communication, April (2008).

[15] K. A. Germina, Personal communication, May (2008).

[16] F. Harary, Graph Theory, Addison-Wesley Publ. Comp., Massachusetts, Reading (1969).

[17] T. Haynes, S. T. Hedetniemi and P. J. Slater, Fundamentals of Domination in Graphs, Marcel-Dekker Inc., New York (1997).

[18] S. M. Hedetniemi, S. T. Hedetniemi, R. C. Laskar, L. Marcus and P. J. Slater, Disjoint dominating sets in graphs, In: Proceedings of International Conference on Discrete Mathematics (Eds.: R. Balakrishnan and C. E. Veni Madhavan), 87–100. Lecture Notes Series in Mathematics, Vol. **7**, Ramanujan Mathematical Society (India) 2008.

[19] O. Ore, Theory of Graphs, American Mathematical Society Colloquium Publications, Volume XXXVIII (1962).

[20] H. B. Walikar, B. D. Acharya and E. Sampathkumar, Recent Developments in the Theory of Domination in Graphs, Lecture Notes in Mathematics, No. 1, Mehta Research Institute of Mathematics and Mathematical Physics, Allahabad (1979).

[21] H. B. Walikar, B. D. Acharya, K. Narayankar and H. G. Shekharappa, Embedding index of nonindominable graphs, In: Graphs, Combinatorics, Algorithms and Applications (Eds.: S. Arumugam, B. D. Acharya and S. B. Rao), pp. 173–179. Narosa Publishing House, New Delhi (2005).

Proc. ICDM 2008, RMS-Lecture Notes Series No. 13, 2010, pp. 19–32.

Chain Decompositions of Normalized Matching Posets of Rank 2

Andrew Pearsall[1] and Shahriar Shahriari[2]

[1]Department of Mathematics, University of Washington Seattle, WA 98195
[2]Department of Mathematics, Pomona College Claremont, CA 91711
e-mail: rcarrot@u.washington.edu; sshahriari@pomona.edu

Abstract. Let P be a graded poset and let μ be a partition of the integer $|P|$. Is it possible to construct a chain partition of P such that the sizes of the chains are given precisely by the (positive) parts of μ? A 20 year old conjecture of Griggs gives the answer when P is a Boolean Lattice. In this paper, after a survey of known results related to the Griggs conjecture, we generalize the conjecture to all normalized matching posets, and provide a proof for all such posets of rank 2.

Keywords. Chain decompositions, nested posets, Griggs nesting conjecture, normalized matching property, LYM property, dominance order.

2000 Subject Classification: Primary 06A07; Secondary 05D05, 05D15

1. Chain partitions of normalized matching posets—a quick survey

Let P be a finite partially ordered set (poset). The number of elements of P is denoted by $|P|$, and a totally ordered subset of P of size k is called a *chain* of size k or length $k-1$. Now let $\mu = (\mu_1, \mu_2, \dots, \mu_k)$ be an integer partition of $|P|$. In other words, $\mu_1 \geq \mu_2 \geq \cdots \geq \mu_k$ are positive integers with $\mu_1 + \mu_2 + \cdots + \mu_k = |P|$. Can we partition the poset P into k chains with sizes $\mu_1, \mu_2, \dots, \mu_k$? Posed at this level, the question is too general, but, surprisingly, much can be said for a broad class of posets—the so-called normalized matching posets defined below—that include the Boolean lattices, the divisor lattices, and the linear lattices.

The case of the Boolean lattices itself is most instructive. For a positive integer n, let $[n] = \{1, \dots, n\}$, and $\mathbf{2}^{[n]}$—called the Boolean or subset lattice of order n—denote the poset of subsets of $[n]$ ordered by inclusion. Sands [30] asked if, for a given k, and large enough n, the poset $\mathbf{2}^{[n]}$ can be partitioned into chains of size 2^k. The answer is trivially yes for $k = 0$ and $k = 1$, and was proved by Griggs, Grinstead, and Yeh [18] to be true for $k = 2$ if and only if $n \geq 9$. Later, generalizing Sands' question, Griggs [17] conjectured that for a given positive integer c, and large enough n, one can partition $\mathbf{2}^{[n]}$ into chains of size c and a single chain of size at most $c - 1$. This conjecture was proved by Lonc [24]. In a variation, Füredi [9] conjectured that $\mathbf{2}^{[n]}$ can be partitioned into $\binom{n}{\lfloor n/2 \rfloor}$ chains—the minimum number of chains possible—such that the sizes of each of the chains is either $a(n)$ or $a(n) + 1$,

The authors thank Pomona College's Summer Undergraduate Research Program for supporting this research.

where $a(n) = \lfloor 2^n / \binom{n}{\lfloor n/2 \rfloor} \rfloor \sim \sqrt{\pi/2}\sqrt{n}$. Maybe surprisingly, Füredi's conjecture remains open after 23 years. Elzobi and Lonc [5] examined Lonc's proof of the conjecture by Griggs to show that, for n sufficiently large, the Boolean lattice $\mathbf{2}^{[n]}$ can be partitioned into chains of size $c = \lfloor \frac{1}{6}\sqrt{\log\log n} \rfloor$, except possibly $c-1$ elements which also form a chain. The best results to date in support of Füredi's conjecture are from Hsu *et al.* [20,21]. It follows from their work that $\mathbf{2}^{[n]}$ has a partition into $\binom{n}{\lfloor n/2 \rfloor}$ chains such that the size of the shortest chains are at least $\frac{1}{2}\sqrt{n}$, and, for $n > 16$, the size of the longest chains is at most $\frac{3}{2}\sqrt{n \log n}$. For the Boolean lattices, in 1988, Griggs [17] gave the most general conjecture for the chain sizes of a chain partition in $\mathbf{2}^{[n]}$. To state it, we need a few definitions.

If all maximal chains of a poset P have the same length n, then the poset is called a *graded poset of rank* n. For such a poset, the *rank* of each element x, is then defined to be the length of the longest chain of the form $x_0 < x_1 \cdots < x_m = x$. The kth *level* and the kth *rank number* of P are, respectively, the set and the number of elements of P of rank k. Let $\mathcal{C} : x_1 < x_2 < \cdots < x_k$ be a chain in P, a graded poset of rank n. The chain $\mathcal{C}$ is *skipless* (also called *saturated* in the literature) if, for $2 \le i \le k$, $\operatorname{rank}(x_i) = \operatorname{rank}(x_{i-1}) + 1$. The chain $\mathcal{C}$ is a *symmetric* chain if it is skipless and $\operatorname{rank}(x_1) + \operatorname{rank}(x_k) = n$. A partition of P into symmetric chains is called a *symmetric chain decomposition* of P.

It is straightforward to inductively construct a symmetric chain decomposition for $\mathbf{2}^{[n]}$ (originally due to de Bruijn, Tengbergen, and Kruyswijk [4], but also see Anderson [2, Chapter 3] or Greene and Kleitman [13]). While a symmetric chain decomposition for $\mathbf{2}^{[n]}$ is far from unique, the set of chain sizes in such a symmetric chain decomposition is indeed unique. In fact, define

$$\sigma = (\sigma_1, \sigma_2, \ldots, \sigma_{\binom{n}{\lfloor n/2 \rfloor}}),$$

where $\sigma_k = n - 2j + 1$, for all k and j such that $\binom{n}{j-1} < k \le \binom{n}{j}$ and $0 \le j \le \lfloor n/2 \rfloor$. Then σ—called the *SCD partition of* 2^n—is a partition of the integer 2^n, and the sizes of the chains in any symmetric chain decomposition of $\mathbf{2}^{[n]}$ are given precisely by the parts of σ.

Let $\mu = (\mu_i)$ and $\nu = (\nu_i)$ be two partitions of an integer m. We say that $\mu \le \nu$ in the *dominance* (or *majorization*) order if and only if, for all j, we have

$$\sum_{i=1}^{j} \mu_i \le \sum_{i=1}^{j} \nu_i .$$

We can now state the elegant conjecture of Griggs [17]—from which Füredi's conjecture follows:

Griggs Dominance Conjecture 1.1. Let n be a positive integer, and let σ and $\mu = (\mu_1, \ldots, \mu_m)$, respectively, be the SCD and an arbitrary partition of 2^n. A partition of the Boolean lattice $\mathbf{2}^{[n]}$ into m chains with sizes $\mu_1, \ldots, \mu_m$ exists if and only if $\mu \le \sigma$ in the dominance order.

It is straightforward to show the "only if" direction of the conjecture, but, in twenty years, no substantial progress has been made on the "if" direction.

What about other posets? We will generalize Griggs' conjecture to normalized matching posets that include Boolean lattices as a special case. A finite graded poset is a *normalized matching* poset (or has the *LYM property*) if, for any two consecutive levels X and Y and every subset $S \subseteq X$, we have

$$\frac{|\Gamma(S)|}{|Y|} \geq \frac{|S|}{|X|}, \tag{1}$$

where $\Gamma(S)$ is the set of elements of Y that are related to some element in S. Any collection of pair-wise incomparable elements in a poset is called an *antichain*. Kleitman [23] showed that a finite graded poset P is normalized matching if and only if every anti-chain $\mathcal{A} \subseteq P$ satisfies the *LYM inequality*:

$$\sum_{x \in \mathcal{A}} \frac{1}{N(x)} \leq 1,$$

where $N(x)$ is the number of elements of P with the same rank as x. In addition to Boolean lattices, subspace lattices (that is, the subspaces of a finite dimensional vector space over a finite field ordered by inclusion), divisor lattices (that is, divisors of a positive integer ordered by divisibility), and regular posets (a graded poset is *regular* if, for each rank i, there are two constants a_i and b_i such that each element of rank i is covered by a_i elements and covers b_i elements) are examples of normalized matching posets. (Normalized Matching was introduced in Graham and Harper [11]. LYM stands for Lubell [26], Yamamoto [32] and Meshalkin [27] who—in addition to Bollobás [3]—independently proved the inequality for the subset lattices. See Anderson [2], Engel [6], and Fredrich *et al.* [8] for more on normalized matching posets.)

Let $r_0, \ldots, r_n$ be an ordered list of positive integers. This list is *symmetric* if, for $0 \leq k \leq n$, we have $r_k = r_{n-k}$. The list is called *unimodal* if there is an index k with $0 \leq k \leq n$ such that

$$r_0 \leq r_1 \leq \cdots \leq r_k \geq r_{k+1} \geq \cdots \geq r_n.$$

A graded poset P is called *rank-unimodal*, respectively *rank-symmetric*, if its rank numbers are unimodal, respectively symmetric.

To generalize Griggs' conjecture, we need to generalize the notion of a symmetric chain decomposition. Two chains in a graded poset are *nested* if the ranks of the elements of one of the chains is a subset of the ranks of the elements of the other chain. A partition of a graded poset into chains is a *nested chain decomposition* (or simply a *nesting*) if every pair of chains in the partition are nested (this vocabulary was first introduced by Gansner [10]). We say that a graded poset is *nested* if it has a nesting. In the case of a rank-symmetric, rank-unimodal poset, a nesting is the same as a symmetric chain decomposition. We note that if the nested poset P is not rank-unimodal, then not all the chains in the nesting will be skipless.

As was mentioned above, in the case of the Boolean lattices, the existence of a symmetric chain decomposition is a theorem. However, in the case of normalized matching posets, the existence of a nested chain decomposition is another long standing conjecture of Griggs [14] (and also [15–17]):

Griggs Nesting Conjecture 1.2. Every normalized matching poset is nested.

This conjecture is a theorem (Theorem 3.6.4 of Anderson [2], proved independently by Anderson [1] and Griggs [14]) in the case of rank-symmetric rank-unimodal posets:

Anderson-Griggs Theorem 1.3. *Every rank-symmetric, rank-unimodal normalized matching poset is nested. In other words, every such poset has a symmetric chain decomposition.*

Surprisingly little progress toward the proof of the 30 year old Griggs Nesting Conjecture has been made. A proof of the conjecture for rank 2 posets is well known (see Corollary 1.8 below and also Wang [31]). Hsu *et al.* [19], and Escamilla *et al.* [7] prove the conjecture for large classes of rank-unimodal rank 3 posets. We summarize some of their results here. Let $\mathcal{NM}(r_0, \dots, r_n)$ denote the set of all normalized matching posets P of rank n whose ith rank number is equal to r_i.

Proposition 1.4 (Hsu et al. [19] and Escamilla et al. [7]). *To prove the Griggs Nesting conjecture for rank-unimodal rank 3 posets, it is enough to prove that every $P \in \mathcal{NM}(r_0, r_1, r_2, r_0)$ is nested, where $r_0 < r_1 < r_2$ are positive integers. Now, fix positive integers $r_0 < r_1 < r_2$, and let $P \in \mathcal{NM}(r_0, r_1, r_2, r_0)$. Assume that at least one of the following conditions are satisfied:*

(1) *r_0 divides r_1; or*
(2) *$r_0 + 1$ divides r_1; or*
(3) *$r_2 > r_0 r_1 - r_0 \gcd(r_1, r_2)$, in particular, this condition is satisfied if $r_2 > r_0 r_1$ or if r_1 divides r_2; or*
(4) *$f(i) \geq 0$ for all $1 \leq i \leq r_1 - r_0 - 1$, where the function f on integers is defined by*

$$f(i) = \left\lceil \frac{r_0(1+i)}{r_2 - r_0} \right\rceil - \left\lfloor \frac{r_0 i}{r_1 - r_0} \right\rfloor,$$

in particular, this condition is satisfied if $r_2 \leq r_1 + \lceil \frac{r_2}{r_0} \rceil - 1$.

Then every $P \in \mathcal{NM}(r_0, r_1, r_2, r_0)$ is nested. As a consequence, every normalized matching rank-unimodal poset of rank 3 and width—width of P is r_2—at most 11 is nested.

In fact, Hsu *et al.* [19], and Escamilla *et al.* [7] prove more. For each of their conditions, they provide an algorithm for finding a nested chain decomposition of P. For example, for $r_0 < r_1 < r_2$, Hsu *et al.* [19] prove that for an arbitrary poset in $\mathcal{NM}(r_0, r_1, r_2, r_0)$, we can start with an *arbitrary* matching of the middle two levels and extend it to a nesting of P if and only if $r_2 \leq r_1 + \lceil \frac{r_2}{r_0} \rceil - 1$.

As an example of the progress made in nesting rank 3 rank-unimodal normalized matching posets, consider $P \in \mathcal{NM}(5, r_1, r_2, 5)$, where $5 < r_1 < r_2$. From Proposition 1.4, it follows that if $r_1 = 15$, then P is nested for all $r_2 \geq 15$. If $r_1 = 16$, then P is nested if $16 \leq r_2 \leq 21$, or if $r_2 = 32, 48, 64, 72$, or if $r_2 \geq 74$. If $r_1 = 17$, then P is nested if $17 \leq r_2 \leq 23$, if $r_2 = 34, 51, 68$, or if $r_2 \geq 81$. Finally, if $r_1 = 18$, then P is nested for all $r_2 \geq 18$. Hence, in general, if r_2 is close to r_1 or much larger than r_1, then we can prove that the poset is nested.

Just as in the case of the Boolean lattices, even though a nested chain decomposition need not be unique, the sizes of the chains in such a chain partition are unique. Let P be a poset of rank n, and let $s_0 \leq s_1 \leq \cdots \leq s_n$ be the rank numbers of P ordered by size. In addition, define $s_{-1} = 0$. For $0 \leq k \leq n$ and $s_{k-1} < i \leq s_k$, define $\sigma_i = n - k + 1$, and let

$$\sigma = (\sigma_1, \sigma_2, \ldots, \sigma_{s_n}).$$

Then σ is a partition of the integer $|P|$, and the sizes of the chains in a nested chain decomposition of P—if such a decomposition exists—will be precisely $\sigma_1, \sigma_2, \ldots, \sigma_{s_n}$. We call σ the *NCD partition of* $|P|$.

We will now state our common generalization of the two conjectures by Griggs (the conjecture was extended to rank-symmetric, rank-unimodal, normalized matching posets by Hsu, Logan, and Shahriari, and to arbitrary normalized matching posets by Shahriari):

Generalized Griggs Conjecture 1.5. Let P be a finite normalized matching poset, σ the NCD partition of $|P|$, and $\mu = (\mu_1, \ldots, \mu_m)$ an arbitrary partition of $|P|$. A partition of the poset P into m chains with sizes $\mu_1, \ldots, \mu_m$ exists if and only if $\mu \leq \sigma$ in the dominance order.

One direction of the conjecture is easy to prove. For completeness, we include an argument here:

Proof (The Proof of the "only if" direction). Define $s_{-1} = 0$, and let $s_0, \ldots, s_n$ be the rank numbers of P ordered by size—that is, s_0 is the size of P's smallest level, while s_n is the size of the largest level of P—and let j be an arbitrary positive integer. Assuming that a partition, $\mathcal{M}$, of the poset P into m chains with sizes $\mu_1 \geq \mu_2 \geq \cdots \geq \mu_m$ exists, the proof will be complete if we show that

$$\sum_{i=1}^{j} \mu_i \leq \sum_{i=1}^{j} \sigma_i.$$

Now, since both σ and μ are partitions of $|P|$, and σ has exactly s_n positive parts, the inequality is clear for $j \geq s_n$. Hence assume that $s_{k-1} < j \leq s_k$, for some k with $0 \leq k \leq n$.

Let A be the elements of the longest j chains—those with sizes $\mu_1, \ldots, \mu_j$—in $\mathcal{M}$ that lie on levels with rank numbers $s_0, \ldots, s_{k-1}$, and let B be the rest of elements of the longest j chains in $\mathcal{M}$. Then clearly $|A| \leq s_0 + \cdots + s_{k-1}$. In addition, since each of the j chains can have at most one element from each of the largest $n - k + 1$ levels, we have $|B| \leq j(n - k + 1)$. We conclude that

$$\sum_{i=1}^{j} \mu_i \leq s_0 + \cdots + s_{k-1} + j(n - k + 1).$$

However, the RHS is exactly the number of elements in the longest j chains in a nested chain decomposition of P, hence equals $\sum_{i=1}^{j} \sigma_i$, and the proof is complete. □

Lonc and Elzobi were the first to consider a generalization of the Griggs conjecture. They proved that the generalized Griggs conjecture 1.5 is true for the product of two chains [25]. A special case of the generalized Griggs conjecture is the *generalized Füredi conjecture* asserting that it is possible to partition any normalized matching poset into w chains, where the sizes of the chains are one of two consecutive integers, and w is the width (the size of the largest antichain) of the poset. While the Füredi conjecture is open for the Boolean lattices, Hsu, Logan, and Shahriari [22] proved the generalized Füredi conjecture for the subspace lattices. In fact, they only considered rank-symmetric, rank-unimodal, normalized matching posets, and within this class, they proved the generalized Füredi conjecture for both rank 2 posets and RDR posets. A rank-symmetric, rank-unimodal poset is an *RDR poset* (RDR stands for *rapidly decreasing rank numbers*) if the ratio between consecutive rank sizes is at least 2 when they are not both equal to the maximum possible rank size. The result for subspace lattices follows since these lattices are rank-symmetric, rank-unimodal, normalized matching, and RDR.

In this paper, extending the techniques of Hsu *et al.* [22], we will prove:

Main Theorem 1.6. *The generalized Griggs conjecture 1.5 is true for all normalized matching posets of rank* 2.

Hence, in addition to proving the Griggs conjecture for this class of posets, we have extended the proof of Hsu *et al.* [22] for the generalized Füredi conjecture for rank-unimodal, rank-symmetric normalized matching posets of rank 2 to include posets that are not rank-unimodal nor rank-symmetric.

Finally, in this introductory section, we mention a result of Greene and Kleitman that is related to the Griggs nesting conjecture. Let k be a positive integer. A collection of elements of a poset that contains no chains of length k is called a *k-family* (hence, 1-family is synonymous with antichain). For any poset P, denote by $d_k(P)$, the size of the largest k-family in P. Let $\mathcal{C} = \{C_1, \ldots, C_\ell\}$ be a partition of P into chains. We say that $\mathcal{C}$ is a *k-saturated partition* of P if

$$\sum_{i=1}^{\ell} \min\{k, |C_i|\} = d_k(P). \qquad (2)$$

We now state the deep Theorem of Greene and Kleitman [12] (for a different proof—presented in Anderson [2, Theorem 13.3.2]—see Saks [29] and Perfect [28]):

Greene-Kleitman Theorem 1.7. *Let k be an arbitrary positive integer. Any poset P has a chain partition that is simultaneously k and $k+1$-saturated.*

It is well known (for example, see Escamilla *et al.* [7, Proposition 2.5]) that a chain partition of a normalized matching poset of rank n is a nested chain partition if and only if it is simultaneously k-saturated for all $1 \leq k \leq n$. (In fact, the full force of normalized matching in not needed. The result remains true for posets that are Strong Sperner. A graded poset of rank n is a *Strong Sperner* poset if, for all $1 \leq k \leq n$, the size of the largest k-family is equal to the sum of the largest k rank numbers. Every normalized matching poset is Strong Sperner—see Lemma 2.6.)

In particular, a normalized matching poset of rank 2 is nested if it has a chain partition that is simultaneously 1 and 2-saturated. Hence an immediate consequence of the Greene-Kleitman Theorem 1.7 is the Griggs Nesting Conjecture for rank 2 posets:

Corollary 1.8. *If P is a normalized matching poset of rank 2 then P is nested.*

2. Preliminaries on normalized matching posets

To verify that a graded poset is normalized matching, we have to check condition (1) for each two consecutive levels (recall that a level of P is the collection of elements of P of some particular fixed rank). Hence it is useful to introduce some notation for normalized matching posets of rank 1, and their Hasse diagrams. The notation $G = (X, \Delta, Y)$ will denote a bipartite graph G, where X and Y are sets that partition the set of vertices of G, the set of edges G is denoted by Δ, and every element of Δ is of the form $\{x, y\}$ with $x \in X$ and $y \in Y$. A bipartite graph $G = (X, \Delta, Y)$ will be called *normalized matching* if for every subset $S \subseteq X$, we have

$$\frac{|\Gamma_Y(S)|}{|Y|} \geq \frac{|S|}{|X|}, \tag{3}$$

where $\Gamma_Y(S)$ is the set of neighbors of S in Y. Given a graded poset P with (not necessarily consecutive) levels X and Y, we use the notation $H_P(X, Y)$ to denote the bipartite graph (X, Δ, Y) with vertex set $X \cup Y$ and edge set $\Delta = \{\{x, y\} \mid x \in X, y \in Y, \text{and } x \leq y \text{ or } y \leq x\}$. The following is straightforward:

Lemma 2.1.

(1) *[6, Proposition 4.5.2] The definition of normalized matching bipartite graphs is symmetric in X and Y. In other words, (X, Δ, Y) is normalized matching if and only if (Y, Δ, X) is normalized matching.*
(2) *A finite graded poset P is normalized matching if and only if, for every two consecutive levels X and Y of P, $H_P(X, Y)$ is a normalized matching bipartite graph.*
(3) *[6, Proposition 4.5.3] Assume a finite graded poset P is normalized matching, and X and Y are two levels (not necessarily consecutive) of P. Then, $H_P(X, Y)$ is a normalized matching bipartite graph.*

We now give a construction of new normalized matching posets from existing ones that generalizes similar results from Hsu, Logan, Shahriar, Towse [21, Theorem 2.2] and Hsu, Logan, Shahriari [22, Lemma 2.5]:

Lemma 2.2. *Let Y_1 and Y_2 be disjoint sets. Assume that (X, Δ_1, Y_1) and (X, Δ_2, Y_2) are normalized matching bipartite graphs. Then so is $(X, \Delta_1 \cup \Delta_2, Y_1 \cup Y_2)$.*

Proof. Let $S \subseteq X$. Then

$$\begin{aligned}
\frac{|\Gamma_{Y_1 \cup Y_2}(S)|}{|Y_1 \cup Y_2|} &= \frac{|\Gamma_{Y_1}(S)| + |\Gamma_{Y_2}(S)|}{|Y_1| + |Y_2|} \\
&= \frac{\frac{|\Gamma_{Y_1}(S)|}{|Y_1|}\frac{1}{|Y_2|} + \frac{|\Gamma_{Y_2}(S)|}{|Y_2|}\frac{1}{|Y_1|}}{\frac{1}{|Y_2|} + \frac{1}{|Y_1|}} \\
&\geq \frac{\frac{|S|}{|X|}\frac{1}{|Y_2|} + \frac{|S|}{|X|}\frac{1}{|Y_1|}}{\frac{1}{|Y_2|} + \frac{1}{|Y_1|}} \\
&\geq \frac{|S|}{|X|},
\end{aligned}$$

and the proof is complete. □

Using the above lemma, we can "move around" levels of a normalized matching poset and construct new normalized matching posets:

Corollary 2.3. *Let P be a normalized matching poset of rank n, and for $0 \leq i \leq n$ let Z_i denote the elements of rank i in P (the i-th level of P). Let r and s be fixed integers with $0 \leq r < s \leq n$.*

(1) *[6, Proposition 4.5.3] Let the poset Q be obtained from P by eliminating level r. In other words, the elements of Q are $P \backslash Z_r$, and for two elements $x, y \in Q$, we have $x \leq y$ in Q if and only if $x \leq y$ in P. Then Q is normalized matching.*
(2) *Let the poset R be obtained from P by "merging" Z_s into Z_r. In other words, the elements of R are the same as the elements of P. For two elements $x, y \in P \backslash Z_s$, we have $x \leq y$ in Q if and only if $x \leq y$ in P. If $x \in Z_s$, and $y \in Z_i$, then, in Q,*

   (a) *if $i = r$, then x is not related to y;*
   (b) *if $i > r$, then $x \leq y$ if $x \leq y$ or $x \geq y$ in P;*
   (c) *if $i < r$, then $x \geq y$ if $x \leq y$ or $x \geq y$ in P.*

   Then R is normalized matching.

Proof.

(1) The levels of the poset Q are $Z_0, \ldots, Z_{r-1}, Z_{r+1}, \ldots, Z_n$. By Lemma 2.1(3), $H_P(Z_{r-1}, Z_{r+1})$ is a normalized matching bipartite graph. Hence, for each two consecutive levels X and Y of Q, we have that $H_Q(X, Y)$ is a normalized matching bipartite graph. By Lemma 2.1(2), Q is normalized matching.
(2) The levels of R are

$$Z_0, \ldots, Z_{r-1}, Z_r \cup Z_s, Z_{r+1}, \ldots, Z_{s-1}, Z_{s+1}, \ldots Z_n.$$

Let $(Z_{r-1}, \Delta, Z_r) = H_P(Z_{r-1}, Z_r)$, and $(Z_{r-1}, \Delta', Z_s) = H_P(Z_{r-1}, Z_s)$. Then, by the definition of R, we have $H_R(Z_{r-1}, Z_r \cup Z_s) = (Z_{r-1}, \Delta \cup \Delta', Z_r \cup Z_s)$ is normalized matching by Lemma 2.2. Similarly, $H_R(Z_r \cup Z_s, Z_{r+1})$ is normalized matching, and so by Lemma 2.1(2), R is normalized matching. □

Let P be a graded poset. For sets $A, B \subseteq P$, a *matching* from A to B is an injective map $\phi : A \to B$ such that, for all $x \in A$, x is comparable to $\phi(x)$. Likewise, in a bipartite graph (X, Δ, Y), a *matching* from X into Y is an injective map $\phi : X \to Y$ such that, for all $x \in X$, $\{x, \phi(x)\} \in \Delta$. The following lemma is immediate and well known:

Lemma 2.4. *Let the bipartite graph (X, Δ, Y) be normalized matching and assume $|X| \leq |Y|$. Then there is a matching from X into Y. In particular, if X and Y are two (not necessarily consecutive) levels in a normalized matching poset P, then there is a matching from the smaller of X and Y into the other.*

Proof. The matching is guaranteed by Hall's marriage condition which follows immediately from the normalized matching condition. □

We need a stronger version of the above lemma.

Corollary 2.5. *Let P be a normalized matching poset, and let $X_1, \ldots, X_k, Y_1, \ldots, Y_\ell$ be some of the levels of P. Let $X = \cup_{i=1}^{k} X_i$ and $Y = \cup_{j=1}^{\ell} Y_j$, and assume $|X| \leq |Y|$. Then there is a matching from X to Y.*

Proof. Let $\Delta = \{\{x, y\} \mid x \in X, y \in Y$, and x and y comparable in $P\}$. Repeatedly apply Corollary 2.3 to get that the bipartite graph (X, Δ, Y) is normalized matching. Now apply Lemma 2.4. □

We also need that normalized matching posets have the Strong Sperner property:

Lemma 2.6 (The Strong Sperner Property). *Let P be a normalized matching poset. Then the size of the largest k-family in P is the sum of its k largest rank numbers.*

Proof. Corollary 4.5.3 in Engel [6]. It follows immediately from Corollary 2.3.3 in Anderson [2]. □

3. Normalized matching posets of rank 2

We now limit ourselves to normalized matching posets of rank 2. We let $(3^a, 2^b, 1^c)$ denote the integer partition $(\underbrace{3, 3, \ldots, 3}_{a}, \underbrace{2, 2, \ldots, 2}_{b}, \underbrace{1, 1, \ldots, 1}_{c})$. If P is poset of rank 2, let s_0, s_1, and s_2 be the rank numbers of P ordered by size—that is, $s_0 \leq s_1 \leq s_2$, but s_i is not necessarily the rank of the ith level. By Corollary 1.8, P has a nested chain decomposition, and, in such a chain partition, the sizes of the chains are given by the parts of the partition $\sigma = (3^{s_0}, 2^{s_1 - s_0}, 1^{s_2 - s_1})$ of the integer $s_0 + s_1 + s_2$. We first characterize the partitions of $s_0 + s_1 + s_2$ that are dominated by σ.

Proposition 3.1. *Let $s_0 \leq s_1 \leq s_2$ be positive integers, and let $\sigma = (3^{s_0}, 2^{s_1 - s_0}, 1^{s_2 - s_1})$. Let $\mu = (3^{t_1}, 2^{t_2}, 1^{t_3})$ where t_1, t_2, t_3 are arbitrary non-negative integers. Then μ is a partition of $s_0 + s_1 + s_2$ and $\mu \leq \sigma$ in the dominance order if and only if*

$$t_1 \leq s_0,$$

$$2t_1 + t_2 \leq s_0 + s_1, \text{ and}$$

$$3t_1 + 2t_2 + t_3 = s_0 + s_1 + s_2.$$

Proof. Let

$$\sigma_i = \begin{cases} 3 & \text{if } 1 \le i \le s_0, \\ 2 & \text{if } s_0 < i \le s_1, \\ 1 & \text{if } s_1 < i \le s_2, \end{cases} \quad \text{and} \quad \mu_i = \begin{cases} 3 & \text{if } 1 \le i \le t_1, \\ 2 & \text{if } t_1 < i \le t_1 + t_2, \\ 1 & \text{if } t_1 + t_2 < i \le t_1 + t_2 + t_3, \end{cases}.$$

Then $\sigma = (\sigma_1, \dots, \sigma_{s_2})$ and $\mu = (\mu_1, \dots, \mu_{t_1+t_2+t_3})$.

The sequence μ is a partition of $s_0+s_1+s_2$ if and only if $\sum_{i=1}^{t_1+t_2+t_3} \mu_i = s_0+s_1+s_2$ and this is equivalent to $3t_1 + 2t_2 + t_3 = s_0 + s_1 + s_2$. Assuming this, then $\mu \le \sigma$ if and only if

$$\sum_{i=1}^{t_1} \mu_i \le \sum_{i=1}^{t_1} \sigma_i, \quad \text{and} \quad \sum_{i=1}^{t_1+t_2} \mu_i \le \sum_{i=1}^{t_1+t_2} \sigma_i.$$

The first inequality is satisfied if and only if $t_1 \le s_0$. Now, if σ has no fewer threes than μ, then the second inequality will be automatically satisfied if $\sigma_{t_1+t_2}$ is 3 or 2. In the remaining case, when $\sigma_{t_1+t_2} = 1$, we have $\sum_{i=1}^{t_1+t_2} \sigma_i = 3s_0+2(s_1-s_0)+(t_1+t_2-s_1)$. Hence—given $3t_1+2t_2+t_3 = s_0+s_1+s_2$ and $t_1 \le s_0$—we have $\mu \le \sigma$ if and only if

$$3t_1 + 2t_2 \le 3s_0 + 2(s_1 - s_0) + (t_1 + t_2 - s_1) \Leftrightarrow 2t_1 + t_2 \le s_0 + s_1.$$

The proof is now complete. □

If $s_0 \le s_1 \le s_2$ are the rank numbers of a normalized matching poset P of rank 2, then, by Lemma 2.6, the size of the largest anti-chain in P is s_2, and hence, by Dilworth's Theorem (see Anderson [2, Theorem 3.2.1]), the minimum number of chains needed to cover P is also s_2. In our proof of the main Theorem 1.6, we will need a characterization of the partitions of $|P|$ that are dominated by the NCD partition of P and that have s_2 non-zero parts.

Proposition 3.2. *Let $s_0 \le s_1 \le s_2$ be positive integers. Let $f = \min(s_0, s_2 - s_1)$. For $0 \le a \le f$, define*

$$\nu_a = (3^{s_0-a}, 2^{s_1-s_0+2a}, 1^{s_2-s_1-a}).$$

Then $\nu_0, \dots, \nu_f$ are the only partitions of $s_0 + s_1 + s_2$ with s_2 non-zero parts that are dominated by ν_0. Furthermore, in the dominance order, we have

$$\nu_0 \ge \nu_1 \ge \cdots \ge \nu_f.$$

Note that if $s_0 \le s_1 \le s_2$ are the rank numbers of a post P. Then ν_0 and ν_f are, respectively, the NCD and the Füredi partition of $|P|$.

Proof. Let $\mu = (3^{t_1}, 2^{t_2}, 1^{t_3})$ be a partition of $s_0 + s_1 + s_2$ with s_2 non-zero parts. Further assume that $\mu \le \nu_0 = (3^{s_0}, 2^{s_1-s_0}, 1^{s_2-s_1})$. By Proposition 3.1, we have

$$\begin{cases} t_1 \le s_0, \\ 2t_1 + t_2 \le s_0 + s_1, \\ 3t_1 + 2t_2 + t_3 = s_0 + s_1 + s_2. \end{cases}$$

Since the number of non-zero parts of μ is s_2, we also have $t_1 + t_2 + t_3 = s_2$. These four equations are equivalent to:

$$\begin{cases} t_1 \leq s_0, \\ 2t_1 + t_2 = s_0 + s_1, \\ t_1 + t_2 + t_3 = s_2. \end{cases}$$

Now let $a = s_0 - t_1 \geq 0$. Since $s_0 - a = t_1 \geq 0$, we have $a \leq s_0$. Also $a = s_0 - t_1 = (2t_1 + t_2 - s_1) - t_1 = t_1 + t_2 - s_1 = s_2 - t_3 - s_1 \leq s_2 - s_1$. Hence $0 \leq a \leq f$.

Now, $t_1 = s_0 - a$, $t_2 = s_0 + s_1 - 2t_1 = s_1 - s_0 + 2a$, and $t_3 = s_2 - t_1 - t_2 = s_2 - s_1 - a$. Hence

$$\mu = (3^{s_0 - a}, 2^{s_1 - s_0 + 2a}, 1^{s_2 - s_1 - a}) = \nu_a.$$

It is clear that the partitions ν_a form a chain, $\nu_0 \geq \nu_1 \geq \cdots \geq \nu_f$ in the dominance order. □

We are now ready to complete the proof of our Main Theorem 1.6. We adapt and generalize the proof in Hsu, Logan, and Shahriari [22] of the generalized Füredi conjecture for rank-symmetric rank-unimodal normalized matching posets of rank 2.

Let P be a poset and $\mu = (\mu_1, \ldots, \mu_m)$ a partition of $|P|$. Assume $\mathcal{C} = \{C_1, \ldots, C_m\}$ is a collection of m chains that partition P. We say that $\mathcal{C}$ *realizes* μ if, for $1 \leq i \leq m$, $|C_i| = \mu_i$.

Proof of Theorem 1.6. We have already proved the "only if" direction of the Theorem, so let P be a normalized matching poset of rank 2 with rank numbers $s_0 \leq s_1 \leq s_2$. Let σ be the NCD partition of $|P|$. Let $\mu = (3^{t_1}, 2^{t_2}, 1^{t_3})$ be a partition of $|P|$ with $\mu \leq \sigma$. We have to construct a chain partition of P that realizes μ.

We first show that it is enough to give a construction in the case when the number of positive parts of μ is s_2, the width of P. In other words, without loss of generality we can focus on chain partitions with the least possible number of chains.

Let $d = (s_0 + s_1) - (2t_1 + t_2)$. By Proposition 3.1, we have $2t_1 + t_2 \leq s_0 + s_1$, and so $d \geq 0$.

Now, let $\mu' = (3^{t_1}, 2^{t_2+d}, 1^{t_3-2d})$. By checking Proposition 3.1, we see that μ' is a partition of $|P|$ with $\mu' \leq \sigma$. Now the number of positive parts in μ' is $t_1 + t_2 + t_3 - d = 3t_1 + 2t_2 + t_3 - s_0 - s_1 = s_2$. Assume that we have constructed a chain partition $\mathcal{C}$ of P that realizes μ'. Now take d of the chains of size 2 in $\mathcal{C}$ and turn them into $2d$ chains of size 1. This gives a chain partition of P that realizes μ.

Hence with no loss of generality, we can assume $t_1 + t_2 + t_3 = s_2$. By Proposition 3.2, for some $0 \leq a \leq f = \min(s_0, s_2 - s_1)$, we have

$$\mu = \nu_a = (3^{s_0 - a}, 2^{s_1 - s_0 + 2a}, 1^{s_2 - s_1 - a}).$$

If $a = 0$, then $\nu_0 = \sigma$, the NCD partition of $|P|$. By Corollary 1.8, P has a nested chain decomposition, and this chain partition realizes ν_0. We now assume that for some $0 \leq a < f$, we have a chain partition $\mathcal{C}$ of P that realizes ν_a. The proof will be complete when we construct a new chain partition $\mathcal{C}'$ of P that realizes ν_{a+1}.

Comparing ν_a and ν_{a+1}, we see that, in $\mathcal{C}$, we have to replace a chain of size 3 and a chain of size 1 with two chains of size 2.

Elements of P are partitioned into three levels. These levels have sizes s_0, s_1, and s_2 (a'priori we do not know which one is the first level, etc.). Following Hsu, *et al.* [22], we refer to the s_2 elements that form the largest level as *children*. Likewise, the s_1 and s_0 elements that form the other two levels will be referred to, respectively, as (biological) *mothers* and *fathers*.

Since the number of chains in $\mathcal{C}$ is s_2, every chain contains exactly one child. A chain of size 3 in $\mathcal{C}$, assigns a (biological) father and mother to a child. A chain of size 2 assigns either a (biological) father or mother to a child. Children on chains of size 0 are *orphans*. Note that, while some children are orphans, each parent has exactly one (biological) child.

Given the chain sizes in $\mathcal{C}$ and the rank numbers of P, we see that $s_0 - a$ children have two (biological) parents, a children have a (biological) father, $s_1 - s_0 - a$ children have a (biological) mother, and $s_2 - s_1 - a$ children are orphans. To construct $\mathcal{C}'$, we need to take one parent from a two-parent family and assign them to an orphan.

Let X be the set of parents (fathers and mothers) and Y be the set of children. By Corollary 2.5, there is a matching from the smaller of X and Y to the other. We will refer to this new matching between parents and children as the *foster* matching.

Case 1. $s_0 + s_1 \geq s_2$ (i.e., there are at least as many parents as children).

In this case, the foster matching assures that every child has exactly one (foster) parent. Take an arbitrary orphan (i.e., an element on a chain of size 0 in $\mathcal{C}$) and match it with the child's foster parent to create a (new) chain of size 2. The parent will be taken away from their biological family. If, in $\mathcal{C}$, this parent was a member of a two parent family, then we have reconfigured a chain of size 3 and a singleton into two chains of size 2 and we are done. However, if the parent was a single parent, we have now created a new orphan. In this case, we go to another orphan (could be the newly created one) and repeat the process. Note that any orphan who has been given a foster parent will never be orphaned again and hence the process will have to stop, and this happens when we get a foster parent from a two parent family.

Case 2. $s_0 + s_1 \leq s_2$ (i.e., there are at least as many children as parents).

In this case, the foster matching assures that every parent has exactly one (foster) child, while exactly $s_2 - s_1 - s_0$ children will not have any foster parents. Since $a < f = \min(s_0, s_2 - s_1) = s_0$, we have $s_2 - s_1 - s_0 < s_2 - s_1 - a$. So there are more orphans than children without foster parents. Choose an orphan who has a foster parent and proceed as in Case 1. Note that unless the process stops (by matching the orphan with their foster parent who is coming from a two parent family) the number of orphans and children without foster parents stays constant, and so we can always find an orphan who has a foster parent to repeat the step.

Since, for $0 \leq a < f$, given a partition $\mathcal{C}$ that realizes ν_a, we can always construct a partition $\mathcal{C}'$ that realizes ν_{a+1}, the proof is complete. □

References

[1] Ian Anderson, Some problems in combinatorial number theory, Ph.D. thesis, University of Nottingham, Nottingham, United Kingdom (1967).

[2] Ian Anderson, Combinatorics of finite sets, Dover Publications, Mineola, NY, 2002, Corrected reprint of the 1989 edition published by Oxford University Press, Oxford.

[3] B. Bollobás, On generalized graphs, *Acta Math. Acad. Sci. Hungar*, **16** (1965) 447–452.

[4] N. G. de Bruijn, Ca. van Ebbenhorst Tengbergen and D. Kruyswijk, On the set of divisors of a number, *Nieuw Arch. Wiskunde (2)*, **23** (1951) 191–193.

[5] Muktar Elzobi and Zbigniew Lonc, On partition of a Boolean lattice into chains of equal sizes, *J. Combin. Math. Combin. Comput.*, **44** (2003) 237–249.

[6] Konrad Engel, Sperner theory, Encyclopedia of Mathematics, Cambridge University Press, Cambridge (1997).

[7] Elinor Gardner Escamilla, Andreea Cristina Nicolae, Paul Russell Salerno, Shahriar Shahriari, and Jordan Olliver Tirrell, On nested chain decompositions of normalized matching posets of rank 3, to appear in Order.

[8] Rae Fredrich, Tim Hsu, Mark J. Logan and Katherine Nolan, Matchwebs, preprint.

[9] Zoltan Füredi, Problem session, Kombinatorik geordneter Mengen, Oberwolfach, B. R. D. (1985).

[10] E. R. Gansner, On the lattice of order ideals of an up-down poset, *Discrete Mathematics*, **39** (1982) 113–122.

[11] R. L. Graham and L. H. Harper, Some results on matching in bipartite graphs, *SIAM J. Appl. Math.*, **17** (1969) 1017–1022.

[12] Curtis Greene and Daniel J. Kleitman, The structure of Sperner k-families, *Journal of Combinatorial Theory Series A*, **20** (1976) 80–88.

[13] Curtis Greene and Daniel J. Kleitman, Proof techniques in the theory of finite sets, Studies in Combinatorics (Gian-Carlo Rota, ed.), MAA studies in Mathematics, vol. 17, The Mathematical Association of America (1978) pp. 22–79.

[14] Jerrold R. Griggs, Sufficient conditions for a symmetric chain order, *Siam Journal of applied Math.*, **32** (1977) 807–9.

[15] Jerrold R. Griggs, Symmetric chain orders, Sperner theorems, and loop matchings, Ph.D. thesis, M. I. T. (1977).

[16] Jerrold R. Griggs, On chains and Sperner k-families in ranked posets, *Journal of Combinatorial Theory Series A*, **28** (1980) 156–168.

[17] Jerrold R. Griggs, Problems on chain partitions, *Discrete Mathematics*, **72** (1988) 157–162.

[18] Jerrold R. Griggs, C. M. Grinstead and R. K. C. Yeh, Partitioning Boolean lattices into chains of subsets, *Order*, **4** (1987) 65–67.

[19] Tim Hsu, Mark Logan and Shahriar Shahriari, Methods for nesting rank 3 normalized matching rank-unimodal posets, *Discrete Math.*, **309** (2009) 521–531.

[20] Tim Hsu, Mark Logan, Shahriar Shahriari and Christopher Towse, Partitioning the Boolean lattice into chains of large minimum size, *Journal of Combinatorial Theory A*, **97** (2002) no. 1, 62–84.

[21] Tim Hsu, Mark Logan, Partitioning the Boolean lattice into a minimal number of chains of relatively uniform size, *European Journal of Combinatorics*, **24** (2003) 219–228.

[22] Tim Hsu, Mark J. Logan and Shahriar Shahriari, The generalized Füredi conjecture holds for finite linear lattices, *Discrete Math.*, **306** (2006) no. 23, 3140–3144.

[23] D. J. Kleitman, On an extremal property of antichains in partial orders. The LYM property and some of its implications and applications, Combinatorics (Proc. NATO Advanced Study Inst., Breukelen, 1974), Part 2: Graph theory; foundations, partitions and combinatorial geometry, Math. Centrum, Amsterdam, 1974, pp. 77–90. Math. Centre Tracts, No. 56.

[24] Zbigniew Lonc, Proof of a conjecture on partitions of a Boolean lattice, *Order*, **8** (1991) 17–27.

[25] Zbigniew Lonc and Muktar E. Elzobi, Chain partitions of products of two chains, *Journal of Combinatorial Theory Series A*, **86** (1999) 140–150.

[26] D. Lubell, A short proof of Sperner's lemma, *J. Combinatorial Theory*, **1** (1966) 299.

[27] L. D. Mešalkin, A generalization of Sperner's theorem on the number of subsets of a finite set, *Teor. Verojatnost. i Primenen*, **8** (1963) 219–220.

[28] Hazel Perfect,Addendum to: "A short proof of the existence of k-saturated partitions of partially ordered sets" [*Adv. in Math.*, **33** (1979), no. 3, 207–211; MR0546293 (82c:06008)] by M. Saks, *Glasgow Math. J.*, **25** (1984) no. 1, 31–33.

[29] Michael Saks, A short proof of the existence of k-saturated partitions of partially ordered sets, *Adv. in Math.*, **33** (1979) no. 3 207–211.

[30] B. Sands, Problem session, Colloquium on Ordered Sets, Szeged, Hungary (1985).

[31] Yi Wang, Nested chain partitions of LYM posets, *Discrete Appl. Math.*, **145** (2005) no. 3, 493–497.

[32] Koichi Yamamoto, Logarithmic order of free distributive lattice, *J. Math. Soc. Japan*, **6** (1954) 343–353.

Proc. ICDM 2008, RMS-Lecture Notes Series No. 13, 2010, pp. 33–44.

On 1-Well-Covered Graphs

Arthur S. Finbow and Bert L. Hartnell

Department of Mathematics and Computing Science, Saint Mary's University, Halifax, Canada B3H 3C3
e-mail: art.finbow@stmarys.ca; bert.hartnell@stmarys.ca

Abstract. We begin with a brief survey of what is known about the structure of well-covered graphs. The theme of removing vertices without destroying well-coveredness leads us to the case of 1-well-covered graphs. A study of triangle free 1-well-covered graphs is then undertaken and a surprising structural characterization is established: that such graphs are precisely the triangle free graphs in which no vertex v satisfies $\{v\} = V(G) - N[S]$ for an independent set $S \subseteq V(G)$. It is also shown that each of their maximal independent sets is *locating* in the sense of Slater [50]. Finally, several methods for constructing such graphs are presented, which simplify and extend previous work.

Keywords. Well-covered graph, 1-well-covered, W_2 graph, triangle free graph, maximal independent set, locating set.

2000 Subject Classification: 05C69, 05C85

1. Introduction

The *independence number* $\alpha(G)$ of a simple loopless graph $G = (V, E)$ is the maximum cardinality of an independent set in $V(G)$. A graph G is said to be *well-covered* provided every maximal independent set $I \subseteq V(G)$ satisfies $|I| = \alpha(G)$. These graphs were introduced by Plummer [41] in 1970. Although the recognition problem of well-covered graphs in general is co-NP-complete [11,48], it is polynomial for certain classes of graphs. For instance, well-covered graphs that are claw-free [53], have girth at least 5 [20], have neither 4-cycles nor 5-cycles (but 3-cycles are permitted) [21], are chordal [43] or are of bounded degree [10] are all recognizable in polynomial time (for further complexity issues see [8,9,53,54]). The reader is referred to Plummer [42] and to Hartnell [29] for their surveys of the work on well-covered graphs up to near the turn of the century.

More recent developments have included some further studies of planar graphs. The complexity of the recognition problem for planar well-covered graphs is unknown but in any case is fraught with difficulties. Campbell and Plummer [6] characterized all the 3-connected, planar, cubic, well-covered graphs (see Section 3). Later King [32,33] was, with considerable effort, able to characterize the 3-connected, planar, claw-free, well-covered graphs. Finbow, Hartnell, Nowakowski and Plummer [23–25] have characterized the well-covered planar triangulations that are 4-connected.

Research supported by the authors' NSERC Discovery Grants.

In a different direction, Dean and Zito [13] characterized perfect well-covered graphs with no induced 4-cycles. The problem of determining if a graph with no 4-cycle (induced or not) is well-covered had been studied in [21,27,30,56]; and has been recently revisited by Brown et al [4]. In a similar vein, Randerath and Vestergaard [46] have characterized the well-covered graphs that contain no 3-, 5- nor 7-cycles.

Additional structural results include the work of Randerath and Vestergaard [47] regarding [1,2]-factors in well-covered graphs, and Yamashita and Kameda's characterization of well-covered graphs G with $\alpha(G) = 2$ [57]. Ramey's 1994 work characterizing minimal non-well-covered graphs [44], was adapted fruitfully in various directions in [9,10,54] (see also Zverovich [58]). Investigations of well-covered Cartesian products, begun in [55] by Topp and Volkmann, have recently been extended by Fradkin [26].

There has been a relatively large number of papers written on both special properties of well-covered graphs and on classes of graphs that are related in some way to the class of well-covered graphs. The following list of references is meant to be a starting point for the interested reader but is in no way complete. [1–3,8,12,14,16–19,22,31,37,39,49].

Turning to matters notational, for a vertex x in G, the *open neighborhood of* x is the set $N_G(x) = \{v \in V(G) | vx \in E(G)\}$. The *closed neighborhood of* x is the set $N_G[x] = N_G(x) \cup \{x\}$, and if $S \subseteq V(G)$ then $N_G[S] = \cup\{N_G[x] | x \in S\}$. If $S \subseteq V(G)$, then $v \in V(G)$ is said to be a *private neighbour of* $x \in V(G)$ *with respect to* S, provided that $vx \in E(G)$ and $v \notin N_G[S - x]$. The graph H is called *complete bipartite with parts* X *and* Y provided $V(H) = X \cup Y$ and $E(H) = \{xy | x \in X$ and $y \in Y\}$.

2. Deleting vertices from well-covered graphs

In the study of well-covered graphs, it is often the case that the effect of deleting a set of vertices from the graph is an important consideration. In this section we discuss the cases in which such a deletion leaves a well-covered graph.

Definition 1. *Let G be a well-covered graph, n a non-negative integer, and let $S \subseteq V(G)$. Then we say that S* is an S_n set in G *provided that the subgraph H induced by $G - S$ is well-covered, and $\alpha(G) = \alpha(H) + n$. If S is an S_n set in G, we say that $G(S)$* is an S_n subgraph of G.

Observation 1. *Let G be a well-covered graph with $\alpha(G) = m$. Then the only S_m subgraph of G is G itself. We further observe that every graph H is an induced S_0 subgraph of its corona F. (That F is well-covered was shown in [15].)*

Clearly then, S_n sets can have arbitrarily large independence numbers. On the other hand we have:

Lemma 2. *Let H be an induced S_n subgraph of a well-covered graph G. Then $\alpha(H) \geq n$.*

The following characterization is immediate:

Proposition 3. *Let G be a well-covered graph with $\alpha(G) > n$ and with induced subgraph H. Then H is an S_n subgraph of G if and only if for each maximal independent subset K of $G - H$, $G - N_G[K]$ induces in H a well-covered graph with independence number n.*

Observation 4. *In Proposition 3, when $n = 0$, this is equivalent to the condition: each maximal independent subset K of $G - H$ satisfies $G - N_G[K] = \emptyset$.*

When $n = 1$, this is equivalent to the condition: for each maximal independent subset K of $G - H$, $G - N_G[K]$ induces in H a nonempty complete graph.

Some version of the next result has been used in virtually every paper exploring well-covered graphs.

Lemma 5 ([5,6,20]). *If G is a well-covered graph and I is an independent set of m vertices in G, then $N[I]$ is an S_m set in G.*

Singleton S_0 sets often play a special role in the theory, and will be of primary importance in the latter sections of this paper.

Definition 2 ([20]). *A vertex v in a well-covered graph G is said to be* extendable in *G if $\{v\}$ is an S_0 set in G.*

Definition 3. *Let G be any graph (not necessarily well-covered). We say that $v \in V(G)$ is* isolatable in *G provided there exists an independent set of vertices K in $V(G)$ such that $G - N_G[K] = \{v\}$. We say that v* is isolated in *G* by *K.*

Lemma 6 ([20]). *Let G be a well-covered graph. Then $v \in V(G)$ is an extendable vertex in G if and only if v is not isolatable in G.*

Proposition 7. *Let G be a well-covered graph and let S be an independent set of extendable vertices in $V(G)$. Then S is an S_0 set.*

Proof. Assume for a contradiction that H, the subgraph induced by $G - S$, has a maximal independent set K of cardinality strictly less than $\alpha(G)$. Note that since G is well-covered, $N_G[K] \not\supseteq S$. Choose $v \in S - N_G[K]$ and set $J = (K \cup (S - N_G[K])) - \{v\}$ and observe that J isolates v in violation of Lemma 6. ■

3. 1-Well-covered graphs

A well-covered graph G is *1-well-covered* if every $v \in V(G)$ is extendable in G. These graphs were introduced by Staples [51,52] where she showed that they were the same as W_2 *graphs* – graphs in which every pair of disjoint independent sets can be extended to a pair of disjoint maximal independent sets. This class was later investigated in more detail by Pinter [35,36,38,40]. Using the characterization of cubic well-covered graphs [7] and girth 5 well-covered graphs [20], Pinter was able to determine the 1-well-covered ones in each class. He then extended this work to 4-regular 1-well-covered if these graphs were also planar and 3-connected [36], and to the planar

1-well-covered graphs of girth 4 [38]. In [40] he also provides several constructions to obtain infinite families of 1-well-covered graphs of girth 4.

More recently, Caro [8] has shown that the recognition problem for 1-well-covered graphs is NP-complete; Hartnell has characterized the 1-well-covered graphs containing no 4-cycles [28]; and King [34] has shown that a graph is 1-well-covered if and only if it is *stable well-covered.* (A graph G is *stable well-covered* provided that G is well-covered and the possibly new graph formed by the addition of any edge uv, where u and v are in $V(G)$ is also well-covered.)

Our primary objective in this section is to obtain some results regarding S_n subsets in 1-well-covered graphs. To begin with, note that Proposition 3, together with the fact that no vertex in a 1-well-covered graph is isolatable imply:

Proposition 8. *Let G be a 1-well-covered graph and S be an S_n subset of $V(G)$ whose induced graph is complete bipartite. Then $n \leq 1$.*

We also observe that Proposition 7 implies:

Proposition 9. *Every independent set of vertices in a 1-well-covered graph is an S_0 set.*

In general, if S is an S_0 set then we may not conclude that $G(G - S)$ is 1-well-covered. Indeed C_5 is 1-well-covered, but the removal of an independent set of cardinality 2 yields a well-covered graph with an isolated vertex, which is thus not 1-well-covered. There are, however, conditions under which 1-well-coveredness is preserved. The next three results are of this character.

Theorem 10. *Let G be a 1-well-covered graph, $n \geq m \geq 1$ be integers, and let S be a subset of $V(G)$ whose induced graph, F, satisfies $\alpha(F) = m$. If S is an S_n set, and if H is the subgraph induced by $G - S$ then H is a 1-well-covered graph.*

Proof. First observe that by Lemma 2, we have $n = m$. Assume for a contradiction that H is not a 1-well-covered graph, and let $v \in V(H)$ be isolatable in H. Let K be an independent set of vertices in H such that $H - N_H[K] = \{v\}$. Now v cannot be isolated in G and thus v is adjacent to at least one vertex, say u, in $S - N_G[K]$. Let L be a maximal independent set in $S - N_G[K]$ that contains u. Set $J = K \cup \{v\}$ and observe that J is a maximal independent set in H and thence has cardinality $\alpha(H)$. On the other hand, $M = K \cup L$ is a maximal independent set in G and hence is of cardinality $\alpha(G)$. But then $\alpha(G) = |M| = |K| + |L| \leq |J| - 1 + m = \alpha(H) + m - 1 \leq \alpha(G) - 1$, a contradiction. ■

Corollary 11. *Let G be a 1-well-covered graph and let S be an S_1 subset of $V(G)$ whose induced graph is complete. Then the subgraph induced by $G - S$ is a 1-well-covered graph.*

The next lemma provides many examples of S_n subsets F of 1-well-covered graphs whose removal preserves 1-well-coveredness, without the requirement that $\alpha(F) = m$.

Lemma 12 ([35,36]). *If G is a 1-well-covered graph and I is an independent set of vertices in G, then $G - N[I]$ is also a 1-well-covered graph.*

4. Triangle free 1-well-covered graphs

We begin this section with a new characterization of triangle free, 1-well-covered graphs in terms of their local structure. Using this, we extend our results regarding the deletion of subgraphs to the triangle free setting. We end this section by showing that maximal independent sets in triangle free 1-well-covered graphs are *locating* in the sense of Slater [50].

Theorem 13. *Let G be any triangle free graph. Then G is a 1-well-covered graph if and only if no $v \in V(G)$ is isolatable in G.*

Proof. If G is a 1-well-covered graph then no $v \in V(G)$ is isolatable in G by Lemma 6. Conversely, suppose that no vertex is isolatable in G. Then it suffices to show that G is well-covered.

Assume for a contradiction that G is not a well-covered graph, and let L be a maximum independent set of vertices in G. Then there exists a maximal independent set K such that

(1) $|K| < |L|$ and
(2) $n = |K - L|$ is minimum for independent sets K satisfying (1).

Note that $K \subseteq L$ would violate the assumption that K is a maximal independent set in G and hence there is a vertex v in $V(G)$ such that

(3) $v \in K - L$.

Now fix such a v and let K_0 and L_0 be maximal independent sets of vertices in G satisfying (1), (2), and (3) such that $m = |(L_0 \cap N_G[v])|$ is minimum. Maximality of L_0 guarantees that m is nonzero. On the other hand, if $u \in L_0 \cap N_G[v]$, and $J = L_0 - \{u\}$, then $G - N_G[J]$ contains a vertex other than u, say x. But then $J \cup \{x\}$ is a maximum independent set that either intersects $N_G[v]$ in $m - 1$ points or contains $n + 1$ points of K dependent upon whether $x = v$. ■

In contrast we have:

Lemma 14. *Let G be a triangle free, 1-well-covered graph. Then each induced subgraph H of G with $H \cong K_2$ is* isolatable *in the sense that there is an independent set $J \subseteq V(G)$ such that $H = G - N_G[J]$.*

Proof. Suppose $V(H) = \{x, y\}$. If $G = H$, then $J = \emptyset$ satisfies the conclusion. Thus we may assume that $G \neq H$.

We observe that $G - N_G[H] = \emptyset$ cannot be true for otherwise, without loss of generality, x would be adjacent to a vertex $u \neq y$ and since G is triangle free, the only neighbours of u would be adjacent to y. But then u is isolated in G by {y} contrary to Theorem 13.

Let J be a maximal independent set in $G - N_G[H]$. Then $G - N_G[J]$ contains no neighbours of x, for if u were a neighbour of x in $G - N_G[J]$ then, since G is triangle free, $J \cup \{y\}$ would isolate u in G. Similarly $G - N_G[J]$ contains no neighbours of y and hence $G - N_G[J] = H$ as required. ■

Next we present a result regarding the deletion of induced bipartite subgraphs that will be useful in the next section.

Proposition 15. *Let G be a triangle free, 1-well-covered graph and let S be a subset of $V(G)$ whose induced graph is complete bipartite with parts X and Y. Then the following are equivalent:*

(a) *S is an S_n set in G for some n.*
(b) *S is an S_1 set in G.*
(c) *$G(G-S)$ is a 1-well-covered graph.*
(d) *$G(G-S)$ is a well-covered graph.*
(e) *No maximal independent subset K of $G-S$ satisfies $G - N_G[K] = \emptyset$.*
(f) *Each maximal independent subset K of $G-S$ satisfies $G - N_G[K] \cong K_2$.*
(g) *No independent subset K of $G-S$ satisfies $X - N_G[K] = \emptyset$.*

Proof.
(a) implies (b): Observe that by Proposition 8, $n \leq 1$ and that Lemma 14 in combination with Observation 4 shows that $n \neq 0$.
(b) implies (c): By Theorem 13, it suffices to show that no vertex in $H = G(G-S)$ is isolatable in H. Assume for a contradiction that $v \in V(H)$ is isolatable in H. Let K be an independent set of vertices in H such that $H - N_H[K] = \{v\}$. Since v cannot be isolated in G, v must be adjacent to at least one element of S, and since G is triangle free, v cannot have neighbours in both X and Y. Assume without loss of generality that $N_G(v) \cap X \neq \emptyset$.

Now observe that $Y \subseteq N_G[K]$, since otherwise, $K \cup (Y - N_G[K])$ would be independent in G and isolate v, violating the fact that G is 1-well-covered. Next, set $J = K \cup \{v\}$ and observe that J is a maximal independent set in H and thence has cardinality $\alpha(H)$. On the other hand, $N_G[J] \supseteq H \cup Y$ and since $|J| < \alpha(G)$, there is at least one vertex x in $X - N_G[J]$. But then the independent set $(J \cup (X - N_G[J])) - \{x\}$ isolates x in G, a contradiction.
(c) implies (d) is clear.
(d) implies (e): Note that Lemma 14 implies that $\alpha(G-S) = \alpha(G) - 1$ and hence (e) follows.
(e) implies (f): Let K be a maximum independent set in $G-S$. Suppose x_1 and x_2 are in $X \cap (N_G[K])$. Then $K \cup \{x_1\}$ isolates x_2 in G contradicting the 1-well-coveredness of G. Hence $G - N_G[K]$ can contain at most one element of X and similarly can contain at most one element of Y. But since $G - N_G[K] \neq \emptyset$ and $|G - N_G[K]| \neq 1$, we must conclude that $G - N_G[K] \cong K_2$.
(f) implies (g) is clear.
(g) implies (a): Let K be any maximal independent set in $G-S$. Then there is a vertex $x \in X - N_G(K)$. If there were another vertex $x_1 \in X - N_G(K)$ with $x_1 \neq x$, then $K \cup \{x_1\}$ would isolate x in G, a contradiction. But then by 1-well-coveredness there would exist a vertex $y \in Y - N_G[K]$, which similarly must be the only vertex in $Y - N_G[K]$. But now $K \cup \{x\}$ is a maximal independent set in G (a well-covered graph) and thus $|K| = \alpha(G) - 1$. Since K was arbitrary, it follows that $G - S$ is well-covered and thence that S is an S_1 set in G. ■

The notion of a locating dominating set was introduced by Slater [50]. He calls a dominating set D of a graph G *locating* if, for every pair of vertices u and v of G, not in D, $N_G(u) \cap D \neq N_G(v) \cap D$.

Definition 4 ([17]). *A graph in which every maximal independent set is a locating dominating set is called* an EIDL graph.

In [17], it was shown that if G is a well-covered graph of girth 5 or more, then G is an EIDL graph. The graph C_4 shows that the girth restriction is sharp. In our next theorem, we show that when G is also 1-well-covered the girth condition can be tightened to 4.

Theorem 16. *If G is a triangle free, 1-well-covered graph then G is an EIDL graph.*

Proof. Assume for a contradiction that G is not an EIDL graph, and let L be a maximal independent set of vertices in G such that there are vertices $u \in V(G) - L$ and $v \in V(G) - L$ satisfying $N_G(u) \cap L = N_G(v) \cap L$. Then the independent set $K = (L - N_G(u)) \cup \{u\}$ isolates v, for otherwise there would be a neighbour x of v with $x \in N_G(L)$, $x \notin N_G(K)$. Then there would be a neighbour y of x with $y \in L$, $y \notin K$ and hence $y \in N_G(u) \cap N_G(v)$. But this would in turn imply that $\{x, y, v\}$ induces a triangle, a contradiction. ■

Remark 1. The converse is false; C_7 for example is an EIDL graph which is not 1-well-covered. (Also see [17])

5. Constructions for triangle free 1-well-covered graphs

In this section we investigate methods of constructing new 1-well-covered graphs from previously known examples. These generalize and extend the constructions described by Pinter in [40]. Our first construction is justified by Proposition 15 (c).

Theorem 17 (Transformation 1). *Suppose that G is a triangle free, 1-well-covered graph, and let S be an S_1 subset of $V(G)$ whose induced graph is complete bipartite. Let H be the subgraph of G induced by $G - S$. Then H is a triangle free, 1-well-covered graph with $\alpha(H) = \alpha(G) - 1$.*

Remark 2. We note that Lemma 12 with $m = 1$, yields a transformation of this type, but that transforms *any* 1-well-covered graph into another 1-well-covered graph.

Theorem 18 (Transformation 2). *Suppose that G is a triangle free, 1-well-covered graph, and let S be an S_1 subset of $V(G)$ whose induced graph is complete bipartite with parts X and Y. Let H be the graph with $V(H) = V(G) \cup \{u, v, w\}$ (where none of u, v, nor w are in $V(G)$) and with $E(H) = E(G) \cup \{uv, uw\} \cup \{xv | x \in X\} \cup \{yw | y \in Y\}$. Then H is a triangle free 1-well-covered graph with $\alpha(H) = \alpha(G) + 1$.*

Proof. Clearly H is triangle free. To see that H is 1-well-covered, it suffices, by Theorem 13, to show that no vertex in H can be isolated. Note first that u cannot be isolated since any such isolating set K must include an element of both X and Y, which would make K non-independent. Next, neither v nor w can be isolated. Indeed

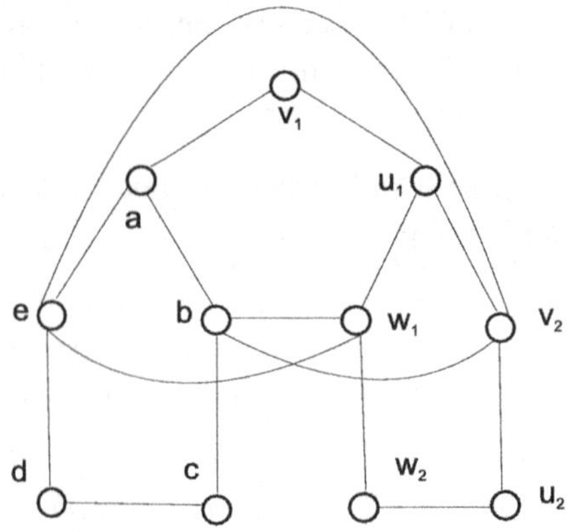

Figure 1.

if w, for example, were isolated by the independent set K, then $v \in K$ (in order to cover u) and $X \cap K = Y \cap K$. But then $K - \{v\}$ would be a maximal independent subset of $G - S$ and $G - N_G[K - \{v\}] = \emptyset$ contradicting Proposition 15 (e). Finally, we show that a vertex in G cannot be isolated in H. Suppose for a contradiction that K isolates a vertex in G. Then K cannot contain u since $K - \{u\}$ would be an isolating set in G. Similarly, K cannot contain both v and w since $K - \{v, w\}$ would be an isolating set in the subgraph of G induced by $G - S$, contradicting Proposition 15 (c). On the other hand, K must include one of v or w (in order that $u \in N_H[K]$). Without loss of generality, assume that $w \in K$, and note that since u is not in K then there must be some $x \in X$ which is also in K (in order to cover v). Then we have $H - N_H[K] = G - N_G[K - \{w\}]$ and hence $K - \{w\}$ would isolate a vertex in G, a contradiction. Thus no vertex in H can be isolated and thence H is a 1-well-covered graph.

To complete the proof, it suffices to exhibit a maximal independent subset K of H with $|K| = \alpha(G) + 1$. Choose any maximal independent subset L of G and set $K = L \cup \{u\}$. ■

Remark 3. After the above construction, S is not an S_1 subset of $V(H)$ in H: indeed $H - S$ has a non well-covered component generated by $\{u, v, w\}$.

We observe that Transformation 2, applied twice to the 5-cycle *abcde*, yields the graph shown in Figure 1 which was not obtainable by previous constructions (see [35,40]).

We also observe that Transformation 2 gives Construction 1 of [38] as well as Construction 1 of [40] (just apply it twice).

Theorem 19 (Transformation 3). *Suppose that G is a triangle free, 1-well-covered graph, and let F be an induced subgraph of G which is both S_1 and complete bipartite with parts X and Y. Let H be the graph with $V(H) = V(G) \cup \{w, v\}$ (where neither w nor v is in $V(G)$) and with $E(H) = (E(G) - E(G(X \cup Y))) \cup \{xv | x \in X\} \cup \{yw | y \in Y\}$. Then H is a triangle free 1-well-covered graph such that $\alpha(H) = \alpha(G) + 1$ and $d_H(v, w) \geq 5$.*

Proof. It is easy to see that $d_H(v, w) \geq 5$ and that H is triangle free. Hence to show that H is a 1-well-covered graph, it suffices, by Theorem 13, to show that no vertex in H can be isolated. First, we note that neither w nor v can be isolated, for if, for instance, K isolates v, then $K - (Y \cup \{w\})$ must be an independent set in G such that $X \subseteq N_G[K - (Y \cup \{w\})]$. This, however would violate Proposition 15 (g) since F is S_1. If a vertex $x \in X$ can be isolated in H by a set K, then by the nature of v, K must contain a vertex in X other than x. Hence, since no neighbour of x can be adjacent to a vertex in Y (G is triangle free), $K - (\{w\} \cup Y)$ must be an independent set in G that isolates x in G, an impossibility since G is 1-well-covered. Similarly, no $y \in Y$ can be isolated.

Lastly, we show that no $z \in G - F$ can be isolated. Assume to the contrary that K isolates z, and is maximal with respect to this property. Now neither v nor w can be in K: if both were, then $K - \{v, w\}$ would be an isolating independent set in $G - F$ which is 1-well-covered (Proposition 15 (c)); if exactly one, say w, were in K, then by the nature of v, K must contain a vertex in X and hence $K - \{w\}$ would be an isolating independent set in G which contradicts 1-well-coveredness.

Such a K then must include elements of both X and Y. Let $T_K = K \cap X$ and without loss of generality assume K is chosen so that $|T_K|$ is minimum. Since, as shown above, no vertex in X can be isolated in H, each vertex $t \in T_K$ must have at least one private neighbour p with respect to K. Moreover, one of these private neighbours must be a neighbour of z for otherwise, $K_p = (K - \{t\}) \cup \{p\}$ would be an independent set isolating z with $|T_{K_p}| < |T_K|$, violating minimality.

Let L be the set of private neighbours of elements of T_K in $N_H(z) = N_G(z)$. Then since H is triangle free, L is independent and, moreover, $J = (K - T_K) \cup L$ is an independent set in $G - X$ with $X \subseteq N_H(J)$. Since no vertex of Y can be adjacent to a neighbour of any vertex in X (G is triangle free), $M = J - Y$ is an independent set in $G - F$ with $X \subseteq N_H(M) = N_G(M)$ which violates Proposition 15 (g) since F is an S_1 set in G. Hence no vertex in H can be isolated and therefore H is 1-well-covered.

Finally, to complete the proof, observe that $H - N_H[\{v, w\}] = G - F$ and thus $\alpha(H) - 2 = \alpha(G) - 1$, since F is an S_1 set in G. ∎

This operation is reversible, since the converse of Theorem 19 is true:

Theorem 20 (Transformation 4). *Suppose that H is a triangle free, 1-well-covered graph, and let v and w be vertices in H with $d_H(v, w) \geq 5$. Let G be the graph with $V(G) = V(H) - \{v, w\}$ and with $E(G) = (E(H) - \{st | s \in \{v, w\}$ and $t \in V(H)\}) \cup \{xy | x \in N_H(v)$ and $y \in N_H(w)\}$. Then G, is a triangle free 1-well-covered graph with $\alpha(G) = \alpha(H) - 1$.*

Proof. By Theorem 13, to show that G is 1-well-covered it suffices to show that no vertex in G can be isolated by an independent set in G. Suppose for a contradiction that K is an independent set in G that isolates a vertex $u \in V(G)$. Now $N_H(v)$ and $N_H(w)$ are disjoint since $d_H(v, w) \geq 5$ and hence, since u can be in only one of them, $N_H[K]$ must contain either $N_H(v)$ or $N_H(w)$, say $N_H(v)$. Therefore, since $N_H[v]$ is an S_1 set in H (by Lemma 5), it follows by Proposition 15 (g) that K contains an element, say z of $N_H(v)$. But then, since $N_H(w) \subseteq N_G(z)$, it follows that both

$u \notin N_H(w)$ and (since K is independent in G), $K \cap N_H(w) = \emptyset$. But then, $\{u\} = G - N_G[K] = H - N_H[K \cup \{w\}]$. Since $K \cup \{w\}$ is independent, this contradicts the 1-well-coveredness of H, and hence G is 1-well-covered.

It remains to exhibit a maximal independent subset J of G with $|J| = \alpha(H) - 1$. Choose any vertex x in $N_H(v)$ and observe that since $N_H[v] \cap N_H[w] = \emptyset$, we can find a maximal independent subset L of H containing both x and w. Now since H is well-covered, $|L| = \alpha(H)$. But by the definition of G, $K = L - \{w\}$ is a maximal independent subset of G. ■

We note that two applications of Transformation 2 followed by an application of Transformation 4 yields Construction 2 given in [40].

References

[1] R. Barbosa and B. Hartnell, Characterizing Z_m-well-covered graphs for some classes of graphs, *Discrete Math.*, **233** (2001) 293–297.

[2] R. C. Brewster and G. MacGillivary, Minimizing $\beta + \Delta$ and well-covered graphs, *Ars Combin.*, **61** (2001) 197–210.

[3] J. I. Brown and R. J. Nowakowski, Well-covered vector spaces of graphs, *SIAM J. Discrete Math.*, **19** (2005) 952–965.

[4] J. I. Brown, R. J. Nowakowski and I. E. Zverovich, The structure of well-covered graphs with no cycles of length 4, *Discrete Math.*, **307** (2007) 2235–2245.

[5] S. R. Campbell, Some results on cubic well-covered graphs, *Ph.D. thesis, Vanderbilt University* (1987).

[6] S. R. Campbell and M. D. Plummer, On well-covered 3-polytopes, *Ars Combin.*, **25-A** (1988) 215–242.

[7] S. R. Campbell, M. N. Ellingham and G. F. Royle, A characterization of well-covered cubic graphs, *J. Combin. Math. Combin. Comput.*, **13** (1993) 193–212.

[8] Y. Caro, Subdivisions, parity and well-covered graphs, *J. Graph Theory*, **25** (1997) 85–94.

[9] Y. Caro, A. Sebö and M. Tarsi, Recognizing greedy structures, *J. Algorithms*, **20(1)** (1996) 137–156.

[10] Y. Caro, M. N. Ellingham and J. Ramey, Local structure when all maximal independent sets have equal weight, *SIAM J. Discrete Math.*, **11(4)** (1998) 644–654.

[11] V. Chvátal and P. J. Slater, A note on well-covered graphs, *Quo Vadis, Graph Theory?, Ann. Discrete Math.*, **55** (1993) 179–182.

[12] J. Currie and R. Nowakowski, A characterization of fractionally well-covered graphs, *Ars Combin.*, **31** (1991) 93–96.

[13] N. Dean and J. Zito, Well covered graphs and extendability, *Discrete Math.*, **126** (1994) 67–80.

[14] O. Favaron, Very well-covered graphs, *Discrete Math.*, **42** (1982) 177–187.

[15] A. Finbow and B. Hartnell, A game related to covering by stars, *Ars Combin.*, **16-A** (1983) 189–198.

[16] A. Finbow and B. Hartnell, A characterization of parity graphs containing no cycle of order five or less, *Ars Combin.*, **140** (1995) 227–234.

[17] A. Finbow and B. Hartnell, On locating dominating sets and well-covered graphs, *Congressus Numerantium*, **65** (1988) 191–200.

[18] A. Finbow and B. Hartnell, Well-located graphs: a collection of well-covered ones, *Discrete Math.*, **276** (2004) 201–209.

[19] A. Finbow, B. Hartnell and R. Nowakowski, Well-dominated graphs: A collection of well-covered ones, *Ars Combin.*, **25A** (1988) 5–10.

[20] A. Finbow, B. Hartnell and R. Nowakowski, A characterization of well-covered graphs of girth 5 or greater, *J. Combin. Theory Ser. B*, **57** (1993) 44–68.
[21] A. Finbow, B. Hartnell and R. Nowakowski, A characterization of well-covered graphs that contain neither 4- nor 5-cycles, *J. Graph Theory*, **18** (1994) 713–721.
[22] A. Finbow, B. Hartnell and C. Whitehead, A characterization of graphs of girth eight or more with exactly two sizes of maximal independent sets, *Discrete Math.*, **125** (1994) 153–167.
[23] A. Finbow, B. Hartnell, R. Nowakowski and M. Plummer, On well-covered triangulations: Part I, *Discrete Appl. Math.*, **132** (2004) 97–108.
[24] A. Finbow, B. Hartnell, R. Nowakowski and M. Plummer, On well-covered triangulations: Part II, *Discrete Appl. Math.*, **157** (2009) 2799–2817.
[25] A. Finbow, B. Hartnell, R. Nowakowski and M. Plummer, On well-covered triangulations: Part III, *Discrete Appl. Math.*, **158** (2010) 894–912.
[26] A. O. Fradkin, On the well-coveredness of Cartesian products of graphs, *Discrete Math.*, **309** (2009) 238–246.
[27] S. Gasquoine, B. Hartnell, R. Nowakowski and C. Whitehead, Techniques for constructing well-covered graphs with no 4-cycles, *J. Combin. Math. Combin. Comput.*, **17** (1995) 65–87.
[28] B. L. Hartnell, A characterization of the 1-well-covered graphs with no 4-cycles, *Graph theory in Paris, Trends Math.*, (2007) 219–224.
[29] B. Hartnell, Well-covered graphs, *J. Combin. Math. Combin. Comput.*, **29** (1999) 107–115.
[30] B. L. Hartnell, On the local structure of well-covered graphs without 4-cycles, *Ars Combin.*, **45** (1997) 77–86.
[31] V. E. Levit and E. Mandrescu, Independence polynomials and the unimodality conjecture for very well-covered, quasi-regularizable, and perfect graphs, *Graph theory in Paris, Trends Math.*, (2007) 243–254.
[32] E. L. C. King, Characterizing and Comparing Some Subclasses of Well-Covered Graphs, Ph.D. thesis, Vanderbilt University (2002).
[33] E. L. C. King, Characterizing a Subclass of Well-Covered Graphs, *Congressus Numerantium*, **160** (2003) 7–31.
[34] E. L. C. King, Stable Well-Covered Graphs, *Ars Combin.*, **93** (2009) 313–319.
[35] M. Pinter, W_2 graphs and strongly well-covered graphs: two well-covered graph subclasses, Ph.D. thesis, Vanderbilt University (1991).
[36] M. Pinter, Planar regular one-well-covered graphs, *Cong. Numer.*, **91** (1992) 159–187.
[37] M. Pinter, Strongly well-covered graphs, *Discrete Math.*, **132** (1994) 231–246.
[38] M. Pinter, A class of planar well-covered graphs with girth four, *J. Graph Theory*, **19** (1995) 69–81.
[39] M. Pinter, Planar strongly well-covered graphs, Vol. 2 *Graph Theory, Combinatorics and Applications*, **2** (1995) 869–879.
[40] M. Pinter, A class of well-covered graphs with girth four, *Ars Combin.*, **45** (1997) 241–255.
[41] M. D. Plummer, Some covering concepts in graphs, *J. Combin. Theory*, **8** (1970) 91–98.
[42] M. D. Plummer, Well-covered graphs: a survey, *Quaestiones Mathematicae*, **16** (1993) 253–287.
[43] E. Prisner, J. Topp and P. D. Vestergaard, Well-covered simplicial, chordal and circular arc graphs, *J. Graph Theory*, **21** (1996) 113–119.
[44] J. E. Ramey, Well-covered graphs with maximum degree three and minimal non-well-covered graphs, Ph.D. thesis, Vanderbilt University (1994).
[45] G. Ravindra, Well-covered graphs, *J. Combin. Inform. System Sci.*, **2** (1977) 20–21.
[46] B. Randerath and P. D. Vestergaard, On well-covered graphs of odd girth 7 or greater, *Discuss. Math. Graph Theory*, **22(1)** (2002) 159–172.

[47] B. Randerath and P. D. Vestergaard, Well-covered graphs and factors, *Discrete Appl. Math.*, **154** (2006) 1416–1428.

[48] R. S. Sankaranarayana and L. K. Stewart, Complexity results for well-covered graphs, *Networks*, **22** (1992) 247–262.

[49] R. S. Sankaranarayana and L. K. Stewart, Recursively decomposable well-covered graphs, *Discrete Math.*, **161** (1996) 243–263.

[50] P. J. Slater, Domination and location in acyclic graphs, *Networks*, **17** (1987) 55–64.

[51] J. Staples, On some subclasses of well-covered graphs, *Ph.D. thesis, Vanderbilt University* (1975).

[52] J. Staples, On some subclasses of well-covered graphs, *J. Graph Theory*, **3** (1979) 197–204.

[53] D. Tankus and M. Tarsi, Well-covered claw-free graphs, *J. Combin. Theory Ser. B*, **66** (1996) 293–302.

[54] D. Tankus and M. Tarsi, The structure of well-covered graphs and the complexity of their recognition problems, *J. Combin. Theory Ser. B*, **69** (1997) 230–233.

[55] J. Topp and L. Volkmann, On the well-coveredness of products of graphs and the complexity of their recognition, *Ars Combin.*, **33** (1992) 199–215.

[56] C. A. Whitehead, A characterization of well-covered claw-free graphs containing no 4-cycles, *Ars Combin.*, **39** (1995) 189–198.

[57] M. Yamashita and T. Kameda, Modeling K-Coteries by Well-Covered Graphs, *Networks*, **34** (1999) 221–228.

[58] I. E. Zverovich, A Characterization of Well-Covered Graphs in Terms of Prohibited Costable Subgraphs, *Mat. Zametki*, **67** (2000) 52–56; English translation in *Math. Notes*, **67** (2000) 41–44.

Proc. ICDM 2008, RMS-Lecture Notes Series No. 13, 2010, pp. 45–57.

New Challenges in the Theory of Hypergraph Coloring

Gábor Bacsó[1], Csilla Bujtás[2], Zsolt Tuza[1,2] and Vitaly Voloshin[3]

[1]Computer and Automation Institute, Hungarian Academy of Sciences,
H–1111 Budapest, Kende u. 13–17, Hungary
[2]Department of Computer Science and Systems Technology, University of Pannonia,
H–8200 Veszprém, Egyetem u. 10, Hungary
[3]Department of Mathematics and Physics, Troy University, Troy, AL 36082, USA
e-mail: bacso@sztaki.hu; bujtas@dcs.vein.hu; tuza@sztaki.hu; vvoloshin@troy.edu

Abstract. Several new fundamental approaches in hypergraph coloring appeared in the last 15 years. They are related to the various types of constraints that are imposed on the edges while coloring the vertices. A significant number of new ideas, results and publications have led to the situation where hypergraph coloring is taking a new shape. We survey some of the trends which in their turn give rise to new challenges in this extremely fast-developing area.

Keywords. Hypergraph coloring, mixed hypergraph, C-perfect hypergraph, uniquely colorable hypergraph, finite geometry, block design, color-bounded hypergraph, stably bounded hypergraph, partition-crossing hypergraph.

1. Introduction

One and a half decades ago, the fourth author introduced a new structure class called mixed hypergraphs [34,35]. This theory of hypergraph coloring offers many types of problems that do not exist in the traditional approach. Despite fast development, still there are lots of interesting fundamental problems in the subject. The questions are not very old, and we think there is hope that some of them can be solved in the near future. In this paper we survey some interesting directions in this promising area of research. For further information, we refer to the research monograph [36] and the regularly updated web site [37]. Some of the problems mentioned here were also collected in the earlier survey [32].

A *mixed hypergraph* is a triple $\mathcal{H} = (X, \mathcal{C}, \mathcal{D})$, where X is the *vertex set* and each of $\mathcal{C}$, $\mathcal{D}$ is a family of subsets of X, the *C-edges* and *D-edges*, respectively. A *proper k-coloring* of a mixed hypergraph is a mapping from the vertex set X into a set of *k colors* so that each C-edge has two vertices with a $\mathcal{C}$ommon color and each D-edge has two vertices with $\mathcal{D}$istinct colors. A mixed hypergraph is *k-colorable* if it has a proper coloring with at most k colors. A *strict k-coloring* is a proper k-coloring using all k colors.

We obtain classical hypergraph coloring in the special case when $\mathcal{H} = (X, \emptyset, \mathcal{D})$, which is denoted by $\mathcal{H}_{\mathcal{D}}$ and called a D-hypergraph. When $\mathcal{H} = (X, \mathcal{C}, \emptyset)$, we denote

Research supported in part by the Hungarian Scientific Research Fund, OTKA grant T-049613 and by Troy University Research Grant.

it by $\mathcal{H}_C$ and call it a C-hypergraph. A mixed hypergraph with $\mathcal{C} = \mathcal{D}$ is called *bi-hypergraph*. A hypergraph is called *r-uniform* if each of its edges has exactly r vertices.

The maximum number of colors in a strict coloring of $\mathcal{H} = (X, \mathcal{C}, \mathcal{D})$ is the *upper chromatic number* $\bar{\chi}(\mathcal{H})$; the minimum number is the *lower chromatic number* $\chi(\mathcal{H})$. Thus, general mixed hypergraphs represent the structures where both the problems on the minimum and the maximum number of colors occur.

Although every C- and D-hypergraph admits a coloring, in general a mixed hypergraph may have no colorings at all. A mixed hypergraph having no colorings is *uncolorable*. Otherwise it is called *colorable*. The general structure of uncolorable mixed hypergraphs is unknown, criteria of uncolorability have been obtained only for some special cases.

2. C-perfect hypergraphs

Every graph G satisfies $\chi(G) \geq \omega(G)$, where $\omega(G)$ is the size of the largest clique. The *perfect graphs* are the graphs such that $\chi(G') = \omega(G')$ for every induced subgraph G'. Many subclasses of perfect graphs have nice structural properties and admit fast algorithms for several optimization problems. The well-known Strong Perfect Graph Conjecture raised by Berge, which stated that a graph G is perfect if and only if no odd cycle of length at least 5 occurs as an induced subgraph of G or its complement $\bar{G}$, has been proved recently by Chudnovsky, Robertson, Seymour and Thomas [13].

Voloshin [35] introduced a natural analogue of perfection for the upper chromatic number. In a mixed hypergraph, a set of vertices is *C-stable* if it contains no C-edge. The *C-stability number* $\alpha_C(\mathcal{H})$ is the maximum cardinality of a C-stable set of $\mathcal{H}$. In contrast to the lower chromatic number of graphs what is bounded from below by the *clique* number, the upper chromatic number of mixed hypergraphs is bounded from above by the *C-stability* number: the inequality

$$\bar{\chi}(\mathcal{H}) \leq \alpha_C(\mathcal{H})$$

always is valid, because a multicolored set with more distinct colors than $\alpha_C(\mathcal{H})$ would assign distinct colors to all vertices of some C-edge. So the C-stability number is a natural upper bound for the upper chromatic number. A mixed hypergraph $\mathcal{H}$ is *perfect* [35] if $\bar{\chi}(\mathcal{H}') = \alpha_C(\mathcal{H}')$ holds for every induced subhypergraph $\mathcal{H}'$; and otherwise it is called *imperfect*. Notice that the perfection of graphs is related to the lower chromatic number, while the perfection of hypergraphs (sometimes called C-perfection) is related to the upper chromatic number.

Analogously to graphs (recall, e.g., the Mycielski graphs), there are examples showing that the difference between the C-stability number and the upper chromatic number may be arbitrary large [36]. Several classes of perfect and minimal imperfect mixed hypergraphs have been found.

For a mixed hypergraph, a *host graph* is a graph G on the same vertex set and such that every C-edge and every D-edge induces a *connected subgraph* of G. Depending on the type of G, we say that $\mathcal{H}$ is an *interval hypergraph* if $\mathcal{H}$ admits a host graph that is a simple path, a *hypertree* if it has a host tree, and a *circular hypergraph* if it has a host cycle.

Several examples of perfect and imperfect C- and mixed hypergraphs were described in [35]. A *cycloid* is an *r-uniform* C-hypergraph, denoted by C_n^r, which has n C-edges and admits a simple cycle on n vertices as a host graph (see example V_5 in Figure 1 for the case $r = 3$). A *polystar* is a mixed hypergraph with at least two C-edges in which the set Y of vertices common to all C-edges (center) is nonempty, and every pair in Y forms a D-edge. When the center consists of one vertex, the polystar is also called *monostar*. In a C-hypergraph, each polystar is a monostar. A *bistar* (called co-bistar in [35]) is a mixed hypergraph in which there exists a pair of distinct vertices that are contained in all C-edges and do not form a D-edge.

Bistars are perfect; polystars are not. Also, cycloids of the form C_{2r-1}^r are not perfect. When $n = 2r - 1$, we have $\alpha_C(C_n^r) = 2r - 3$ and $\bar{\chi}(C_n^r) = 2r - 4$. These cycloids are analogous to the known minimal imperfect graphs. Polystars and cycloids of the form C_{2r-1}^r, $r \geq 3$, are *minimal imperfect* mixed hypergraphs in the sense that every proper induced subhypergraph of such a cycloid is perfect, and every subhypergraph of a polystar that is not a polystar is perfect.

Voloshin conjectured [35] that an r-uniform C-hypergraph is perfect if and only if it has no monostar or cycloid of the form C_{2r-1}^r, $r \geq 3$, as an induced subhypergraph. For $r = 3$ these conditions mean the examples V_1–V_5 shown in Figure 1. There are some classes of C-hypergraphs for which the conjecture is true, namely interval hypergraphs [12], C-hypertrees [8], and also circular C-hypergraphs without edges containing each other [9].

However, the situation is more complex than in case of graphs. Král' [21] has disproved the mixed hypergraph perfection conjecture for each $r \geq 3$ by constructing a new family of minimal imperfect C-hypergraphs different from monostars and cycloids. His example K_1 for the case $r = 3$ is shown in Figure 1: it consists of 6 C-edges obtained by shifting (clockwise rotation) the edge shown on the left plus two more edges shown on the right (they should be on the same vertex set). This C-hypergraph does not contain any of the monostars $V_1 - V_4$, is not the cycloid $V_5 = C_5^3$, and it is not perfect, but each of its induced subhypergraphs is perfect.

There may be other types of minimal imperfect mixed hypergraphs. In fact, for $r \geq 4$, an increasing family of them has been constructed in [9], and a necessary and sufficient condition is described for minimal imperfect C-hypergraphs with C-transversal number 2 in the same paper, but no examples different from those in Figure 1 are known for $r = 3$. (The *C-transversal number*, denoted by $\tau_C(\mathcal{H})$, is the minimum cardinality of a vertex set that meets all C-edges of $\mathcal{H}$; the Gallai-type equality $\alpha_C + \tau_C = |X|$ is valid for every mixed hypergraph.) A condition can also be given for the minimal imperfect C-hypergraphs with $\tau_C = 3$, but it is more complicated than the one for $\tau_C = 2$.

Polystars generally are not uniform and they already suggested that the family of non-uniform minimal imperfect mixed hypergraphs may be complex. All these results

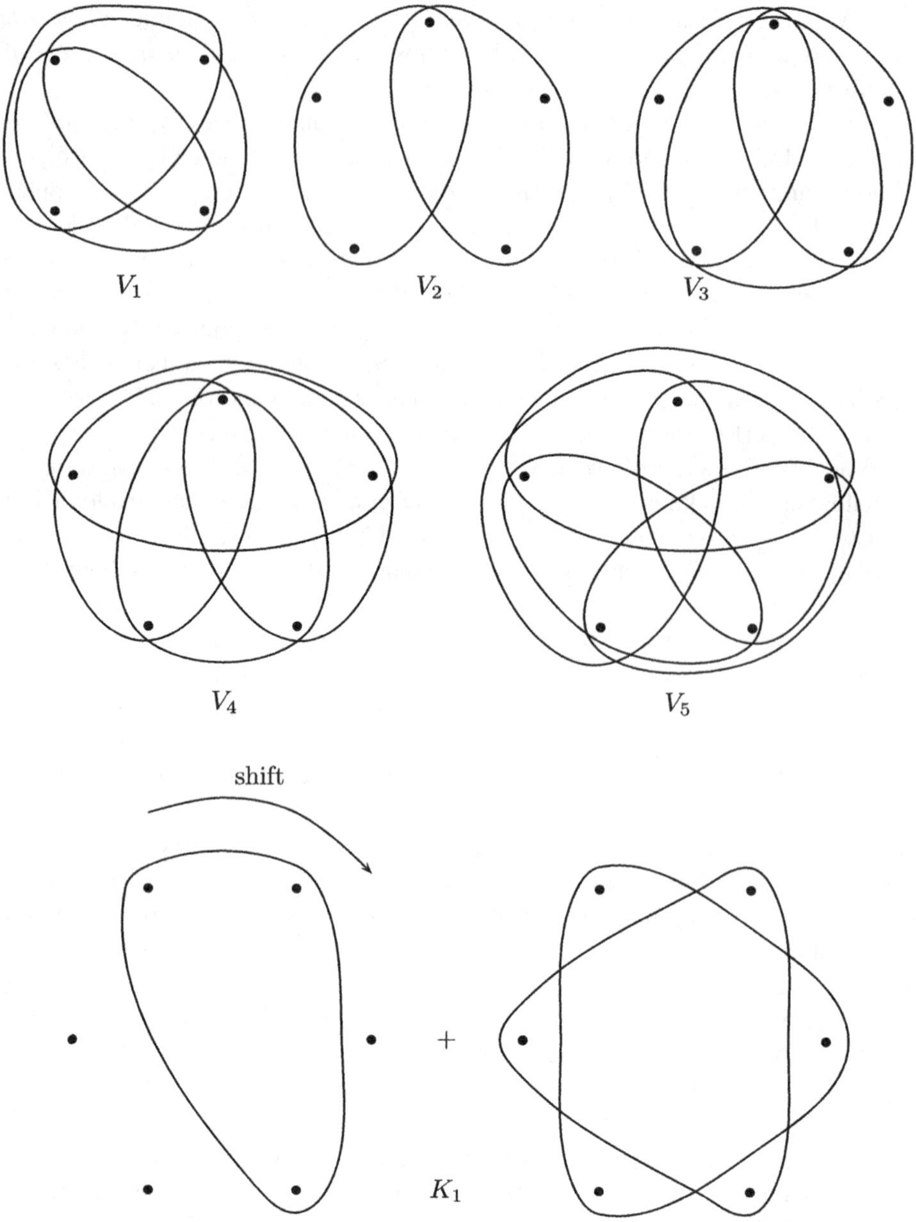

Figure 1. The six known examples of minimal non-perfect 3-uniform hypergraphs.

and investigations will definitely lead to a more general conjecture about perfect mixed hypergraphs.

Concerning time complexity it is well-known that the perfection of graphs has led to efficient polynomial- and linear-time algorithms for solving some problems that are $\mathcal{NP}$-complete in general. One can already see that this is true to some extent for the perfection of hypergraphs, too. For example, the perfect C-hypertrees can be α_C-colored in polynomial time [8]. Actually, all classes of perfect mixed hypergraphs, that are known so far, can be upper-colored efficiently. When monostars are allowed in C-hypertrees, however, the problem is already $\mathcal{NP}$-complete. It is $\mathcal{NP}$-complete even for monostars themselves. More explicitly, it is $\mathcal{NP}$-complete to determine the upper chromatic number of C-hypertrees [22], and also to test whether a generic C-hypertree is perfect [8]. We should also note that some perfect C-hypertrees do contain (non-induced) monostars. These results indicate that the situation is more complicated than in case of graph perfectness.

One may expect in general that hypergraph perfection will lead to efficient polynomial-time algorithms for finding the upper chromatic number and respective colorings. In addition, perfection may serve as a hint for the search of efficient polynomial algorithms for other hard combinatorial problems on discrete structures.

Open problems.

2.1 Search for new classes of perfect and minimal imperfect mixed hypergraphs.
2.2 Describe further classes of C-hypergraphs in which all imperfect uniform C-hypergraphs contain induced monostars or cycloids.
2.3 Develop efficient algorithms to compute the upper chromatic number, and find maximum colorings for various classes of perfect mixed hypergraphs.
2.4 Prove or disprove: A 3-uniform C-hypergraph is perfect if and only if it does not contain any of the monostars V_1, V_2, V_3, V_4, cycloid V_5, and Král's construction K_1 (shown in Figure 1), as induced subhypergraphs
2.5 Do there exist minimal imperfect C-hypergraphs with transversal number greater than three?
2.6 Characterize perfect *planar* mixed hypergraphs ([36], cf. Section 5).
2.7 Characterize perfect uniquely colorable mixed hypergraphs (see the next section).

3. Uniquely colorable hypergraphs

A mixed hypergraph is *uniquely colorable* (UC, for short) if it admits precisely one proper color partition[1] [33]. This new class of "absolutely rigid" combinatorial objects is a generalization of complete graphs (cliques) in the sense that exactly the latter ones are the uniquely colorable D-hypergraphs. Contrary to cliques, however, the structure of UC mixed hypergraphs is rather general, since constructions show that they contain every colorable mixed hypergraph as an induced subhypergraph. Nevertheless, *uniquely colorable separators* of mixed hypergraphs — vertex sets that induce a UC subhypergraph and whose removal makes $\mathcal{H}$ disconnected — behave similarly as *clique-cutsets* of graphs, and can be applied for deriving recursive formulas for various functions (e.g., chromatic polynomial, lower and upper chromatic number) [39]. In this way, new structures with interesting coloring properties can be constructed, analogously to the *chordal graphs* built from cliques.

[1]The requirement is much stronger than the equality $\chi(\mathcal{H}) = \bar{\chi}(\mathcal{H})$; moreover, conceptually it differs from the notion of uniquely colorable graphs in the standard sense. In the latter it is assumed that a graph G has just one color partition *with* $\chi(G)$ *colors*. This would lead to "weakly uniquely colorable" mixed hypergraphs, a wider class that has also been studied in [33] but we do not consider it here.

It is co-$\mathcal{NP}$-complete to decide whether a generic input $\mathcal{H}$ given together with a proper coloring is uniquely colorable [33]. But some UC subclasses can be characterized and recognized efficiently, e.g. those with $\chi = n - 1$ and $\chi = n - 2$ [30], UC mixed hypertrees [29], and UC circular mixed hypergraphs [38]. It has also been characterized, which *size distributions* of color classes can occur in uniquely colorable *r-uniform bi-hypergraphs* [2].

On the other hand, some natural restrictions still yield complex classes. We call a mixed hypergraph $\mathcal{H} = (X, \mathcal{C}, \mathcal{D})$ *UC-orderable* [33] if there exists an ordering $x_1, x_2, \ldots, x_n$ of the vertex set — called *UC-order* — such that every subhypergraph induced by an initial segment $\{x_1, \ldots, x_i\}$ $(i = 2, 3, \ldots, n = |X|)$ of this ordering is uniquely colorable. Although this looks like a strong condition, unexpectedly it is $\mathcal{NP}$-complete to recognize the mixed hypergraphs that are UC-orderable [4]. On the positive side, the UC-orderable mixed *hypertrees* have been characterized [29].

If $x_1, x_2, \ldots, x_n$ is a UC-order of $\mathcal{H}$, then so is $x_2, x_1, \ldots, x_n$ as well. We call $\mathcal{H}$ *uniquely UC-orderable* if it has a unique UC-order apart from the transposition of the first two vertices. Although the recognition problem of uniquely UC-orderable mixed hypergraphs is still open, their possible *color sequences* are characterized and can be recognized in linear time [4].

Open problems.

3.1 Describe subclasses of uniquely colorable and UC-orderable mixed hypergraphs that can be characterized and recognized efficiently.

3.2 Characterize UC mixed hypergraphs with vertex degree k, for $k \geq 2$ fixed.

3.3 Prove or disprove: For every constant c, it can be decided in polynomial time whether a generic input mixed hypergraph on n vertices is uniquely colorable with $n - c$ colors.

3.4 Determine the complexity of recognizing uniquely UC-orderable mixed hypergraphs, and characterize their structure.

4. Finite geometries and block designs

Symmetric configurations, that are of great interest in various fields including e.g. geometry and statistics, have been widely studied within combinatorics. Since colorings of block designs have been surveyed in [28], here we only briefly consider two important types, namely finite projective planes and Steiner systems (balanced incomplete block designs, BIBDs) with small block size and index 1, putting emphasis on recent developments and open problems.

A *bi-Steiner triple system* of order n is a *3-uniform* mixed hypergraph $\mathcal{H} = (X, \mathcal{C}, \mathcal{D})$ with $\mathcal{C} = \mathcal{D} = \mathcal{B}$, i.e. a bi-hypergraph over the n-element vertex set X, where $\mathcal{B}$ is a family of 3-element subsets (also called blocks) such that every *pair* of vertices is contained in precisely one block $B \in \mathcal{B}$. Similarly, a *bi-Steiner quadruple system* of order n is a *4-uniform* bi-hypergraph $\mathcal{H} = (X, \mathcal{C}, \mathcal{D})$ with $\mathcal{C} = \mathcal{D} = \mathcal{B}$ where each *triple* of vertices is contained in precisely one block $B \in \mathcal{B}$. We write $BSTS(n)$ and $BSQS(n)$ to denote a generic bi-Steiner triple system and bi-Steiner quadruple system, respectively, of order n. The systems $CSTS(n)$ and $CSQS(n)$ are defined in a similar way, except that $\mathcal{D} = \emptyset$ in this case, i.e. each $B \in \mathcal{B}$ is assumed to be a C-edge only.

The extremal behavior of the upper chromatic number of Steiner triple systems is understood fairly well [25,26]. If $n \le 2^t - 1$, then $\bar{\chi}(BSTS(n)) \le \bar{\chi}(CSTS(n)) \le t$, this upper bound is tight for both types of systems, moreover if $\bar{\chi} = t$ then the cardinalities of color classes in any t-coloring form the sequence $2^{t-1}, 2^{t-2}, \ldots, 2, 1$, and also the blocks are constructed in a well-defined recursive way.

On the other hand, very little is known about the colorability properties of quadruple systems. The bound $\bar{\chi}(BSQS(n)) \le \bar{\chi}(CSQS(n)) = O(\ln n)$ has been proved[2] for $n \to \infty$ [25], but its tightness for $BSQS$ is not at all clear. Also, while there exist uncolorable bi-Steiner triple systems [18,25] for all orders $n \ge 15$, all $BSQS(n)$ on $n \le 16$ vertices are colorable [24], and no uncolorable $BSQS$ has been found yet.

Analogously to the above, we can view every block design as a bi-hypergraph or a C-hypergraph. It seems to depend mostly on the parameters of the BIBD in question whether there is a significant difference between the two alternatives (bi- or C-) or their coloring properties are similar. Finite projective planes (see e.g. [19] for general information) are relatively sparse structures, and this is expressed also by the fact that their upper chromatic number is nearly the number of points. For instance, we can assign color 1 to a specified point x, color 2 to all points of $L' \backslash \{x\}$ and $L'' \backslash \{x\}$ for two lines L', L'' containing x, color 1 again to all points of $L \backslash (L' \cup L'')$ where L is a line not containing x, and then a distinct new color to each point not in $L \cup L' \cup L''$. If the plane has order q, we obtain a coloring with $q^2 - 2q + 3$ colors in this way, where neither monochromatic nor multicolored lines occur. In this example we have roughly q^2 vertices and $q^2 - cq$ colors for some constant c; i.e., almost all vertices have their dedicated colors.

Motivated by this, we define the *decrement* of a mixed hypergraph $\mathcal{H} = (X, \mathcal{C}, \mathcal{D})$ as $dec(\mathcal{H}) = |X| - \bar{\chi}(\mathcal{H})$. The decrement becomes informative and interesting to study if it is significantly smaller than the number of vertices. The above example shows that

$$dec(\Pi(q)) \le 3q - 2$$

for every projective plane $\Pi(q)$ of order q. This is the only general upper bound on the decrement known so far. It was an open problem since the mid-1990's whether or not $3q$ is a nearly tight estimate. Recently, it has been proved by Bacsó and Tuza [1] that

$$dec(\Pi(q)) \ge 2q + \sqrt{q}/2 - o(\sqrt{q})$$

for every $\Pi(q)$, as $q \to \infty$, both when viewed as a C- or bi-hypergraph. If $\Pi(q)$ is a *Galois plane*, the bound can be improved to $dec(PG(2,q)) \ge 2q + \sqrt{q} - o(\sqrt{q})$. In general, these bounds are tight up to the additive term $\Theta(\sqrt{q})$, for the following reason. If $q = p^{2a}$ is a prime power with *even* exponent and $\Pi(q) = PG(2,q)$, then

---

[2] This upper bound remains valid for Steiner systems $CS(t, t+1, n)$ and also for nearly complete partial ones, too, for every fixed t [27]. (An $S(t, t+1, n)$ system is a block design of order n and block size $t+1$, containing each t-tuple of vertices in precisely one block.)

the union of two disjoint *Baer subplanes* of order $\sqrt{q}$ (which are well-known to exist) meet each line in at least two points. A set with this property is called *2-transversal*. Assigning a dedicated color to each of the remaining points, $dec(PG(2,q)) \le 2q + 2\sqrt{q}+1$ follows. On the other hand, if $q = p^{2\alpha+1}$ is a prime power with *odd* exponent, then the stronger lower bound $dec(PG(2,q)) \ge 2q + \Theta(q^{2/3})$ is valid as $q \to \infty$. One can formulate an estimate on $dec(\Pi(q))$ that involves 2-transversals. In this way, the lower bounds on $dec(PG(2,q))$ can be derived from the results of [3].

Clearly, one can raise very natural generalizations of these problems for C- and bi-hypergraphs that are derived from subspaces of finite geometries (both projective and affine ones), but practically nothing is known so far about their colorability properties.

Open problems.

4.1 Prove or disprove that almost all bi-Steiner triple systems are uncolorable.
4.2 Do there exist uncolorable $BSQS$ systems? If the answer is affirmative, find the order of the smallest one.
4.3 Can $\bar{\chi}(BSQS(n))$ be arbitrarily large? How fast can it tend to infinity as $n \to \infty$?
4.4 Can $\chi(BSQS(n))$ be arbitrarily large? How fast can it tend to infinity as $n \to \infty$?
4.5 Determine the exact or asymptotic value of $\bar{\chi}(PG(2,q))$, the upper chromatic number of the Galois plane of order q.
4.6 Is $\bar{\chi} \ge q^2 - q - o(q)$ for every projective plane of order q as $q \to \infty$?
4.7 Study the analogous problems for subspaces of projective and affine spaces of higher dimension.

5. Color-bounded and stably bounded hypergraphs

Color-bounded and stably bounded hypergraphs were introduced recently by Bujtás and Tuza [5–7], generalizing the concept of mixed hypergraph. Instead of distinguishing between two sets of edges, $\mathcal{C}$ and $\mathcal{D}$, throughout this section we consider hypergraphs as pairs $\mathcal{H} = (X, \mathcal{E})$ where X is the vertex set and $\mathcal{E} = \{E_1, \dots, E_m\}$ is the *edge set*. What makes this model more general is that edge types are replaced with *color-bounds*, which will impose stronger conditions.

Color-bounded hypergraph means that each edge E_i is associated with a *lower color-bound* s_i and with an *upper color-bound* t_i, whereby a vertex coloring is considered proper if each E_i gets at least s_i and most t_i different colors.

In classical hypergraph coloring and also for C-hypergraphs, every hypergraph has a strict k-coloring for all integers k between its lower and upper chromatic number. Surprisingly, this property holds neither for mixed nor for color-bounded hypergraphs, what made it necessary to introduce the notion of *feasible set* $\Phi(\mathcal{H})$, whose elements are those values k for which the hypergraph $\mathcal{H}$ admits a strict k-coloring. We say that $\mathcal{H}$ has a *gap* at an integer ℓ if $\chi(\mathcal{H}) < \ell < \bar{\chi}(\mathcal{H})$ but $\ell \notin \Phi(\mathcal{H})$. In the study of a restricted subclass of color-bounded hypergraphs, the first fundamental question arising is whether or not all of its colorable members have a gap-free feasible set.

As proved in [6], every color-bounded *interval hypergraph* has gap-free feasible set and its lower chromatic number equals the maximum of lower color-bounds s_i, for which we shall write s. Moreover, the proof of this result contains a polynomial-time recoloring algorithm that transforms a strict k-coloring to a strict $(k-1)$-coloring whenever $k > s$. It is worth noting that the procedure does not require the hyperedges as input. On the other hand, at present only few results are known concerning the upper chromatic number and conditions for the colorability of color-bounded interval hypergraphs.

The widest known class of *mixed* hypergraphs with gap-free feasible sets appears to be the one whose members admit host graphs in which any two cycles are vertex-disjoint [23]. In particular, mixed hypertrees always have gap-free feasible sets and their lower chromatic number equals one or two [22]; but this is very far from being valid in the color-bounded model. As a matter of fact, color-bounded hypertrees play central role regarding colorability properties and feasible sets [6]. Although they have quite restricted structure, for every finite subset S of integers greater than two, there can be constructed a color-bounded hypertree with feasible set S. Compared to the general case, this characterization for the possible sets S contains only one additional restriction: the feasible set of every 2-colorable color-bounded hypertree necessarily is gap-free. It is even true that for any finite set $S \subset \mathbb{N}$ satisfying $\min(S) \geq 3$, we can arbitrarily prescribe the number r_k of proper color partitions with k nonempty classes for each $k \in S$, and then there exists a color-bounded hypertree with feasible set S and exactly r_k proper k-colorings (apart from renumbering of colors).

Significant differences between mixed and color-bounded hypertrees appear also concerning the time complexity properties. For mixed hypertrees the decision problem of colorability can be solved in linear time [31] and, if it is colorable, a coloring with χ colors can be obtained easily. In sharp contrast with this, the decision problem of colorability is $\mathcal{NP}$-complete already for 3-uniform color-bounded hypertrees.

As regards circular color-bounded hypergraphs, the lower chromatic number is known to be at most $2s-1$ and there cannot be any gaps in the feasible set above this value [6].

The notion of planar hypergraph was introduced analogously to the classical idea of planar graphs. In general, the *incidence graph* representation of a hypergraph $\mathcal{H} = (X, \mathcal{E})$ means the *bipartite* graph $B(\mathcal{H})$ with vertex set $X \cup \mathcal{E}'$ where the vertices in $\mathcal{E}'$ are in one-to-one correspondence with the edges in $\mathcal{E}$, furthermore $x \in X$ and $e'_i \in \mathcal{E}'$ are adjacent in $B(\mathcal{H})$ if and only if the incidence relation $x \in E_i$ holds for the corresponding vertex and edge in $\mathcal{H}$. For mixed hypergraphs, $\mathcal{E}$ is considered as the union $\mathcal{C} \cup \mathcal{D}$. By definition, a hypergraph is *planar* if its incidence graph is a planar graph. The possible feasible sets of planar mixed hypergraphs are fairly restricted [20] since either they are gap-free intervals with minimum value $1 \leq \chi \leq 4$ or they have lower chromatic number $\chi = 2$ and the only gap occurs at 3. On the other hand, a characterization for the feasible sets of color-bounded planar hypergraphs has not been proposed yet.

Coloring constraints used in the class of color-bounded hypergraphs yield a stronger model than mixed hypergraphs. But the structure class of *stably bounded hypergraphs* [7] is even more general than the former ones. It offers a common frame also to express several types of non-classical graph coloring constraints. The heart of the matter is that each edge E_i is associated not only with bounds s_i and t_i on the cardinality of largest *polychromatic* subset of the edge, but also there is a prescribed lower bound a_i and an upper bound b_i for the cardinality of the largest *monochromatic* subset of E_i. That is, there must exist a color occurring on at least a_i vertices of E_i, whilst no color is allowed to occur more than b_i times in E_i.

If the lower color-bound function s_i or a_i has constant value 1, or the upper color-bound function t_i or b_i assigns value $|E_i|$ to each edge E_i, it has no effect on the proper colorings and can be omitted. Taking different subsets of the four color-bound functions by disregarding the non-restrictive conditions, the obtained structure classes are denoted with capital letters corresponding to the restrictive functions. Especially, (S, T)-hypergraph means a color-bounded hypergraph.

The upper color-bounds t_i and b_i can always be reduced (by inserting new edges) to a_i and s_i, respectively, if structural conditions (e.g., hypertree or uniformity) are not imposed. But the lower-bounds s_i and a_i generally cannot be replaced by the other ones. In this sense, the model (S, A) is the only universal pair.

The (T, A)- and also the (S, B)-hypergraphs are always colorable and have gap-free feasible sets. All the other four pairs admit uncolorability and also gaps in the feasible sets of colorable hypergraphs.

One way to compare the structure classes obtained by choosing different color-bound functions as restrictive ones is to consider the sets of possible chromatic polynomials. The *chromatic polynomial* of a mixed, color-bounded, or stably bounded hypergraph $\mathcal{H}$ is the polynomial $P(\mathcal{H}, \lambda)$ whose value at $\lambda = k \in \mathbb{N}$ is the number of proper colorings $\varphi : X \to \{1, \dots, k\}$ with at most k colors. That is, the colors are distinguished and renumbering of colors is counted to be a different coloring. The concept of $P(\mathcal{H}, \lambda)$ for mixed hypergraphs was introduced in [34]. There can be gaps in the coefficients of the chromatic polynomials of hypergraphs; e.g., $\mathcal{H} = (\{1, 2, 3\}, \emptyset, \{\{1, 2, 3\}\})$ has $P(\mathcal{H}, \lambda) = \lambda^3 - \lambda$. Extending partly the results of [16], some properties of $P(\mathcal{H}, \lambda)$ for S-hypergraphs were described in [17]. Moreover, the possible chromatic polynomials of non-1-colorable hypergraphs were characterized in [5] in terms of linear inequalities on the coefficients.

In Figure 2 we exhibit the Hasse-diagram of classes of chromatic polynomials depending on the subsets of $\{S, T, A, B\}$; moreover, also the positions of the classes of C-, D-, and mixed hypergraphs are indicated with the notation $\mathcal{P}_\mathcal{C}$, $\mathcal{P}_\mathcal{D}$, and $\mathcal{P}_\mathsf{M}$, respectively. We should note, however, that the hierarchy of hypergraph classes is not stable: taking different aspects for the comparison, the order may change partially.

Open problems.

5.1 Are there polynomial algorithms to decide whether a generic color-bounded interval hypergraph is colorable, and to determine its $\bar{\chi}$?
5.2 Are the feasible sets of color-bounded circular hypergraphs gap-free?

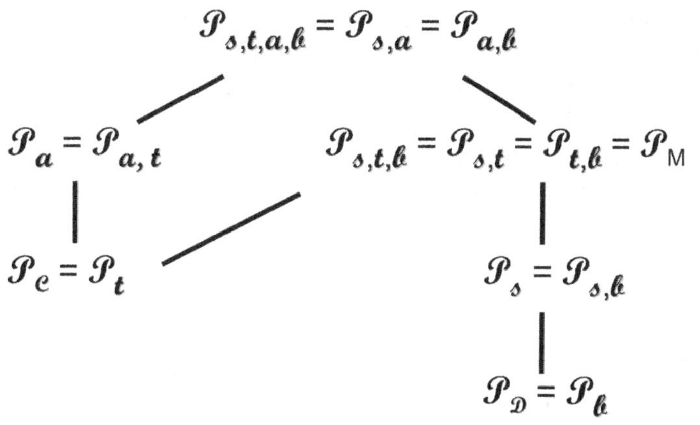

Figure 2. Hasse-diagram of possible chromatic polynomials belonging to different structure classes.

5.3 Characterize the feasible sets of color-bounded and stably bounded *planar* hypergraphs; and solve the analogous problem for other subsets of the conditions S, T, A, B.
5.4 Characterize the chromatic polynomials of mixed, color-bounded, and stably bounded hypergraphs admitting a proper 1-coloring.
5.5 What are the combinatorial meanings of the coefficients in the chromatic polynomials $P(\mathcal{H}, \lambda)$ of mixed, color-bounded, and stably bounded hypergraphs? (Cf. [16] for D-hypergraphs and [17] for S-hypergraphs.)
5.6 Characterize the chromatic polynomials of mixed interval hypergraphs and other well-structured classes.

6. C-hypergraphs and partition crossing

It follows immediately by the pigeon-hole principle that if $\mathcal{H} = (X, \mathcal{C}, \emptyset)$ is an r-uniform C-hypergraph, then every vertex partition into fewer than r classes yields a proper coloring of $\mathcal{H}$. As a direct consequence, we have $\bar{\chi}(\mathcal{H}) \geq r - 1$. This observation leads to the very natural extremal problem of determining the minimum number of edges in an r-uniform C-hypergraph on n vertices whose upper chromatic number is equal to $r - 1$. We denote this minimum by $f(n, r)$.

Diao, Zhao and Zhou [15] observed that a certain kind of connectivity condition implies the inequality

$$f(n, r) \geq \frac{2}{n - r + 2} \binom{n}{r}$$

for every $n \geq r \geq 3$. This lower bound turns out to be best possible for all n if $r = 3$ [14] — then it simplifies to the formula $f(n, 3) = \lceil \frac{n(n-2)}{3} \rceil$ — what is not the case if $r \geq 4$ [10]. Nevertheless, it is asymptotically tight for every fixed r as $n \to \infty$, as proved in [10]. What is more, the estimates given in [10] are within ratio $(1 + o(1))$ from optimum even when r is allowed to grow at any speed not faster than $o(n^{1/3})$.

In the recent manuscript [11], the problem is treated in a more general setting, by defining a three-parameter function and showing that it is related to several extensively studied concepts of graph theory and combinatorics, including connectivity, Turán-type functions on graphs and hypergraphs, and the exact and asymptotic variants of balanced incomplete block designs. For this reason, such extremal functions related to the upper chromatic number may play an important role in future research in different contexts.

Open problems.

5.1 Determine the exact value of $f(n, k)$ for $k \geq 4$ fixed and $n > n_0(k)$.
5.2 Determine the asymptotic value of $f(n, k)$ for $k = \Omega(n^{1/3})$ as $n \to \infty$.

References

[1] G. Bacsó and Zs. Tuza, Upper chromatic number of finite projective planes, *J. Combin. Designs*, **16** (2008) 221–230.
[2] G. Bacsó, Zs. Tuza and V. Voloshin, Unique colorings of bi-hypergraphs, *Australasian J. Combin.*, **27** (2003) 33–45.
[3] A. Blokhuis, L. Storme and T. Szőnyi, Lacunary polynomials, multiple blocking sets and Baer subplanes, *J. London Math. Soc. (2)*, **60** (1999) 321–332.
[4] Cs. Bujtás and Zs. Tuza, Orderings of uniquely colorable mixed hypergraphs, *Discrete Applied Math.*, **155** (2007) 1395–1407.
[5] Cs. Bujtás and Zs. Tuza, Color-bounded hypergraphs, I: General results, *Discrete Math.*, **309(15)**, (2009) 4890–4902.
[6] Cs. Bujtás and Zs. Tuza, Color-bounded hypergraphs, II: Interval hypergraphs and hypertrees, *Discrete Math.*, **309(22)**, (2009) 6391–6401.
[7] Cs. Bujtás and Zs. Tuza, Color-bounded hypergraphs, III: Model comparison, *Applicable Analysis and Discrete Math.*, **1(1)** (2007) 36–55.
[8] Cs. Bujtás and Zs. Tuza, Voloshin's conjecture for C-perfect hypertrees, *Australasian J. Combin.*, to appear.
[9] Cs. Bujtás and Zs. Tuza, C-perfect hypergraphs, *J. Graph Theory*, **64(2)** (2010) 132–149.
[10] Cs. Bujtás and Zs. Tuza, Smallest set-transversals of k-partitions, *Graphs and Combinatorics*, **25(6)** (2009) 807–816.
[11] Cs. Bujtás and Zs. Tuza, Partition-crossing hypergraphs. Manuscript (2008).
[12] E. Bulgaru and V. Voloshin, Mixed interval hypergraphs, *Discrete Applied Math.*, **77(1)** (1997) 24–41.
[13] M. Chudnovsky, N. Robertson, P. Seymour and R. Thomas, The Strong Perfect Graph Theorem, *Annals of Math.*, **164(1)** (2006) 51–229.
[14] K. Diao, G. Liu, D. Rautenbach and P. Zhao, A note on the least number of edges of 3-uniform hypergraphs with upper chromatic number 2, *Discrete Math.*, **306** (2006) 670–672.
[15] K. Diao, P. Zhao and H. Zhou, About the upper chromatic number of a C-hypergraph, *Discrete Math.*, **220** (2000) 67–73.
[16] K. Dohmen, A broken-circuits-theorem for hypergraphs, *Archives of Math.*, **64** (1995) 159–162.
[17] E. Drgas-Burchardt and E. Łazuka, On chromatic polynomials of hypergraphs, *Applied Math. Letters*, **20(12)** (2007) 1250–1254.
[18] B. Ganter, private communication with V. Voloshin (1997).
[19] D. R. Hughes and F. C. Piper, Projective Planes, Springer-Verlag (1973)

[20] D. Kobler and A. Kündgen, Gaps in the chromatic spectrum of face-constrained plane graphs, *Electronic J. Combin.*, **8(1)** (2001) #N3.

[21] D. Král', A counter-example to Voloshin's hypergraph co-perfectness conjecture, *Australasian J. Combin.*, **27** (2003) 25–41.

[22] D. Král', J. Kratochvíl, A. Proskurowski and H.-J. Voss, Coloring mixed hypertrees, *Discrete Applied Math.*, **154(4)** (2006) 660–672.

[23] D. Král', J. Kratochvíl and H.-J. Voss, Mixed hypercacti, *Discrete Math.*, 286(1–2) (2004) 99–113.

[24] G. Lo Faro, L. Milazzo and A. Tripodi, The first $BSTS$ with different lower and upper chromatic numbers, *Australasian J. Combin.*, **22** (2000) 123–133.

[25] L. Milazzo and Zs. Tuza, Upper chromatic number of Steiner triple and quadruple systems, *Discrete Math.*, **174** (1997) 247–259.

[26] L. Milazzo and Zs. Tuza, Strict colourings for classes of Steiner triple systems, *Discrete Math.*, **182** (1998) 233–243.

[27] L. Milazzo and Zs. Tuza, Logarithmic upper bound for the upper chromatic number of $S(t, t+1, v)$ systems, *Ars Combinatoria*, **92** (2009) 213–223.

[28] L. Milazzo, Zs. Tuza and V. I. Voloshin, Strict colorings of Steiner triple and quadruple systems: A survey, *Discrete Math.*, **261** (2003) 399–411.

[29] A. Niculitsa and V. Voloshin, About uniquely colorable mixed hypertrees, *Discuss. Math. Graph Theory*, **20(1)** (2000) 81–91.

[30] A. Niculitsa and H.-J. Voss, A characterization of uniquely colorable mixed hypergraphs of order n with upper chromatic numbers $n-1$ and $n-2$, *Australasian J. Combin.*, **21** (2000) 167–177.

[31] Zs. Tuza and V. Voloshin, Uncolorable mixed hypergraphs, *Discrete Applied Math.*, **99** (2000) 209–227.

[32] Zs. Tuza and V. Voloshin, Problems and results on colorings of mixed hypergraphs, Horizons of Combinatorics (E. Győri *et al.*, eds.), *Bolyai Society Mathematical Studies*, **17** Springer-Verlag (2008) 235–255.

[33] Zs. Tuza, V. I. Voloshin and H. Zhou, Uniquely colorable mixed hypergraphs, *Discrete Math.*, **248** (2002) 221–236.

[34] V. I. Voloshin, The mixed hypergraphs, *Computer Sci. J. Moldova*, **1** (1993) 45–52.

[35] V. I. Voloshin, On the upper chromatic number of a hypergraph, *Australasian J. Combin.*, **11** (1995) 25–45.

[36] V. I. Voloshin, Coloring Mixed Hypergraphs: Theory, Algorithms and Applications, Fields Institute Monographs 17, *Amer. Math. Soc.*, (2002).

[37] V. Voloshin, Mixed Hypergraph Coloring Web Site:
http://spectrum.troy.edu/~voloshin/mh.html

[38] V. Voloshin and H.-J. Voss, Circular mixed hypergraphs I: colorability and unique colorability, Proc. Thirty-first Southeastern International Conference on Combinatorics, Graph Theory and Computing, Boca Raton, FL, (2000), *Congr. Numer.*, **144** (2000) 207–219.

[39] V. I. Voloshin and H. Zhou, Pseudo-chordal mixed hypergraphs, *Discrete Math.*, **202** (1999) 239–248.

Proc. ICDM 2008, RMS-Lecture Notes Series No. 13, 2010, pp. 59–67.

A General Approach to Distance-Two Colouring of Graphs on Surfaces

Jan van den Heuvel

Department of Mathematics, London School of Economics, London, U.K.
e-mail: jan@maths.lse.ac.uk

Abstract. The square of a graph G, denoted G^2, is the graph with the same vertex set as G and with an edge between any two different vertices that have distance at most 2 in G. Finding the chromatic number of squares of graphs has been an area of intensive research, in particular for planar graphs.

An alternative way to look at a vertex colouring of the square of a graph G is as a colouring of the vertices of G in such a way that (i) adjacent pairs get different colours and (ii) pairs with a common neighbour get different colours. In this note we introduce a more general concept in which we assume there are two vertex sets (not necessarily disjoint) A and B. And the requirement is to colour the vertices of B so that (i) adjacent pairs get different colours, and (ii) non-adjacent pairs with a common neighbour from A get different colours. We show that this concept not only generalises colouring of the square, but other types of colourings as well.

For graphs embedded in a fixed surface we can give asymptotically best possible upper bounds on the number of colours required for such colourings (in terms of the natural degree condition).

Keywords. Graph colouring, square of a graph, cyclic colouring, graphs on surfaces.

2000 Subject Classification: 05C15, 05C10

1. Introduction

Most of the terminology and notation we use in this note is standard and can be found in any textbook on graph theory (such as [3] or [8]). All our graphs and multigraphs will be finite. A *multigraph* can have multiple edges; a *graph* is supposed to be simple. We use V for the set of vertices of a (multi)graph, and E for the set of edges.

A *vertex colouring* of a graph G is an assignment of colours to the vertices so that adjacent vertices receive different colours. The minimum number of colours required is the *chromatic number*, denoted $\chi(G)$. For a *list vertex colouring* we assume that each vertex v has its own list $L(v)$ of colours, and when colouring the vertices, each vertex should get a colour from its own list, but so that adjacent vertices still get a different colour. The *list chromatic number* $\chi_\ell(G)$ is the minimum integer k so that for each assignment of a list of colours $L(v)$ with at least k colours to the vertices v, there exists a vertex colouring in which vertices receive colours from their own lists.

This note is based on recent research done jointly with Omid Amini (Max-Planck-Institut für Informatik, Saarbrücken, Germany) and Louis Esperet (LaBRI, Bordeaux, France) [2], and with Frédéric Havet (INRIA, Sophia-Antipolis, France), Colin McDiarmid (University of Oxford, Oxford, U.K.) and Bruce Reed (INRIA, Sophia-Antipolis, France and McGill University, Montreal, Canada) [9].

All the concepts above can be defined for edges as well. We talk about an *edge colouring*, the *edge chromatic number* (or *chromatic index*) $\chi'(G)$, *list edge colouring*, and the *list edge chromatic number* (or *list chromatic index*) $\chi'_\ell(G)$.

For a graph G, its *square* G is the graph with the same vertex set as G and with an edge between any two different vertices that have distance at most 2 in G. A proper vertex colouring of the square of a graph can also be seen as a vertex colouring of the original graph satisfying:

- vertices that are adjacent must receive different colours, and
- non-adjacent vertices that have a common neighbour must receive different colours.

Another way to formulate these conditions is as "vertices at distance 1 or 2 must receive different colours". This is why the name *distance-two colouring* is also used in the literature.

In this note we consider a colouring concept that generalises the concept of colouring the square of a graph, but that also can be used to study different concepts such as *cyclic colouring of plane graphs* (definition will be given later).

Assume that a graph G with vertex set V has two specially assigned subsets A, B of V. (It is not required for A and B to be disjoint.) An (A, B)*-colouring* of G is an assignment of colours to the vertices in B so that:

- vertices that are adjacent must receive different colours, and
- vertices that have a common neighbour from A must receive different colours.

When each vertex $v \in B$ has its own list $L(v)$ of colours from which its colour must be chosen, we talk about a *list* (A, B)*-colouring*.

We denote by $\chi(G; A, B)$ the minimum number of colours required for an (A, B)-colouring to exist. Its list variant is denoted by $\chi_\ell(G; A, B)$, and is defined as the minimum integer k so that for each assignment of a list $L(v)$ with at least k colours to the vertices $v \in B$, there exists a proper (A, B)-colouring of G in which the vertices in B are assigned colours from their own lists.

Notice that we trivially have:

$$\chi(G) = \chi(G; \varnothing, V), \qquad \chi(G^2) = \chi(G; V, V),$$
$$\chi_\ell(G) = \chi_\ell(G; \varnothing, V), \quad \chi_\ell(G^2) = \chi_\ell(G; V, V).$$

For a vertex $v \in V$, let $N(v)$ be the set of vertices adjacent to v, and define $d(v) = |N(v)|$, $N_B(v) = N(v) \cap B$, and $d_B(v) = |N_B(v)|$. If we define $\Delta(G; A, B) = \max\{d_B(v) | v \in A\}$, then it is clear that we always need at least $\Delta(G; A, B)$ colours in a proper (A, B)-colouring.

In the case $A = B = V$, there exist many graphs G that require $O(\Delta(G)^2)$ colours (where $\Delta(G) = \Delta(G; V, V)$ is the normal maximum degree of a graph). But for planar graphs, it is known that a constant times $\Delta(G)$ colours is enough (even for list colouring). We'll take a closer look at this in the next subsection.

Following Wegner's Conjecture on colouring the square of planar graphs, we propose the following conjecture.

Conjecture 1.1. Let G be a planar graph and $A, B \subseteq V$. Then there exist constants c_1, c_2, c_3 such that

$$\chi(G; A, B) \leq \left\lfloor \frac{3}{2}\,\Delta(G; A, B) \right\rfloor + c_1;$$

$$\chi_\ell(G; A, B) \leq \left\lfloor \frac{3}{2}\,\Delta(G; A, B) \right\rfloor + c_2;$$

$$\chi(G; A, B) \leq \left\lfloor \frac{3}{2}\,\Delta(G; A, B) \right\rfloor + 1, \quad \text{if } \Delta(G; A, B) \geq c_3.$$

If $A = \varnothing$ (hence $\Delta(G; A.B) = 0$) and $B = V$, then the Four Colour Theorem means that the smallest possible value for c_1 is 4; while the fact that planar graphs are always 5-list colourable but not always 4-list colourable, shows the smallest possible value for c_2 is 5.

Our main result is that Conjecture 1.1 is asymptotically correct. In fact, we can prove a more general asymptotic version, which holds for general surfaces.

Theorem 1.2 (Amini, Esperet & van den Heuvel [2]). *Let S be a fixed surface, G a graph embeddable in S, and $A, B \subseteq V$. Then $\chi_\ell(G; A, B) \leq (1 + o(1))\,\frac{3}{2}\,\Delta(G; A, B)$.*

In other words, for all $\varepsilon > 0$, there exists $D_{S,\varepsilon}$, so that for all $D \geq D_{S,\varepsilon}$ we have: If G is a graph embeddable in S, with $A, B \subseteq V$ so that $\Delta(G; A, B) \leq D$, and L is a list assignment so that each vertex v in B gets a list $L(v)$ of at least $\left(\frac{3}{2}+\varepsilon\right) D$ colours, then there exists an (A, B)-colouring of G in which the vertices in B are assigned colours from their own lists.

In the next subsections we discuss two special consequences of this result. In the final section of this note we give an overview of the proof of the theorem.

1.1 *Colouring the square of graphs*

If G has maximum degree Δ, then a vertex colouring of its square will need at least $\Delta + 1$ colours, but the greedy algorithm shows it is always possible with $\Delta^2 + 1$ colours. Diameter 2 cages such as the 5-cycle, the Petersen graph and the Hoffman-Singleton graph (see [3, page 239]) show that there exist graphs that in fact require $\Delta^2 + 1$ colours.

Regarding the chromatic number of the square of a planar graph, Wegner [19] posed the following conjecture (see also the book of Jensen and Toft [11, Section 2.18]), suggesting that for planar graphs far less than $\Delta^2 + 1$ colours suffice.

Conjecture 1.3 (Wegner [19]). For a planar graph G of maximum degree Δ:

$$\chi(G^2) \leq \begin{cases} 7, & \text{if } \Delta = 3, \\ \Delta + 5, & \text{if } 4 \leq \Delta \leq 7, \\ \left\lfloor \frac{3}{2}\,\Delta \right\rfloor + 1, & \text{if } \Delta \geq 8. \end{cases}$$

Wegner also gave examples showing that these bounds would be tight. For even $\Delta \geq 8$, these examples are sketched below. The graph in the picture has maximum degree $2k$ and yet all the vertices except z are adjacent in its square. Hence to colour these $3k+1$ vertices, we need at least $3k+1 = \frac{3}{2}\Delta + 1$ colours.

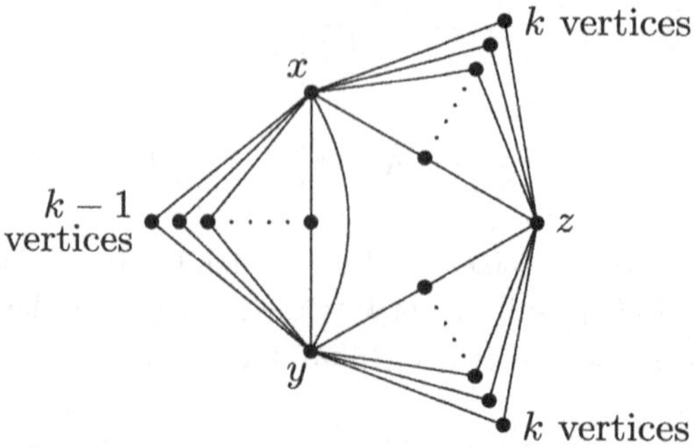

Figure 1.

Kostochka and Woodall [14] conjectured that for the square of a graph the chromatic number equals the list chromatic number. This conjecture and Wegner's together imply the following conjecture for planar graphs G:

$$\chi_\ell(G^2) \leq \begin{cases} 7, & \text{if } \Delta = 3,\\ \Delta + 5, & \text{if } 4 \leq \Delta \leq 7,\\ \lfloor \frac{3}{2}\Delta \rfloor + 1, & \text{if } \Delta \geq 8. \end{cases}$$

Wegner also showed that if G is a planar graph with $\Delta = 3$, then G^2 can be 8-coloured. Very recently, Thomassen [18] solved Wegner's Conjecture for $\Delta = 3$, and Cranston and Kim [7] showed that the square of every connected graph (not necessarily planar) which is subcubic (i.e., with $\Delta \leq 3$) has list chromatic number at most 8, except for the Petersen graph. However, the best possible upper bound (7 or 8) of the list chromatic number of the square of subcubic planar graphs is still open.

The first upper bound on $\chi(G^2)$ for planar graphs in terms of Δ, $\chi(G^2) \leq 8\Delta - 22$, was implicit in the work of Jonas [12]. This bound was later improved by Wong [20] to $\chi(G^2) \leq 3\Delta + 5$ and then by Van den Heuvel and McGuinness [10] to $\chi(G^2) \leq 2\Delta + 25$. Better bounds were then obtained for large values of Δ. It was shown that $\chi(G^2) \leq \lceil \frac{9}{5}\Delta \rceil + 1$ for $\Delta \geq 750$ by Agnarsson and Halldórsson [1], and the same bound for $\Delta \geq 47$ by Borodin *et al.* [5]. Finally, the best known upper bound so far has been obtained by Molloy and Salavatipour [15]: $\chi(G^2) \leq \lceil \frac{5}{3}\Delta \rceil + 78$. As mentioned in [15], the constant 78 can be reduced for sufficiently large Δ. For example, it was improved to 24 when $\Delta \geq 241$.

Since $\chi_\ell(G^2) = \chi_\ell(G; V, V)$, as an immediate corollary of Theorem 1.2 we obtain.

Corollary 1.4. *Let S be a fixed surface. Then the square of every graph G embeddable in S and of maximum degree Δ has list chromatic number at most* $(1 + o(1))\,\frac{3}{2}\,\Delta$.

In fact, the same asymptotic upper bound can be proved for much larger classes of graphs, and with a stronger conclusion on the colouring. For the following result we assume that colours are integers, which allows us to talk about the "distance" $|c_1 - c_2|$ between two colours c_1, c_2.

Theorem 1.5 (Havet, van den Heuvel, McDiarmid & Reed [9]). *Let k be a fixed positive integer and let $\mathcal{K}_k$ be the class of graphs that don't have $K_{3,k}$ as a minor. Then the square of every graph $G \in \mathcal{K}_k$ of maximum degree Δ has list chromatic number at most $(1 + o(1)) \frac{3}{2} \Delta$. Moreover, given lists of this size, there is a proper colouring in which the colours on every pair of adjacent vertices of G differ by at least $\Delta^{1/4}$.*

Note that planar graphs do not have a $K_{3,3}$ minor. In fact, for every surface $\mathcal{S}$, there is a constant k so that all graphs embeddable on $\mathcal{S}$ don't have $K_{3,k}$ as a minor. This proves that the first order term $\frac{3}{2} \Delta$ in Wegner's Conjecture is asymptotically achievable for graphs embeddable on any fixed surface.

1.2 *Cyclic colourings of embedded graphs*

Given a surface S and a graph G embeddable in S, we denote by G^S that graph with a prescribed embedding in S. If the surface S is the sphere, we talk about a *plane graph* G^P. The size (number of vertices in its boundary) of a largest face of G^S is denoted by $\Delta^*(G^S)$.

A *cyclic colouring* of an embedded graph G^S is a vertex colouring of G such that any two vertices incident to the same face have distinct colours. The minimum number of colours required in a cyclic colouring of an embedded graph is called the *cyclic chromatic number* $\chi^*(G^S)$. This concept was introduced for plane graphs by Ore and Plummer [16], who also proved that for a plane graph G^P we have $\chi^*(G^P) \leq 2\,\Delta^*$. Borodin [4] (see also Jensen and Toft [11, page 37]) conjectured the following.

Conjecture 1.6 (Borodin [4]). For a plane graph G of maximum face degree Δ^* we have $\chi^*(G^P) \leq \left\lfloor \frac{3}{2}\,\Delta^* \right\rfloor$.

The bound in this conjecture is best possible. For even Δ^*, this can be seen because of the following graphs. Take two disjoint triangles $x_1y_1z_1$ and $x_2y_2z_2$, and connect each of the pairs $\{x_1, x_2\}$, $\{y_1, y_2\}$ and $\{z_1, z_2\}$ with disjoint paths with $k - 2$ internal vertices. The resulting graph has $3\,k$ vertices, is planar, and has two faces of size 3 and three faces of size $\Delta^* = 2\,k$. Since all pairs of vertices have a face they are incident with, we need $3\,k = \frac{3}{2}\,\Delta^*$ colours in a cyclic colouring.

Borodin [4] also proved Conjecture 1.6 for $\Delta^* = 4$. For general values of Δ^*, the original bound $\chi^*(G^P) \leq 2\,\Delta^*$ of Ore and Plummer [16] was improved by Borodin *et al.* [6] to $\chi^*(G^P) \leq \left\lfloor \frac{9}{5}\,\Delta^* \right\rfloor$. The best known upper bound in the general case is due to Sanders and Zhao [17]: $\chi^*(G^P) \leq \left\lceil \frac{5}{3}\,\Delta^* \right\rceil$.

Although Wegner's and Borodin's Conjectures seem to be closely related, nobody has ever been able to bring to light a direct connection between them. Most of the results approaching these conjectures use the same ideas, but up until this point (as far as we know) no one had proved a general theorem implying both a result on the colouring of the square and a result an the cyclic colouring of plane graphs.

In order to show that our Theorem 1.2 provides an asymptotically best possible upper bound for the cyclic chromatic number for a graph G with some fixed embedding G^S, we need some extra notation. For each face f of G^S, add a vertex x_f and call X_F the set of vertices that were added to G. For any face f of G^S, and any vertex v incident with f, add an edge between v and x_f. We denote by G_F the graph obtained from G^S by this construction, so $V(G_F) = V(G) \cup X_F$. Observe that a (list) $(X_F, V(G))$-colouring of G_F is exactly a cyclic (list) colouring of G^S and that $\Delta(G_F; X_F, V(G)) = \Delta^*(G^S)$. We get the following corollary of Theorem 1.2.

Corollary 1.7. *Let S be a fixed surface. Every embedded graph G^S of maximum face size Δ^* has cyclic list chromatic number at most* $(1 + o(1))\,\frac{3}{2}\,\Delta^*$.

2. Sketch of the proof of theorem 1.2 for planar graphs

To simplify the remainder of this note we only give the main steps of the proof of Theorem 1.2 for the case that G is a planar graph. Apart from some technical details, no essential new ideas are required for the proof for general surfaces.

First note that to prove Theorem 1.2 we may as well assume that A contains all vertices of degree at most $\Delta(G; A, B)$. We introduce some extra notation. For β a non-negative integer, define $V^\beta = \{v \in V | d(v) \leq \beta\}$. So to prove Theorem 1.2 for planar graphs, it is enough to prove the following theorem.

Theorem 2.1 ([2]). *For all real $\varepsilon > 0$, there exists a β_ε so that the following holds for all $\beta \geq \beta_\varepsilon$. Let G be a planar graph, with $B \subseteq V$ a set of vertices, and suppose every vertex $v \in B$ has a list $L(v)$ of at least $\left(\frac{3}{2} + \varepsilon\right)\beta$ colours. Then a list (V^β, B)-colouring of G with those colours exist.*

The proof of this theorem is done by induction on the number of vertices of G.

A *β-neighbour* of a vertex $v \in B$ is a vertex $u \in B$, $u \neq v$, so that u and v are adjacent, or u and v have a common neighbour in V^β. Denote the set of β-neighbours of v by $N^\beta(v)$, and its number by $d^\beta(v)$. Note that we have $d^\beta(v) \leq d(v) + \sum_{u \in N(v) \cap V^\beta}(d(u) - 1)$. Also, when colouring a vertex $v \in B$, we only need to make sure it receives a colour difference from its β-neighbours. In particular, a vertex v can have at most $d^\beta(v)$ "forbidden" colours.

The main tool in our proof of Theorem 2.1 is the following structural result.

Lemma 2.2 ([2]). *There exist integer constants C_1, C_2 so that the following holds. Let $\beta \geq C_1$ and let G be a plane graph. Then G contains one of the following:*

(S1) *a vertex with degree 0 or 1;*

(S2) *a face f and two vertices u, v on the boundary of f with $d(u) + d(v) \leq \beta$ and $d^\beta(u) \leq \frac{3}{2}\beta$;*

(S3) *two disjoint non-empty sets $X, Y \subseteq V^\beta$ with the following properties:*

 (i) *Every vertex $y \in Y$ has degree at most 4. Moreover, y is adjacent to exactly two vertices of X and the other neighbours of y have degree at most 4 as well.*

For $y \in Y$, let X^y be the set of its two neighbours in X. And for $W \subseteq X$, let Y^W be the set of vertices $y \in Y$ with $X^y \subseteq W$ (that is, the set of vertices of Y having their two neighbours from X in W).

(ii) *For all pairs of vertices $y, z \in Y$, if y and z are adjacent or have a common neighbour $w \notin X$, then $X^y = X^z$.*

(iii) *For all non-empty subsets $W \subseteq X$, we have the following inequality:*

$$|E(W, V \setminus Y)| \leq |E(W, Y \setminus Y^W)| + C_2 |W|.$$

If G contains a vertex v with degree 0 or 1, then we can just remove this vertex and use induction on the smaller graph. Since v is in conflict with at most β other vertices, we can easily find a colour for v.

If G contains a face f and two vertices u, v on the boundary of f according to (S2), then we do the following. Construct a new planar graph G_0 by contracting uv into a new vertex w. We also give w a list of colours $L(w)$ with $L(w) = L(v)$ (if $v \in B$). Notice that w has degree at most $d_G(u) + d_G(v) - 2 \leq \beta$ in G_0. In other words, if V_0 denotes the vertex set of G_0, then $w \in V_0^\beta$. By induction there exists a list (V_0^β, B)-colouring of G_0. We define a colouring of G as follows: every vertex different from u and v keeps its colour from the colouring of G_0. We colour v with the colour given to w in G_0. By assumption, $d_G^\beta(u) \leq \frac{3}{2}\beta$, and so there exists a colour for u different from the colours of all the vertices in conflict with u. We colour u with one of these colours. It is easy to verify that this provides a list (V^β, B)-colouring of G.

It remains to deal with the case when G has a structure according to (S3). This is the only non-trivial case, and in this note we can only sketch the approach taken.

First we essentially remove the vertices of Y from G (adding appropriate edges to avoid increasing distances between the remaining vertices). We colour the smaller graph and need to extend this colouring to the original graph G. For each vertex y in Y, let us remove from $L(y)$ the colours which are forbidden for y. These colours are exactly the colours of the vertices of $V \setminus Y$ at distance at most 2 from y.

Let us define the multigraph H as follows: H has vertex set X. And for each vertex $y \in Y$ we add an edge e_y between the two neighbours of y in X (in other words, between the two vertices from X^y). We associate a list $L(e_y)$ to e_y in H by taking the list of y obtained after removing the set of forbidden colours for y from the original list $L(y)$. We also associate an arbitrarily chosen list $L(e)$ of at least $\left(\frac{3}{2} + \varepsilon\right)\beta$ colours to all edges e of $G[X]$ in H. Using property (S3)(ii) in Lemma 2.2 it is not difficult to prove that a list edge colouring of H, using the edge list assignment L defined as above, provides a proper extension of the list (V^β, B)-colouring of G to Y by giving each vertex $y \in Y$ the colour of the edge e_y in H.

So we need to prove that there is a list edge colouring of H using the lists $L(e)$ defined above. From its definition, H "inherits" properties related to the structural properties in (S3). Together with the observations above, that means that the multigraph H and the list assignment L of the edges of H satisfies the following properties.

(H1) For all vertices v in H we have $d_H(v) \le d_G(v) \le \beta$.
(H2) For all edges $e = uv$ in H: $|L(e)| \ge \left(\frac{3}{2} + \varepsilon\right)\beta - (d_G(u) - d_H(u)) - (d_G(v) - d_H(v)) - 10$.
(H3) For all non-empty subsets $W \subseteq V(H)$: $\sum_{w \in W}(d_G(w) - d_H(w)) \le |E_H(W, X \setminus W)| + C_2\,|W|$.

The fact that a multigraph H with edge list assignment L satisfying (H1)–(H3) has a proper edge colouring can be proved using a deep result essentially due to Kahn [13].

Full details can be found in [2].

The proof of Theorem 1.5 in [9] follows a similar line to the approach sketched above. Again, a structural result on graphs without $K_{3,k}$-minor leads to an inductive argument. And again we find ourselves in the situation that the non-trivial case is more or less equivalent to finding a list edge colouring of a multigraph. But for that proof we have a couple of extra elements: The structural results are not so clean as Lemma 2.2 (since we are dealing with graphs that are more general than planar graphs), and the required distance between the colours of adjacent vertices means that the list colouring of the edges of the multigraph is "more" than just a normal edge colouring. There are some extra elements involving both the colours on adjacent edges, and between edges that are not adjacent in the multigraph. But all these extra conditions appear to be of smaller order than the main requirement. Repeating and extending the arguments from Kahn [13], we can prove that a suitable colouring of the edges is possible. And this list edge colouring results can be used to prove Theorem 1.5, in a way very much like the sketch of Theorem 1.2 above.

The details of that proof are given in [9].

References

[1] G. Agnarsson and M. M. Halldórsson, Coloring powers of planar graphs, *SIAM J. Discrete Math.*, **16** (2003) 651–662.

[2] O. Amini, L. Esperet and J. van den Heuvel, A unified approach toe distance-two colouring of graphs on surfaces, Submitted (2008). A preprint is available at www.cdam.lse.ac.uk/Reports/Files/cdam-2008-23.pdf.

[3] J. A. Bondy and U. S. R. Murty, Graph Theory with Applications, Macmillan, London and Elsevier, New York (1976).

[4] O. V. Borodin, Solution of the Ringel problem on vertex-face coloring of planar graphs and coloring of 1-planar graphs (in Russian), *Metody Diskret. Analyz.*, **41** (1984) 12–26.

[5] O. V. Borodin, H. J. Broersma, A. Glebov and J. van den Heuvel, Minimal degrees and chromatic numbers of squares of planar graphs (in Russian), *Diskretn. Anal. Issled. Oper. Ser. 1*, **8(4)** (2001) 9–33.

[6] O. V. Borodin, D. P. Sanders and Y. Zhao, Cyclic colorings and their generalizations, *Discrete Math.*, **203** (1999) 23–40.

[7] D. W. Cranston and S.-J. Kim, List-coloring the square of a subcubic graph, Manuscript (2006).

[8] R. Diestel, Graph Theory, 3rd edition. Springer-Verlag, Berlin and Heidelberg (2005).

[9] F. Havet, J. van den Heuvel, C. McDiarmid and B. Reed, List colouring squares of planar graphs, Submitted (2008), A preprint is available at www.cdam.lse.ac.uk/Reports/Files/cdam-2008-09.pdf.

[10] J. van den Heuvel and S. McGuinness, Coloring the square of a planar graph, *J. Graph Theory*, **42** (2003) 110–124.

[11] T. R. Jensen and B. Toft, Graph Coloring Problems, John-Wiley & Sons, New York (1995).

[12] T. K. Jonas, Graph coloring analogues with a condition at distance two: $L(2, 1)$-labelings and list λ-labelings, Ph.D. Thesis, University of South Carolina (1993).

[13] J. Kahn, Asymptotics of the list-chromatic index for multigraphs Random Structures Algorithms, **17** (2000) 117–156.

[14] A. V. Kostochka and D. R. Woodall, Choosability conjectures and multicircuits, *Discrete Math.*, **240** (2001) 123–143.

[15] M. Molloy and M. R. Salavatipour, A bound on the chromatic number of the square of a planar graph, *J. Combin. Theory Ser. B*, **94** (2005) 189–213.

[16] O. Ore and M. D. Plummer, Cyclic coloration of plane graphs, In: Recent Progress in Combinatorics, Proceedings of the Third Waterloo Conference on Combinatorics, Academic Press, San Diego (1969) 287–293.

[17] D. P. Sanders and Y. Zhao, A new bound on the cyclic chromatic number, *J. Combin. Theory Ser. B*, **83** (2001) 102–111.

[18] C. Thomassen, The square of a planar cubic graph is 7-colorable, Manuscript (2007).

[19] G. Wegner, Graphs with given diameter and a coloring problem, Technical Report, University of Dortmund (1977).

[20] S. A. Wong, Colouring graphs with respect to distance, M.Sc. Thesis, Department of Combinatorics and Optimization, University of Waterloo (1996).

Proc. ICDM 2008, RMS-Lecture Notes Series No. 13, 2010, pp. 69–79.

Graceful Labelings of Unicyclic Graphs

Jay Bagga[1] and Adrian Heinz[2]

[1]Ball State University, Muncie, Indiana, USA
[2]Georgia Gwinnett College, Lawrenceville, GA, USA
e-mail: jbagga@bsu.edu; aheinz@ggc.edu

Abstract. Graceful graphs were first studied by Rosa in 1966. In this paper we present an algorithm to generate graceful labelings of a special class of unicyclic graphs. These graphs $Sun(n, k)$ are obtained by adding k pendant edges at each vertex of a cycle C_n. We also present some properties of graceful labelings of $Sun(n, k)$.

1. Introduction

Graceful labelings of graphs were first defined and studied in a paper by Rosa [11]. He defined several types of graphs labelings in order to study decompositions of complete graphs into trees. One of these labelings was called a β-valuation in Rosa's paper. See below for definitions. In 1972 Golomb [9] used the term graceful labeling for a β-valuation. Since then graceful labelings have been studied extensively and several variations have also been introduced. For an up-to-date survey on graceful and other graph labelings, we refer the reader to the excellent survey by Gallian [8].

In this paper we study properties of graceful labelings of a special class of unicyclic graphs called $Sun(n, k)$, which is the class of unicyclic graphs obtained by adding k pendant edges to each vertex of the cycle C_n ($n \geq 3$, $k \geq 0$). In [7] Frucht and Harary defined the corona $G \odot H$ of two graphs G and H. If G has n vertices then the corona $G \odot H$ is the graph obtained by taking n copies of H (one for each vertex of G) and then by adding edges from each vertex of G to all vertices in its copy of H. It follows that $Sun(n, k) = C_n \odot kK_1$. We shall call the k pendant edges at each vertex of the cycle as the *rays*.

In particular, we present an algorithm to determine all graceful labelings of $Sun(n, k)$. In an earlier paper [2] we presented a graceful labeling enumeration algorithm for cycles C_n. The algorithm presented in this paper generalizes the earlier algorithm to $Sun(n, k)$. The data generated by our algorithms have led to the discovery of several properties of graceful labelings of cycles and $Sun(n, k)$. We presented several such properties for cycles in [3]. In this paper we obtain some new results on graceful labelings of $Sun(n, k)$.

This paper is organized as follows. In this section we give a short historical account of progress in this field, introduce the definitions and list several known results and questions. In Section 2 we present some properties of graceful labelings of $Sun(n, k)$.

In Section 3 we describe our algorithm for determining all graceful labelings of $Sun(n,k)$. We close with a summary of our conclusions and describe our proposed future work.

For a graph G, we are interested in labeling the vertices and/or edges of G with nonnegative integers and study labelings which have some special properties. More formally, a *labeling* of a graph $G = (V, E)$ is a one-one mapping $f : V \to X$, where X is a (finite) set of nonnegative integers. Such a vertex labeling f induces an edge labeling, where an edge $e = uv$ gets the label $|f(u) - f(v)|$. Suppose that G has q edges. We are interested in labelings where the set of possible labels X is $\{0, 1, 2, \ldots, q\}$ or $\{0, 1, 2, \ldots, 2q\}$. Rosa [11] called such a labeling a *valuation*. He considered four special types of valuations.

Let f be a valuation of a graph G with q edges.

1. f is called an *α-valuation* if $X \subseteq \{0, 1, 2, \ldots, q\}$, the induced edge labels are $1, 2, \ldots, q$ and there exists an $x \in \{0, 1, 2, \ldots, q\}$ such that for any edge uv of G, either $f(u) \leq x < f(v)$ or $f(v) \leq x < f(u)$.
2. f is called a *β-valuation* if $X \subseteq \{0, 1, 2, \ldots, q\}$ and the induced edge labels are $1, 2, \ldots, q$.
3. f is called a *σ-valuation* if $X \subseteq \{0, 1, 2, \ldots, 2q\}$ and the induced edge labels are $1, 2, \ldots, q$.
4. f is called a *ρ-valuation* if $X \subseteq \{0, 1, 2, \ldots, 2q\}$ and the induced edge labels are $x_1, x_2, \ldots, x_q$ where, $x_i = i$ or $x_i = 2q + 1 - i$ for each $1 \leq i \leq q$.

Observe that the above definitions are from the strongest to weakest in the sense that every α-valuation is also a β-valuation, every β-valuation is also a σ-valuation, and so on. As mentioned earlier, the term graceful labeling was used by Golomb [9] for a β-valuation. Figure 1 shows a graceful labeling of $Sun(8, 3)$.

Rosa considered the above valuations in order to investigate decompositions of graphs into trees. For the sake of completeness, we include a brief description here. A *decomposition* of a graph $G = (V, E)$ can be considered to be a partition of the edge set E into subsets $E_1, E_2, \ldots, E_r$. If H_i is the subgraph of G induced by E_i, then we also say the G decomposes into subgraphs $\{H_i\}$. If the subgraphs H_i are all isomorphic to a single graph (say) H then we say that G is $H-decomposable$ and we write $H|G$. In 1963, Ringel [10] made the following conjecture.

Conjecture 1.1. If T is a tree with m edges, then K_{2m+1} decomposes into $2m + 1$ copies of T.

Rosa [11] proved the following theorem.

Theorem 1.1. *If a tree T with m edges has a graceful labeling, then K_{2m+1} decomposes into $2m + 1$ copies of T.*

In fact, Rosa proved more. A decomposition of K_n by r copies of a subgraph H is called a *cyclic decomposition* if, when K_n is drawn appropriately with its vertices on a regular polygon, the r copies of H can be obtained by rotations of an appropriate copy of H. Rosa [11] proved the following results.

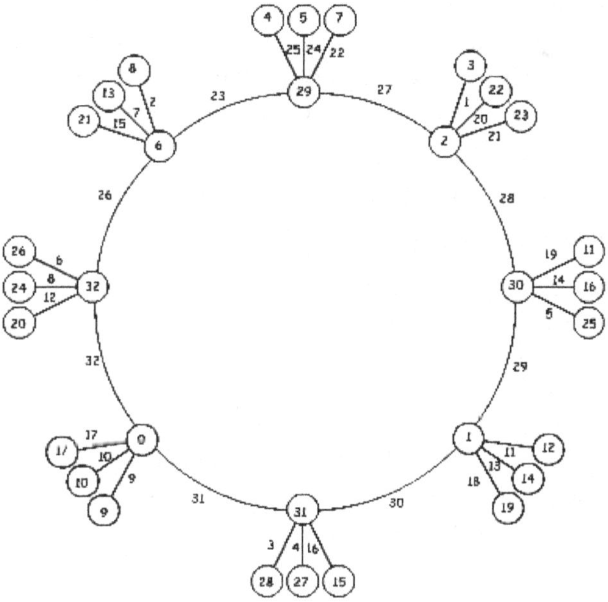

Figure 1. A gracefully labeled *Sun*(8, 3)

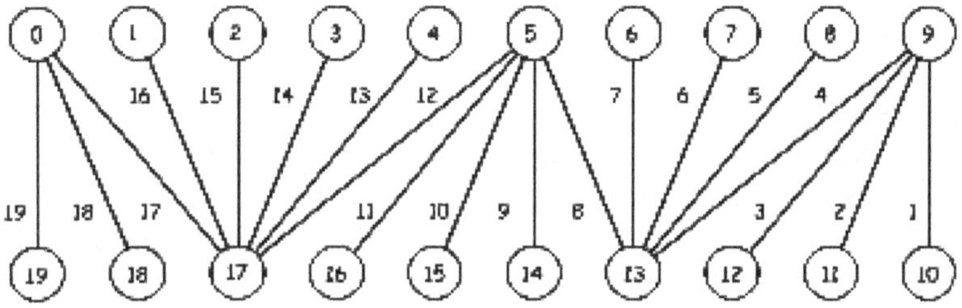

Figure 2. An *α-valuation* of a caterpillar

Theorem 1.2. *Let G be a graph with m edges. Then G has a ρ-valuation if and only if* K_{2m+1} *has a cyclic decomposition into* $2m + 1$ *copies of G.*

Theorem 1.3. *Let G be a graph with m edges, and* $k \geq 1$ *be an integer. If G has an α-valuation then* K_{2km+1} *has a cyclic decomposition into* $k(2km + 1)$ *copies of G.*

Conjecture 1.1 was followed by the following stronger conjecture, which is known as the Kotzig–Ringel graceful tree conjecture.

Conjecture 1.2. Every tree has a graceful labeling.

A substantial amount of research effort has been spent towards finding a proof of this conjecture. Aldred and McKay [1] used a computer program to prove that every tree of order up to 27 is graceful. A number of special classes of trees have been shown to be graceful. These include paths, caterpillars, trees with up to four leaves, and several others. Figure 2 shows an *α-valuation* of a caterpillar as given by Rosa [11].

The survey by Edwards and Howard [6] gives more details of results related to the graceful tree conjecture. A more general survey by Gallian [8] covers graceful labelings of trees but also includes details of the extensive work done on graceful labelings of other types of graphs. In particular, we have the following result of Rosa [11].

Theorem 1.4. *The cycle C_n is graceful iff $n \equiv 0(mod\ 4)$ or $n \equiv 3(mod\ 4)$. Moreover, C_n has an α-valuation iff $n \equiv 0(mod\ 4)$.*

We observe that for a given graceful labeling of a cycle or a unicyclic graph with q edges, the unique vertex labels belong to the set $\{0, 1, 2, \ldots, q\}$. Since there are q vertices, it follows that exactly one element of the set $\{0, 1, 2, \ldots, q\}$ is missing from the set of labels. We denote this missing label by M.

In [2] we presented an algorithm to generate all graceful labelings of cycles C_n for $n \equiv 0(mod\ 4)$ and $n \equiv 3(mod\ 4)$. The data generated by our algorithm led to the discovery of several properties of such graceful labelings. In particular, we proved the following result. See [3] for more results.

Theorem 1.5. *For every graceful labeling of C_n,*

$$\left\lceil \frac{n}{4} \right\rceil \leq M \leq \left\lfloor \frac{3n}{4} \right\rfloor.$$

The bounds are sharp.

Cycles are special cases of unicyclic graphs. Truszczyński [12] conjectured in 1984 that all unicyclic graphs, except cycles C_n with $n \equiv 1(mod\ 4)$ or $n \equiv 2(mod\ 4)$, are graceful. Doma [5] studied unicyclic graphs in which the cycle length is 3 through 9, and showed that all unicyclic graphs on up to nine edges (except C_5, C_6 and, C_9) are graceful. Barrientos [4] studied graceful labelings of a special class of unicyclic graphs. He defined a *hairy cycle* to be a unicyclic graph in which the deletion of any edge in the cycle results in a caterpillar. Barrientos [4] showed that all hairy cycles are graceful. His proof uses an appropriate modification of an α-*valuation* of a caterpillar (as shown in Figure 2).

Our algorithm (as presented in Section 3) generates all graceful labelings of $Sun(n, k)$. We next present some properties of graceful labelings of $Sun(n, k)$.

2. Graceful labelings of *Sun(n, k)*

In this section we present some properties of graceful labelings of $Sun(n, k)$. Observe that our definition of $Sun(n, k)$ includes cycles ($k = 0$). Rosa's Theorem 1.4 showed which cycles are graceful. The next theorem addresses the case $k \geq 1$.

Theorem 2.1. *For $n \geq 3$ and $k \geq 1$, $Sun(n, k)$ is graceful.*

Proof. This follows immediately from the result of Barrientos [4] mentioned above since $Sun(n, k)$ is a hairy cycle.

Proposition 2.1. *For a graceful labeling of Sun(n, k), the number of odd labeled edges on the cycle is even.*

Proof. If there are no odd edges on the cycle, we are done. Otherwise, an odd labeled edge has vertex labels of different parity for its ends. Consider the subgraph of the cycle of *Sun*(n, k) induced by vertices of odd labels on the cycle. This is a union of disjoint paths. If there are p such paths, then there are $2p$ odd edges on the cycle, one at each end of each of these paths.

Theorem 2.2. *For a graceful labeling of Sun(n, k), if k is odd then the missing label M is even.*

Proof. *Sun*(n, k) has $n(k+1)$ vertices and the same number of edges, which is even. Suppose the cycle has a even vertices. We denote these vertices by $v_1, v_2, \ldots, v_a$. We denote the $n-a$ odd vertices of the cycle by $v_{a+1}, v_{a+2}, \ldots, v_n$. Note that all the v_i's occur on the cycle in some order. For each such i, let o_i denote the number odd labeled end vertices adjacent to the vertex v_i.

Hence the number of odd rays in this *Sun*(n, k) is

$$o_1 + o_2 + \cdots + o_a + (k - o_{a+1}) + (k - o_{a+2}) + \cdots + (k - o_n).$$

It follows that the number of odd cycle edges is

$$\frac{n(k+1)}{2} - (o_1 + o_2 + \cdots + o_a) - (k - o_{a+1}) - (k - o_{a+2}) - \cdots - (k - o_n).$$

Now suppose that M is odd. Then the graceful labeling has $\frac{n(k+1)}{2} + 1$ even vertex labels $0, 2, 4, \ldots, n(k+1)$ and $\frac{n(k+1)}{2} - 1$ odd labels $1, 3, \ldots, \hat{M}, \ldots, n(k+1) - 1$, where $\hat{M}$ denotes the missing odd label. On the other hand, the number of odd vertices is $(n-a) + o_1 + o_2 + \cdots + o_n$. This gives $\frac{n(k+1)}{2} = 1 + (n-a) + o_1 + o_2 + \cdots + o_n$.
Thus the number of odd edges on the cycle is

$$1 + (n-a) + o_1 + o_2 + \cdots + o_n - (o_1 + o_2 + \cdots + o_a) - (k - o_{a+1})$$
$$- (k - o_{a+2}) - \cdots - (k - o_n) = 1 + (n-a)(1-k) + 2(o_{a+1} + \cdots + o_n),$$

which is odd, a contradiction to Proposition 2.1.

Theorem 2.3. *For a graceful labeling of Sun(n, k), if k is even, and n and the number of even vertices on the cycle have the same parity, then the missing label M is even.*

Proof. This follows by an argument similar that in Theorem 2.2 and we omit the details.

We observe that M can be odd in general. See Figure 8 for an example of a graceful labeling of *Sun*$(4, 2)$ with $M = 9$.

3. Graceful labeling algorithm for *Sun*(n, k)

We now describe our algorithm for generating graceful labelings of *Sun*(n, k). This is an extension of the backtrack algorithm we gave in our paper [2] for cycles. For

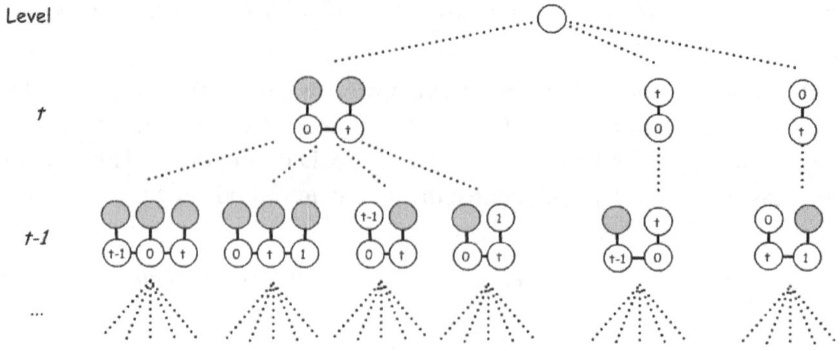

Figure 3. Execution Tree

convenience, we denote the highest edge label $n(k+1)$ by t. Our algorithm finds graceful labelings of $Sun(n,k)$ by generating edge labels in decreasing order as it traverses a tree. Note that the highest edge label t can only be generated from the vertex labels 0 and t and therefore, these vertex labels are adjacent. At the root of this tree (level t), there are three branches, corresponding to the three cases where the edge labeled t is (1) a cycle edge, or (2) a ray with the vertex labeled 0 on the cycle, or (3) a ray with the vertex labeled t on the cycle. Successive edge labels in decreasing order are generated at different levels of the tree as shown in Figure 3. For levels $k < t$, there are several possibilities for edge label k.

3.1 *Description*

The algorithm starts the computation at level $L = t$, where the level indicates that it is necessary to find a labeling containing two labels v_i and v_j such that $|v_i - v_j| = L$. At level $L = t$ there exists only two labels, namely 0 and t and hence the initial labeling consists of two adjacent vertices with these labels. The next step is to find labelings for $L = t-1$. The two alternatives available consist of adding the label $t-1$ adjacent to 0 or the label 1 adjacent to t. For each case the algorithm splits into a new computation branch. The algorithm continues computing for $L = t-2$ by splitting into several computation branches each time and recursively calling each branch for other values of L. The computation for a particular branch continues until a graceful labeling has been found or no graceful labeling is possible for the current branch. Figure 3 illustrates the execution tree.

3.2 *General Step*

To describe the general step of our algorithm, we define $c_1, c_2, \ldots, c_n$ as vertex labels in the cycle of $Sun(n,k)$ and r_{ij} where $1 \le j \le k$ as the jth pendant vertex label adjacent to c_i. We denote by $\overline{r_i}$ the set $\{r_{i1}, r_{i2}, \ldots, r_{ik}\}$ of pendant vertex labels adjacent to c_i.

A given labeling $f = <(c_1, \overline{r_i}), \ldots, (c_n, \overline{r_n})>$ of $Sun(n,k)$ can be considered an ordered sequence of labels. A *sublabeling* is an ordered union of disjoint subsequences

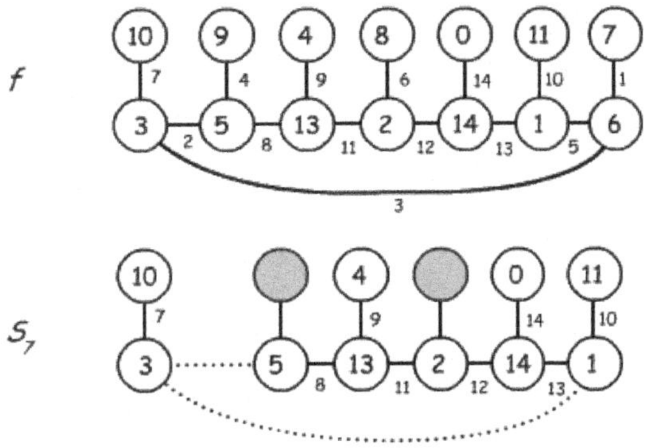

Figure 4. Sublabeling S_7 of f

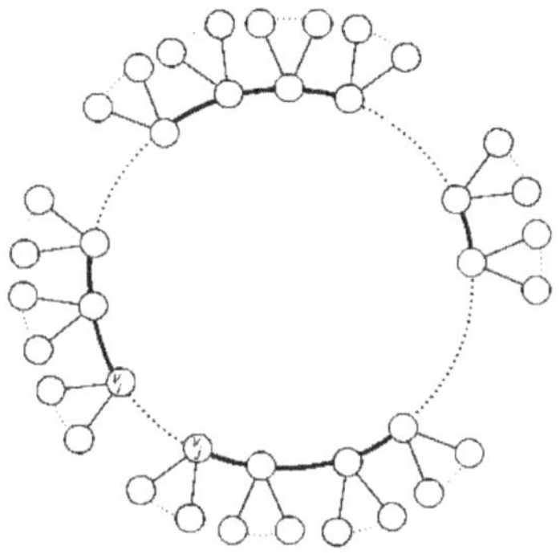

Figure 5. Case (i)

of f. With $t = n(k+1)$ as the highest label, we denote by S_q with $1 \leq q \leq t$ a sublabeling containing edge labels $q, q+1, \ldots, t$.

For instance, for the graceful labeling

$$f = \langle(3, (10)), (5, (9)), (13, (4)), (2, (8)), (14, (0)), (1, (11)), (6, (7))\rangle$$

of $Sun(7, 1)$ displayed in Figure 4 at the top, the sublabeling S_7 of f displayed at the bottom consists of $S_7 = \langle(3, (10))\rangle\langle(5, (-)), (13, (4)), (2, (-)), (14, (0)), (1, (11))\rangle$ and therefore S_7 is the ordered union of two subsequences. Note that $f = S_1$.

As our algorithm computes graceful labelings by traversing a tree and splitting into computation branches, it is essential to determine at each step how to find the next edge label. We proceed to explain how this process is achieved.

For a given level l, the previous sublabeling is S_{l+1}. At every step, it is necessary to find two vertex labels v_i and v_j such that $|v_i - v_j| = l$. There are three cases to consider.

(i) $v_i \in S_{l+1}$ *and* $v_j \in S_{l+1}$. Subsequences are merged into a single subsequence. Note that in Figure 5, v_i and V_j are at the extremity of their corresponding subsequences before merging.

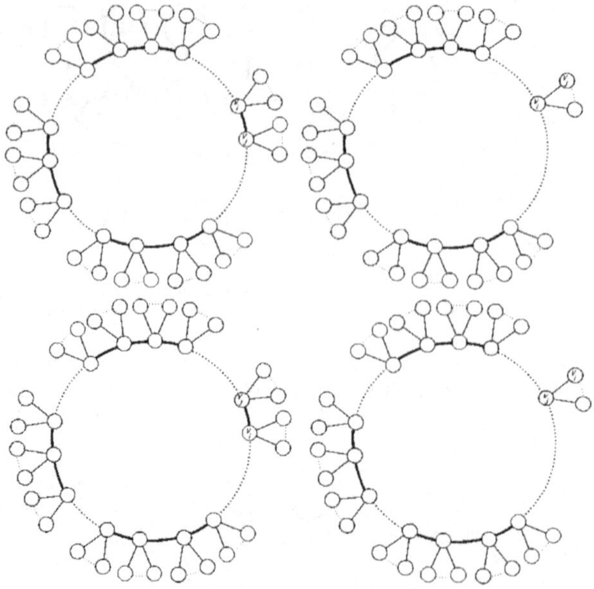

Figure 6. Case (ii)

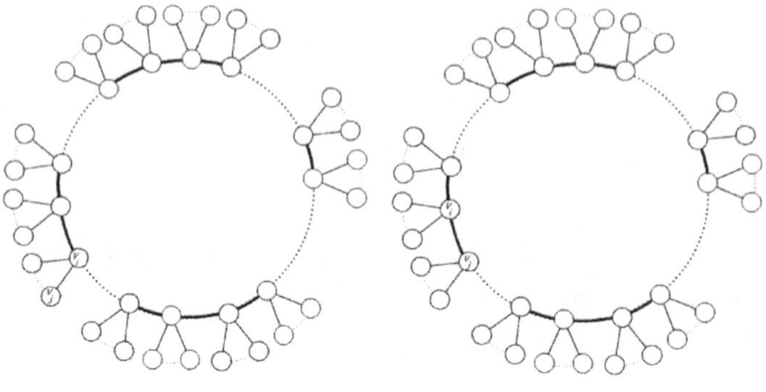

Figure 7. Case (iii)

(ii) $v_i \notin S_{l+1}$ *and* $v_j \notin S_{l+1}$. In this instance, v_i and v_j are added adjacent to each other. Figure 6 illustrates all possible scenarios. Observe that in the sublabelings displayed on the left, both v_i and v_j are added in the cycle while in the sublabelings shown on the right, v_i is added in the cycle and v_j as ray (top) or v_j in the cycle and v_i as ray (bottom).

(iii) $v_i \in S_{l+1}$ *and* $v_j \notin S_{l+1}$. Then v_j is added adjacent to v_i. Notice that in Figure 7, v_j can be added as a ray of v_i (right) or as neighbor in the cycle (left).

If none of these cases occur at a given branch, then that computation branch dies. We have also verified that the algorithm is correct in that it generates each graceful labeling precisely once. We omit the proof.

3.3 *Proof of correctness*

Theorem 3.1. *Suppose* $f = \langle(c_1, (\overline{r_1})), \ldots, (c_n, (\overline{r_n}))\rangle$ *is a graceful labeling of* $Sun(n, k)$. *The algorithm achieves* f *exactly once.*

Proof. Let $t = n(k + 1)$ be the highest vertex label. The proof is by induction on the edge labels $l = t, t - 1, \ldots, 2, 1$.

Base Case: $S_t = \langle(0, (t))\rangle$ or $S_t = \langle(t, (0))\rangle$ or $S_t = \langle(0, (-)), (t, (-))\rangle$ since for any f, the edge label t can only be generated by taking 0 and t together. The algorithm does this at the very first step by creating independent execution branches for each instance. For $t - 1$, six branches are created as illustrated in Figure 3.

Notice that on each branch, either vertex label $t - 1$ or 1 was added adjacent to 0 or t respectively to obtain edge label $t - 1$. The algorithm considers all these cases and hence it achieves S_{t-1}. Also, since f is given exactly one of these alternatives will match with S_{t-1} since they are distinct options. This indicates that at edge label $t - 1$ the algorithm achieves S_{t-1} uniquely.

Now, for a general label l with $1 \leq l \leq t$, assume that the algorithm achieves S_{l+1} uniquely. Suppose that in S_l the edge label l is obtained by vertex labels v_i and v_j in f, so that $|v_i - v_j| = l$. We consider S_l an order union of subsequences. There are three cases:

(i) l is an interior edge of the cycle.
(ii) l is the only edge of a subsequence.
(iii) l is a pendant edge of a subsequence either as a ray or in the cycle.

As f is given, precisely one of these will match S_l. Since the algorithm achieves S_{l+1} uniquely, exactly one of the three cases described in Section 3.2 applies to achieve S_l. Since $S_1 = f$ the theorem follows by induction.

3.4 *Sample output*

Figure 8 displays partial output for the computation of graceful labelings of $Sun(4, 2)$. On each row, the number to the left corresponds to the current level. It is followed by the vertex labels in the cycle. The next set of four numbers are labels of the end vertices of the first ray at each of the four cycle vertices. Similarly, the last set of four numbers are labels of the end vertices of the second ray at each of the four cycle vertices. The character E denotes an empty ray and M indicates the missing label. When the algorithm finds a graceful labeling, it displays the symbol $*$ to the left of it. In this example, the graceful labeling found is $f = \langle(11, (3, 4)), (0, (1, 6)), (12, (8, 10)), (2, (5, 7))\rangle$.

4. Summary

In this paper we presented an algorithm for generating all graceful labelings of $Sun(n, k)$, a class of unicyclic graphs. The data generated by this algorithm led us to

```
12 | 0 12 |->|E E |E E |
11 | 11 0 12 |->|E E E |E E E |
10 | 11 0 12 2 -2 |->|E E E E E |E E E E E |
9 | 11 0 12 2 |->|E E E E |E E E E |
8 | 11 0 12 2 |->|3 E E E |E E E E |
7 | 11 0 12 2 |->|3 E E E |4 E E E |
6 | 11 0 12 2 |->|3 6 E E |4 E E E |
5 | 11 0 12 2 |->|3 5 E E |4 6 E E |
4 | 11 0 12 2 |->|3 5 8 E |4 6 E E |
3 | 11 0 12 2 |->|3 5 8 E |4 6 9 E |
5 | 11 0 12 2 |->|3 6 7 E |4 E E E |
4 | 11 0 12 2 |->|3 6 7 E |4 E 8 E |
3 | 11 0 12 2 |->|3 6 7 5 |4 E 8 E |
5 | 11 0 12 2 |->|3 6 E 7 |4 E E E |
4 | 11 0 12 2 |->|3 6 8 7 |4 E E E |
3 | 11 0 12 2 |->|3 6 8 7 |4 E 9 E |
3 | 11 0 12 2 |->|3 6 8 5 |4 E E 7 |
2 | 11 0 12 2 |->|3 6 8 5 |4 E 10 7 |
* 1 | 11 0 12 2 |->|3 1 8 5 |4 6 10 7 |   M=9   [1]
6 | 11 0 12 2 |->|3 E 6 E |4 E E E |
5 | 11 0 12 2 |->|3 5 6 E |4 E E E |
4 | 11 0 12 2 |->|3 5 6 E |4 E 8 E |
5 | 11 0 12 2 |->|3 E 6 E |4 E 7 E |
5 | 11 0 12 2 |->|3 E 6 7 |4 E E E |
4 | 11 0 12 2 |->|3 E 6 7 |4 E 8 E |
3 | 11 0 12 2 |->|3 E 6 5 |4 E 8 7 |
6 | 11 0 12 2 |->|3 E E 8 |4 E E E |
5 | 11 0 12 2 |->|3 5 E 8 |4 E E E |
```

Figure 8. Sample output for *Sun*(4, 2)

the discovery of some properties of such labelings. This research is ongoing. We also propose to investigate algorithms for other classes of graphs such as caterpillars and other unicyclic graphs. We thank the referees for their helpful comments.

References

[1] R. E. L. Aldred and B. D. McKay, Graceful and harmonius labelings of trees, *Bull. Inst. Combin. Appl.*, **23** (1998) 69–72.

[2] J. Bagga, A. Heinz and M. M. Majumder, An Algorithm for Computing All Graceful labelings of Cycles, *Congressus Numerantium*, **186** (2007) 57–63.

[3] J. Bagga, A. Heinz and M. M. Majumder, Properties of Graceful Labelings of Cycles, *Congressus Numerantium*, **188** (2007) 109–115.

[4] C. Barrientos, Graceful graphs with pendant edges, *Autralasian Journal of Combinatorics*, **33** (2005) 99–107.

[5] J. Doma, Unicyclic Graceful Graphs, M. S. Thesis, McMaster Univ. (1991).

[6] M. Edwards and L. Howard, A Survey of Graceful Trees, *Atlantic Electronic Journal of Mathematics*, **1(1)** (2006) 5–30.

[7] R. Frucht and F. Harary, On the corona of two graphs, *Aequationes Mathematicae*, **4(3)** (1970) 322–325.

[8] J. A. Gallian, A Dynamic Survey of Graph Labeling, *The Electronic Journal of Combinatorics*, **16** (2009) #DS6.

[9] S. W. Golomb, How to number a graph, in Graph Theory and Computing, R. C. Read, ed., Academic Press, New York (1972) 23–37.

[10] G. Ringel, Problem 25. In Theory of Graphs and Its Applications (Proc. Symp. Smolenice 1963), Czech. Acad. Sci. (1964) 162.

[11] A. Rosa, On certain valuations of the vertices of a graph, Theory of Graphs (Internat. Symposium, Rome, July 1966), Gordon and Breach, N. Y. and Dunod Paris (1967) 349–355.

[12] M. Truszczyński, Graceful unicyclic graphs, *Demonstatio Mathematica*, **17** (1984) 377–387.

Proc. ICDM 2008, RMS-Lecture Notes Series No. 13, 2010, pp. 81–89.

The Spectra of Super Line Multigraphs

Jay Bagga[1], Robert B. Ellis[2] and Daniela Ferrero[3]

[1]Department of Computer Science, Ball State University, Muncie, IN
[2]Department of Applied Mathematics, Illinois Institute of Technology, Chicago, IL
[3]Department of Mathematics, Texas State University, San Marcos, TX
e-mail: jbagga@bsu.edu; rellis@math.iit.edu; dferrero@txstate.edu

Abstract. For an arbitrary simple graph G and a positive integer r, the super line multigraph of index r of G, denoted $\mathcal{M}_r(G)$, has for vertices all the r-subsets of edges. Two vertices S and T are joined by as many edges as pairs of distinct edges $s \in S$ and $t \in T$ share a common vertex in G. We present spectral properties of $\mathcal{M}_r(G)$ and particularly, if G is a regular graph, we calculate all the eigenvalues of $\mathcal{M}_r(G)$ and their multiplicities in terms of those of G.

Keywords. Super line graph, adjacency matrix, graph spectrum.

Subject Classification: 05C50, 05C12, 05C20

1. Introduction

Bagga, Beineke, and Varma introduced the concept of super line graph in [3]. Given a simple graph (i.e. without loops or multiple edges) $G = (V, E)$, let $p = |V|$ be the number of vertices and $q = |E|$ be the number of edges. For a given positive integer $r \leq q$, the *super line graph of index r* of G is the graph $\mathcal{L}_r(G)$ which has for vertices all of the r-subsets of E, and two vertices S and T are adjacent whenever there exist distinct $s \in S$ and $t \in T$ such that s and t share a common vertex. Super line graphs are a generalization of line graphs. Indeed, from the definition it follows that $\mathcal{L}_1(G)$ coincides with the line graph $L(G)$. Properties of super line graphs were presented in [2,4], and a good and concise summary can be found in [15]. A recent survey on several generalizations of line graphs can be found in [1]. More specifically, some results regarding the super line graph of index 2 were presented in [6] and [2].

This paper deals with super line multigraphs, a variation of the super line graph that was first defined and studied in [2]. Intuitively, the super line multigraph of a simple graph has the same set of vertices as the super line graph, but the number of edges between any two vertices S and T is given by the number of pairs $s \in S$ and $t \in T$ such that s and t are distinct and share a common vertex. Formally, for a given graph $G = (V, E)$ and a positive integer $r \leq q$, the *super line multigraph of index r* of G is the multigraph $\mathcal{M}_r(G)$ whose vertices are the r-subsets of $E(G)$, and two vertices S and T are joined by as many edges as ordered pairs of edges $(s, t) \in S \times T$ such that s and t are distinct and share a common vertex in G. Let $n = \binom{q}{r}$ be the number of vertices of $\mathcal{M}_r(G)$ and let m be the number of edges of $\mathcal{M}_r(G)$. In this paper, the underlying graph G is simple.

A simple graph G on p vertices labeled as $v_1, \ldots, v_p$ can be associated with an *adjacency matrix* $A = A(G)$, which is the $p \times p$ matrix whose entries $a_{i,j}$ are given by $a_{i,j} = 1$ if there is an edge joining v_i and v_j in G and $a_{i,j} = 0$ otherwise. If G has q edges labeled as $e_1, \ldots, e_q$, the *incidence matrix* $B = B(G)$ is the $p \times q$ matrix whose entries $b_{i,j}$ are given by $b_{i,j} = 1$ if e_j is incident to v_i, and 0 otherwise. Analogously, the entry $a_{i,j}$ of the adjacency matrix of a multigraph is the number of edges between v_i and v_j when $i \neq j$, and is the number of loops at v_i when $i = j$. The *characteristic polynomial* of the graph G, denoted by $\chi(G; \lambda)$, is defined as $\det(A - \lambda I)$. The *eigenvalues* of the graph G are those of A. The *algebraic multiplicity* of an eigenvalue α is the multiplicity of α as a root of the characteristic polynomial of A and is denoted by $m_a(\alpha, A)$ or $m_a(\alpha, G)$. The *geometric multiplicity* of an eigenvalue α is the dimension of $\ker(A - \alpha I)$ and is denoted by $m_g(\alpha, A)$ or $m_g(\alpha, G)$. If A is a real symmetric matrix, every eigenvalue α satisfies $m_a(\alpha, A) = m_g(\alpha, A)$. The *spectrum* of G is the set of eigenvalues of G together with their multiplicities as eigenvalues of A. The spectrum of a graph provides valuable information on its structure. More information on this topic can be found in [7,9,14].

The authors first investigated the spectra of super line multigraphs in [5]. In this paper we continue that study and give several new results as well extensions of some results in [5]. It is known that the eigenvalues of a line graph are greater than or equal to -2 (see [10] for properties of graphs with least eigenvalue -2). In this paper we generalize that result by giving a lower bound for the eigenvalues of any super line multigraph. Moreover, we characterize the graphs whose super line multigraph has the lower bound as an eigenvalue. This is done through the study of the relationships between the adjacency matrix and characteristic polynomial of a graph and the adjacency matrix and characteristic polynomial of its super line multigraph. Furthermore, those relationships allow us to give a complete description of the spectrum of the super line multigraph of a regular graph in terms of the spectrum of the original graph.

We refer the reader to [8,12] for background on terminology or concepts not included here.

2. General results

Given a super line multigraph $\mathcal{M}_r(G)$ of index $r \leq q$ where $n = \binom{q}{r}$ represents its number of vertices, we denote by $U_{q,r}$ the $n \times q$ binary matrix whose rows are the binary strings of length q with exactly r entries equal to 1. Therefore, if G is a graph with q edges $e_1, \ldots, e_q$, the rows of the matrix $U_{q,r}$ represent all the possible r-subsets of edges, or the vertices of the super line multigraph $\mathcal{M}_r(G)$. We assume that the rows of $U_{q,r}$ are ordered "lexicographically" in the order of the edges $e_1, \ldots, e_q$. Thus, the first row (with ones in the first r entries) corresponds to the vertex $\{e_1, \ldots, e_r\}$ in $\mathcal{M}_r(G)$, while the last row (with ones in the last r entries) corresponds to the vertex $\{e_{q-r+1}, \ldots, e_q\}$. Whenever the parameters q and r are apparent from context, we will use U in place of $U_{q,r}$. Henceforth, for any matrix X, let $(X)_{ij}$ denote the entry in row i and column j of X, and let X^t denote the transpose of X. Before considering the spectrum of $\mathcal{M}_r(G)$, we first obtain a formula for the number of its edges.

Lemma 2.1. *Let $1 \leq r < q$. If G is a simple graph with q edges and p vertices whose degrees are given by the sequence $(d_1, \ldots, d_p)$, then the numbers of non-loop edges and loop edges, respectively, of $\mathcal{M}_r(G)$ are*

$$\left[\binom{q-1}{r-1}^2 - \binom{q-2}{r-2}\right]\sum_{i=1}^{p}\binom{d_i}{2}, \quad \text{and} \quad 2\binom{q-2}{r-2}\sum_{i=1}^{p}\binom{d_i}{2}.$$

Proof. The proof is a straightforward extension of the proof for the case $r = 2$ of [2, Theorem 4.1]. Given a vertex v of degree d, consider a pair of edges $e, f \in \{e_1 \ldots, e_q\}$ in G that are incident at v. By renumbering if necessary, assume $e = e_1$ and $f = e_2$.

We count the number of non-loop edges of $\mathcal{M}_r(G)$ that arise due to the co-incidence of e_1, e_2 at v. These edges will have distinct end-points of the form $S = \{e_1, f_2, f_3, \ldots, f_r\}$ and $T = \{e_2, f_2', \ldots, f_r'\}$, When $\{e_1, e_2\} \not\subseteq S \cap T$, we may view S and T as being ordered, distinguished by $e_1 \in S$ and $e_2 \in T$. When $\{e_1, e_2\} \subseteq S \cap T$, the incidence of e_1 and e_2 gives rise to two edges of the form $\{S, T\}$: one from $e_1 \in S$ and $e_2 \in T$, and the other from $e_2 \in S$ and $e_1 \in T$. The total number of non-loop edges arising from the incidence of e_1 and e_2 is therefore obtained by selecting $\{f_2, \ldots, f_r\} \subseteq E(G)\backslash\{e_1\}$ and $\{f_2', \ldots, f_r'\} \subseteq E(G)\backslash\{e_2\}$, ignoring only the case $S = T$, and is therefore

$$\binom{q-1}{r-1}^2 - \binom{q-2}{r-2}.$$

This quantity multiplies by the $\binom{d}{2}$ pairs of edges $\{e, f\}$ incident at v.

The loop edges arising from the co-incidence of e_1 and e_2 are of the form $\{\{e_1, e_2, f_3, \ldots, f_r\}, \{e_1, e_2, f_3, \ldots, f_r\}\}$, where $f_3, \ldots, f_r$ are distinct edges in $E(G)\backslash\{e_1, e_2\}$. There are $\binom{q-2}{r-2}$ ways to select $f_3, \ldots, f_r$, and by definition, one instance of this form gives rise to two loops. The formula for loops follows since there are $\binom{d}{2}$ pairs of edges e, f incident at v. ■

Lemma 2.1 also holds for $r = q$, when $\mathcal{M}_r(G)$ is a 1-vertex graph with no non-loop edges. We will need the following lemma and its corollary. These were first presented without proof in [5]. For the sake of completeness we give the proofs here.

Lemma 2.2. *For a simple graph G, the adjacency matrix of the super line multigraph $\mathcal{M}_r(G)$ is given by*

$$A(\mathcal{M}_r(G)) = UA_LU^t$$

where A_L is the adjacency matrix of $L(G)$.

Proof. As above, let the edges of G be $e_1, \ldots, e_q$. Given two vertices i and j in $\mathcal{M}_r(G)$, we count the edges between i and j that arise due to an adjacency of e_k in j with edges of i, for $1 \leq k \leq q$. Now the $(i, j)^{\text{th}}$ entry of UA_LU^t is

$$(UA_LU^t)_{ij} = \sum_{k=1}^{q}(UA_L)_{ik}(U^t)_{kj} = \sum_{k=1}^{q}(UA_L)_{ik}(U)_{jk}.$$

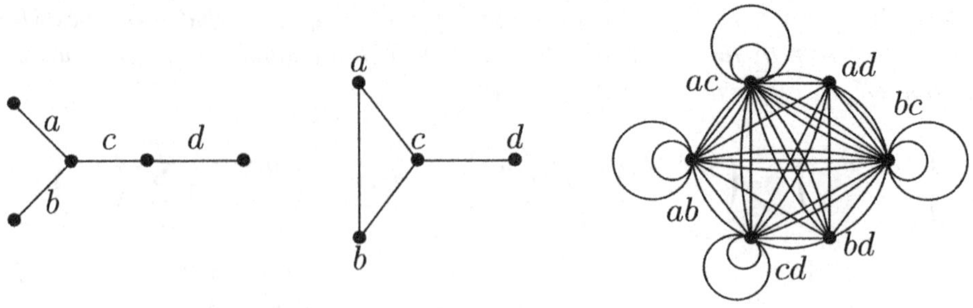

Figure 1. A graph G, its line graph $L(G)$ and super line graph $\mathcal{M}_2(G)$.

Moreover, $(UA_L)_{ik} = \sum_{l=1}^{q}(U)_{il}(A_L)_{lk}$ counts the number of adjacencies of vertex i that arise due to the adjacency of an edge in this vertex with the edge e_k. On the other hand, $(U)_{jk} = 1$ iff the edge e_k is contained in vertex j. The result follows. ■

Note that for a simple graph G, $\mathcal{M}_r(G)$ may have loops, and consequently, $A(\mathcal{M}_r(G))$ may have nonzero entries in the diagonal. The next matrix computation illustrates how the Lemma 2.2 works in the example presented in Figure 1. The edge labels of G, in alphabetical order, are the indices of A_L.

$$A_L = \begin{pmatrix} 0 & 1 & 1 & 0 \\ 1 & 0 & 1 & 0 \\ 1 & 1 & 0 & 1 \\ 0 & 0 & 1 & 0 \end{pmatrix} \qquad q = 4, r = 2;$$

$$U_{4,2}A_L U_{4,2}^t = \begin{pmatrix} 1 & 1 & 0 & 0 \\ 1 & 0 & 1 & 0 \\ 1 & 0 & 0 & 1 \\ 0 & 1 & 1 & 0 \\ 0 & 1 & 0 & 1 \\ 0 & 0 & 1 & 1 \end{pmatrix} \begin{pmatrix} 0 & 1 & 1 & 0 \\ 1 & 0 & 1 & 0 \\ 1 & 1 & 0 & 1 \\ 0 & 0 & 1 & 0 \end{pmatrix} \begin{pmatrix} 1 & 1 & 1 & 0 & 0 & 0 \\ 1 & 0 & 0 & 1 & 1 & 0 \\ 0 & 1 & 0 & 1 & 0 & 1 \\ 0 & 0 & 1 & 0 & 1 & 1 \end{pmatrix};$$

$$A(\mathcal{M}_r(G)) = U_{4,2}A_L U_{4,2}^t = \begin{pmatrix} 2 & 3 & 1 & 3 & 1 & 2 \\ 3 & 2 & 2 & 3 & 3 & 2 \\ 1 & 2 & 0 & 3 & 1 & 2 \\ 3 & 3 & 3 & 2 & 2 & 2 \\ 1 & 3 & 1 & 2 & 0 & 2 \\ 2 & 2 & 2 & 2 & 2 & 2 \end{pmatrix}.$$

Corollary 2.3. *Under the same assumptions as in Lemma 2.2,*

$$A(\mathcal{M}_r(G)) = U_{q,r}B^tBU_{q,r}^t - 2U_{q,r}U_{q,r}^t,$$

where B is the incidence matrix of G.

Proof. We observe that $B^t B = A_L + 2I$. The proof now follows from Lemma 2.2. ■

Observe that $U_{q,1}^t U_{q,1} = I$. Besides, if $2 \le r < q$, it was shown in [5] that the product $U_{q,r}^t U_{q,r}$ still has a particular pattern. We will give an expression for $U_{q,r}^t U_{q,r}$, for any r, $1 \le r < q$, that will be very useful in proving other results in this work. We need the following standard result from linear algebra, which we will apply to the eigensystems of UU^t and $U^t U$.

Lemma 2.4. *Let M and N be two (real or complex) matrices of sizes $m \times n$ and $n \times m$, respectively. Then* $\det(NM - xI_n) = (-x)^{n-m} \det(MN - xI_m)$. ■

Lemma 2.5. *For an integer $q \ge 2$ and an integer r, $1 \le r < q$, let b and c be defined by*

$$b = \binom{q-2}{r-2} \quad \text{and} \quad c = \binom{q-1}{r-1}, \quad \text{if } r > 1$$

$$b = 0 \qquad \text{and} \quad c = 1, \qquad \text{if } r = 1$$

Then

$$U_{q,r}{}^t U_{q,r} = bJ + (c-b)I,$$

where J is the all-ones matrix. Furthermore, both $U_{q,r}{}^t U_{q,r}$ and $U_{q,r} U_{q,r}{}^t$ have eigenvalue rc with multiplicity 1 corresponding to eigenvectors $\vec{1}_{n\times 1}$ and $\vec{1}_{q\times 1}$, respectively; and eigenvalue $c - b$ with multiplicity $q - 1$; $U_{q,r} U_{q,r}^t$ also has eigenvalue 0 with multiplicity $n - q$.

Proof. $U^t U$ is a $q \times q$ matrix whose entries are

$$(U^t U)_{ij} = \sum_{k=1}^{\binom{q}{r}} (U^t)_{ik} (U)_{kj} = \sum_{k=1}^{\binom{q}{r}} (U)_{ki} (U)_{kj}.$$

Therefore, for $i = j$, $(U)_{ki}$ and $(U)_{kj}$ coincide in exactly $\binom{q-1}{r-1}$ positions, and for $i \ne j$, those coincide in exactly $\binom{q-2}{r-2}$ positions. Thus the entries of the matrix $U^t U$ are all $b = \binom{q-2}{r-2}$, except in the diagonal where all are $c = \binom{q-1}{r-1}$. For $U^t U$, $\vec{1}$ has eigenvalue rc, and all vectors orthogonal to $\vec{1}$ have eigenvalue $c - b$. Also for UU^t, $\vec{1}$ has eigenvalue rc, because for all $1 \le i \le q$,

$$UU^t \vec{1}(i) = \sum_{j=1}^{n} \sum_{k=1}^{q} (U)_{ik} (U)_{jk} = \sum_{k=1}^{q} (U)_{ik} \sum_{j=1}^{n} (U)_{jk}.$$

This last sum is rc, because for each of r 1's of row i of U, there are $\binom{q-1}{r-1}$ total rows j of U with a 1 in the same column. From Lemma 2.4, the nonzero portions of the spectra of $U^t U$ and UU^t coincide. ■

Next we generalize the result [10, p. 5] that establishes that the eigenvalues of a line graph are greater than or equal to -2, and characterize the cases where the lower bound is attained in terms of a specific eigenvector.

Theorem 2.6. *Let G be a simple graph with q edges and incidence matrix $B = B(G)$, let r be an integer, where $1 \leq r < q$, and set $U = U_{q,r}$. If λ is an eigenvalue of $\mathcal{M}_r(G)$, then*

$$\lambda \geq -2\binom{q-2}{r-1}.$$

Furthermore, the lower bound is achieved iff there exists a nonzero vector $\phi \in \mathbb{R}^n$, where $n = \binom{q}{r}$, such that

(i) $\sum_{i=1}^{n} \phi(i) = 0$,
(ii) $BU^t\phi = \vec{0}$, *and*
(iii) $UU^t\phi = \binom{q-2}{r-1}\phi$.

Proof. The spectrum of UU^t is given by Lemma 2.5. Let $\{(\mu_i, \psi_i) : 1 \leq i \leq n\}$ be an orthonormal eigensystem for UU^t with eigenvalues $\mu_1 = r\binom{q-1}{r-1}$, $\mu_2 = \cdots = \mu_q = \binom{q-2}{r-1}$, and $\mu_{q+1} = \cdots = \mu_n = 0$, so that $\psi_1 = \vec{1}/\|\vec{1}\| = n^{-1/2}\vec{1}$. Note that $A(\mathcal{M}_r(G))$ is real symmetric and thus Hermitian. If λ is the smallest eigenvalue of $A(\mathcal{M}_r(G))$, then by the Rayleigh-Ritz Theorem ([13, p. 176]), we have $\lambda = \min_{\phi\in\mathbb{R}^n, \|\phi\|=1}\langle\phi, A(\mathcal{M}_r(G))\phi\rangle$. Since $\{\phi \in \mathbb{R}^n : \|\phi\| = 1\}$ is compact, let ϕ_0 be a vector achieving the minimum. Write $\phi_0 = \sum_{i=1}^{n} \alpha_i\psi_i$, so that

$$\begin{aligned}\lambda &= \left\langle \sum_{i=1}^{n} \alpha_i\psi_i, A(\mathcal{M}_r(G))\sum_{i=1}^{n} \alpha_i\psi_i \right\rangle \\ &= \langle \alpha_1 n^{-1/2}\vec{1}, A(\mathcal{M}_r(G))\alpha_1 n^{-1/2}\vec{1}\rangle \\ &\quad + \left\langle \sum_{i=2}^{n} \alpha_i\psi_i, A(\mathcal{M}_r(G))\sum_{i=2}^{n} \alpha_i\psi_i \right\rangle. \qquad (1)\end{aligned}$$

Since $\alpha_1^2 \geq 0$ and $A(\mathcal{M}_r(G))$ is a nonnegative matrix, the first term of (1) is zero, when $\phi_0 \perp \vec{1}$ or $A(\mathcal{M}_r(G)) = 0$, and otherwise positive. If $\sum_{i=2}^{n} \alpha_i\psi_i = \vec{0}$, the second term of (1) is zero and the lower bound follows. Otherwise, using Corollary 2.3 and the fact that $UU^t\psi_i = \mu_i\psi_i$, the second term is

$$\left\langle \sum_{i=2}^{n} \alpha_i\psi_i, UB^tBU^t\sum_{i=2}^{n} \alpha_i\psi_i \right\rangle - 2\left\langle \sum_{i=2}^{n} \alpha_i\psi_i, UU^t\sum_{i=2}^{n} \alpha_i\psi_i \right\rangle \qquad (2)$$

$$= \left\| BU^t\sum_{i=2}^{n} \alpha_i\psi_i \right\|^2 - 2\binom{q-2}{r-1}\sum_{i=2}^{q} \alpha_i^2 \qquad (3)$$

The lower bound follows since the first term of (2) is nonnegative, and the condition $\|\phi_0\| = 1$ implies that $\sum_{i=2}^{q} \alpha_i^2 \leq 1$.

The lower bound is achieved if and only if $\|BU^t\sum_{i=2}^n \alpha_i \psi_i\|^2 = 0$ and $\sum_{i=2}^q \alpha_i^2 = 1$. Since $\sum_{i=2}^q \alpha_i^2 = 1$ iff $\alpha_1 = \alpha_{q+1} = \cdots \alpha_n = 0$, the lower bound is achieved iff $\phi_0 = \sum_{i=2}^q \alpha_i \psi_i$ and $BU^t\phi_0 = 0$, which is equivalent to conditions (i)–(iii). ■

By using the following result from [11] concerning the kernel of B we can characterize when $\mathcal{M}_r(G)$ achieves the lower bound in the theorem in terms of the structure of G.

Proposition 2.7. *Let G be a simple graph. There exists a nonzero vector $\theta \in \mathbb{R}^q$ such that $\sum_{i=1}^q \theta(i) = 0$ and $B\theta = \vec{0}$ iff G contains an even cycle or two edge-disjoint odd cycles in the same connected component.* ■

Corollary 2.8. *The super line multigraph $\mathcal{M}_r(G)$ of a simple graph G achieves the eigenvalue lower bound in Theorem 2.6 iff G contains an even cycle or two edge-disjoint odd cycles in the same connected component.*

Proof. Suppose that $\phi \in \mathbb{R}^n$ is a nonzero vector that achieves the lower bound and thus satisfies conditions (i)–(iii) of Theorem 2.6. Set $\theta = U^t\phi$. Then from (iii) we have $\theta \neq \vec{0}$; from (i), $\sum_{i=1}^q \theta(i) = \sum_{i=1}^q \sum_{j=1}^n (U^t)_{ij}\phi(j) = \sum_{j=1}^n \phi(j) \sum_{i=1}^q (U)_{ji} = \sum_{j=1}^n \phi(j)r = 0$; and, from (ii), $B\theta = \vec{0}$. Thus the forward direction follows from Proposition 2.7.

For the converse, by Proposition 2.7 suppose there exists a nonzero $\theta \in \mathbb{R}^q$ such that $\sum_{i=1}^q \theta(i) = 0$ and $B\theta = \vec{0}$. Using the same orthonormal eigensystem for UU^t as in the proof of Theorem 2.6, let $\mathcal{U} = \{U^t\psi_1, \ldots, U^t\psi_n\}$, so that $\mathcal{U}$ spans $\mathbb{R}^q$, as the eigenvalues $\mu_1, \ldots, \mu_q$ are nonzero by Lemma 2.5. $\mathcal{U}$ is an orthogonal basis for $\mathbb{R}^q$ since $\langle U^t\psi_i, U^t\psi_j\rangle = \langle UU^t\psi_i, \psi_j\rangle = \mu_i\langle\psi_i, \psi_j\rangle$ for $1 \le i, j \le q$. Now write $\theta = \sum_{i=1}^q \beta_i U^t\psi_i$ and note that $\beta_1 = 0$ because $\langle U^t\psi_1, \theta\rangle = n^{-1/2}\binom{q-1}{r-1}\langle\vec{1}, \theta\rangle = 0$. Set $\phi = \sum_{i=2}^q \beta_i\psi_i$ so that ϕ satisfies conditions (i)–(iii). ■

The following result, which originally appeared in [5], is a direct consequence of Lemma 2.4. We will need this in the next section.

Corollary 2.9. *For a simple graph G, and with notation as above, $\chi(UA_LU^t; \lambda) = (-\lambda)^{n-q}\chi(U^tUA_L; \lambda)$.* ■

Note that if one follows the definition of characteristic polynomial given in [7] the above proposition becomes $\chi(UAU^t; \lambda) = \lambda^{n-q}\chi(U^tUA_L; \lambda)$.

3. Regular graphs

In this section we consider a d-regular graph G with q edges, which has a $2(d-1)$-regular line graph $L(G)$. Let A_L be the adjacency matrix of $L(G)$, and let $\alpha_1 = 2(d-1), \alpha_2, \ldots, \alpha_q$ be the eigenvalues of A_L with corresponding eigenvectors $\phi_1 = \vec{1}, \phi_2, \ldots, \phi_q$, respectively. (Notice that $\vec{1}$ is an eigenvector because G, and therefore $L(G)$, are regular graphs.) Proposition 3.1 and Theorem 3.2 originally appeared in [5], but are included here for completeness, with simplified proof and determination of sign in Theorem 3.2.

Proposition 3.1. *Let G be a simple d-regular graph with q edges, and let A_L be the adjacency matrix of $L(G)$. Suppose that r is an integer, with $1 \le r < q$, and define b and c as in Lemma 2.5. Let $\alpha_1, \dots, \alpha_q$ be the eigenvalues of A_L corresponding to eigenvectors $\phi_1 = \vec{1}, \phi_2, \dots, \phi_q$, respectively. Then the eigenvalues of $U^t U A_L$ are $\lambda_1 = 2bq(d-1) + 2(c-b)(d-1)$ with eigenvector $\phi_1 = \vec{1}$; and, for $i = 2, \dots, q$, $\lambda_i = (c-b)\alpha_i$ with eigenvector ϕ_i; and*

$$\chi(U^t U A_L; \lambda) = \prod_{i=1}^{q} (\lambda_i - \lambda).$$

Proof. From Lemma 2.5, $U^t U = bJ + (c-b)I$. Therefore, since $L(G)$ is $2(d-1)$-regular, $U^t U A_L = 2b(d-1)J + (c-b)A_L$. Furthermore, A_L and J have eigenvector $\vec{1}$ simultaneously, and so all other eigenvectors of A_L are also eigenvectors, and are in the null space, of J. ■

Theorem 3.2. *Let G be a simple d-regular graph with q edges, r an integer, $1 \le r < q$, and A_L the adjacency matrix of $L(G)$. Let $\alpha_1, \dots, \alpha_q$ be the eigenvalues of A_L, where α_1 corresponds to the eigenvector $\vec{1}$. Let $n = \binom{q}{r}$. Then*

$$\chi(\mathcal{M}_r(G); \lambda) = (-\lambda)^{n-q} \prod_{i=1}^{q} (\lambda_i - \lambda),$$

where $\lambda_1 = 2bq(d-1) + 2(c-b)(d-1)$, and $\lambda_i = (c-b)\alpha_i$, for $i = 2, \dots, q$. ■

The previous theorem is a direct consequence of Corollary 2.9 and Proposition 3.1. It gives the spectrum of the super line multigraph $\mathcal{M}_r(G)$ for any r, $1 \le r < q$, in terms of the spectrum of $L(G)$ in the case that G is d-regular.

Moreover, we shall give a formula for the spectrum of the super line multigraph $\mathcal{M}_r(G)$ in terms of the spectrum of G for some graphs. This is based on the following result by Sachs ([7, p. 19]).

Theorem 3.3 (Sachs). *Let G be a simple d-regular graph with p vertices and q edges and let $L(G)$ be the line graph of G. Then*

$$\chi(L(G); \lambda) = (-2-\lambda)^{q-p} \chi(G; \lambda + 2 - d).$$ ■

In other words, if $\beta \ne d$ is an eigenvalue of G, then $d - 2 + \beta$ is an eigenvalue of $L(G)$. This observation yields the following corollary of Theorem 3.2.

Corollary 3.4. *Let G be a simple d-regular graph with p vertices and q edges. Let the eigenvalues of the adjacency matrix of G be $d = \beta_1 \ge \cdots \ge \beta_p$, where β_1 corresponds to the eigenvector $\vec{1}$. Let r be an integer, $1 \le r < q$, and let $n = \binom{q}{r}$. Then*

$$\chi(\mathcal{M}_r(G); \lambda) = (-\lambda)^{n-q} \prod_{i=1}^{q} (\lambda_i - \lambda),$$

where $\lambda_1 = 2bq(d-1) + 2(c-b)(d-1)$, $\lambda_i = (c-b)(d-2+\beta_i)$ for $i = 2, \dots, p$, and $\lambda_{p+1} = \cdots = \lambda_q = -2(c-b)$. ■

The authors wish to thank the referees for their helpful comments.

References

[1] Jay Bagga, Old and new generalizations of line graphs, *International Journal of Mathematics and Mathematical Sciences*, **2004** (2004) no. 29, 1509–1521.

[2] J. S. Bagga, L. W. Beineke and B. N. Varma, The super line graph L_2, *Discrete Math.*, **206** (1999) no. 1–3, 51–61.

[3] K. S. Bagga, L. W. Beineke and B. N. Varma, Super line graphs. Graph theory, combinatorics, and algorithms, Vol. 1, 2 (Kalamazoo MI, 1992), 35–46, Wiley-Intersci. Publ., Wiley, New York (1995).

[4] J. S. Bagga, L. W. Beineke and B. N. Varma, Super line graphs and their properties. Combinatorics, graph theory, algorithms and applications (Beijing 1993), 1–6, World Sci. Publishing, River Edge, NJ (1994).

[5] J. Bagga, R. B. Ellis and D. Ferrero, The structure of super line graphs, Proc. 8th. Int. Symposium on Parallel Architectures, Algorithms and networks (ISPAN'05) (2005), 468–471.

[6] K. J. Bagga and M. R. Vasquez, The super line graph L_2 for hypercubes, *Congr. Numer.*, **93** (1993) 111–113.

[7] N. Biggs, Algebraic Graph Theory, Cambridge University Press (1974).

[8] G. Chartrand and L. Lesniak, Graphs and Digraphs, Chapman and Hall (1996).

[9] D. Cvetković, M. Doob and H. Sachs, Spectra of Graphs, Academic Press (1980).

[10] D. Cvetković, P. Rowlinson and S. Simić, Spectral Generalizations of Line Graphs: On Graphs with Least Eigenvalue-2, Cambridge University Press (2004).

[11] M. Doob, An interrelation between line graphs, eigenvalues and matroids, *J. Combin. Theory Ser. B*, **15** (1973) 40–50.

[12] F. Harary, Graph Theory, Addison-Wesley, Reading MA (1969).

[13] R. A. Horn and C. R. Johnson, Matrix analysis, Cambridge Univ. Press (1990).

[14] C. Godsil and G. Royle, Algebraic Graph Theory, Graduate Texts in Mathematics, Springer (2001).

[15] E. Prisner, Graph Dynamics, Pitman Research Notes in Mathematics Series. Longman (1995).

Proc. ICDM 2008, RMS-Lecture Notes Series No. 13, 2010, pp. 91–99.

Out-Branchings with Extremal Number of Leaves

Jørgen Bang-Jensen[1] and Gregory Gutin[2]

[1]Department of Mathematics and Computer Science, University of Southern Denmark, Odense M, DK-5230, Denmark
[2]Department of Computer Science, Royal Holloway, University of London, Egham, Surrey TW20 0EX, UK
e-mail: jbj@imada.sdu.dk; gutin@cs.rhul.ac.uk

Abstract. A subdigraph T of a digraph D is called an *out-tree* if T is an oriented tree with just one vertex s of in-degree zero. A spanning *out-tree* is called an out-branching. A vertex x of an out-branching B is called a leaf if $d_B^+(x) = 0$.

This is mainly a survey paper on out-branchings with minimum and maximum number of leaves. We give short proofs of some well-known theorems.

Keywords. Digraphs, out-trees, out-branchings, leaves.

1. Introduction

This is mainly a survey paper on out-branchings with minimum and maximum number of leaves. Nevertheless, we prove some new results: Theorems 3.5 and Lemma 4.3. We also give short proofs to some well-known theorems.

The reader will see, in what follows, that out-branchings with minimum and maximum number of leaves are of great interest to graph theory, algorithms and applications.

A subdigraph T of a digraph D is called an *out-tree* if T is an oriented tree with just one vertex s of in-degree zero. The vertex s is the **root** of T. If an *out-tree* B is a spanning subdigraph of D, B is called an **out-branching**. A vertex x of an out-branching B is called a **leaf** if $d_B^+(x) = 0$. Figure 1 shows out-branchings with minimum and maximum number of leaves.

The problem of finding an out-branching with extremal number of leaves is of interest in applications; e.g., the problem of finding an out-branching with minimum number of leaves was considered in the US patent [10] by Demers and Downing, where its application to the area of database systems was described.

For general digraphs, the problems of finding an out-branching with minimum/maximum number of leaves are $\mathcal{NP}$-hard: the problem of verifying the existence of an out-branching with just one leaf is the same as the hamiltonian path problem and the problem of finding a spanning tree with maximum number of leaves in an undirected graph is $\mathcal{NP}$-hard [13] (we may transform a undirected graph to the corresponding directed graph by replacing each edge xy by two arcs xy and yx). Thus, it is natural to consider parameterized complexities of the two problems. Let k be a parameter. The problem of checking whether a digraph D has an out-branching with at least k leaves

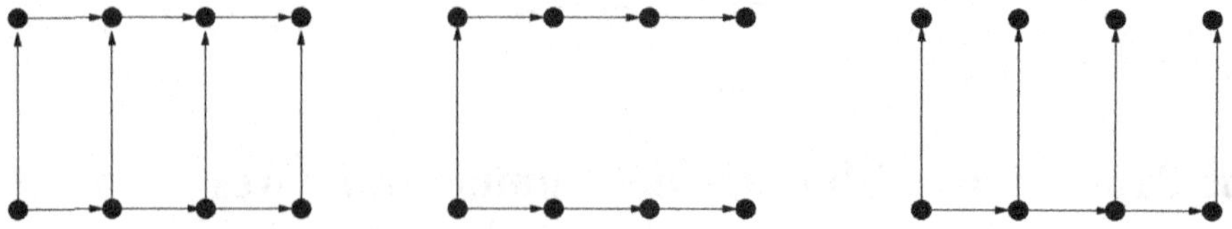

Figure 1. A digraph D and its out-branchings with minimum and maximum number of leaves (Q and R, respectively).

(k non-leaves) is fixed-parameter tractable which was proved by Bonsma and Dorn [6] (by Gutin *et al.* [15]). This means that each of the two parameterized problems can be solved in time $O(f(k) \cdot n^{O(1)})$, where $f(k)$ is a computable function dependent on k but not on n and n is the order of D. We would like to note, in passing, that the problem of checking whether a digraph D has an *out-tree* with at least k leaves is also fixed-parameter tractable [1].

Note that restricted to acyclic digraphs the problems of finding an out-branching with minimum and maximum number of leaves are of different complexity (provided $\mathcal{P} \neq \mathcal{NP}$): while the former is polynomial time solvable (see Section 3), the latter is $\mathcal{NP}$-hard (see Section 4).

Our paper is organized as follows. In the next section, we will provide additional terminology and notation. Sections 3 and 4 are devoted to out-branchings with minimum and maximum number of leaves, respectively.

2. Terminology and notation

For an out-branching B, let $\boldsymbol{L(B)}$ denote the set of leaves of B. For a digraph D containing an out-branching, let $\boldsymbol{\ell_{\min}(D)}$ and $\boldsymbol{\ell_{\max}(D)}$ denote the minimum and maximum number of leaves in an out-branching of D. If D has no out-branching we will write $\ell_{\min}(D) = 0$ or $\ell_{\max}(D) = 0$; it is well-known and easy to prove that a connected digraph D contains an out-branching if and only if D has only one initial strong component [4].

For a digraph D, $\boldsymbol{\alpha(D)}$ denotes the **independence number** of D, i.e., the maximum size of a vertex set X of D such that there is no arc between any pair of vertices of X. A vertex x of a digraph D is called a **source**, if the in-degree of x equals zero. The **path covering number pc($\boldsymbol{D}$)** of a digraph D is the minimum number of disjoint directed paths needed to cover $V(D)$. A digraph D is called **transitive** if the existence of arcs xy, yz implies the existence of the arc xz, where x, y and z are distinct vertices of D. The underlying (undirected) graph of a digraph D will be denoted by $\boldsymbol{UG(D)}$.

For more terminology and notation on digraphs, see Chapter 1 of [4].

3. Minimum leaf out-branchings

In this subsection, we give upper bounds on $\ell_{\min}(D)$ for general and strong digraphs D (Subsection 3.1) and a polynomial algorithm for computing $\ell_{\min}(D)$ for acyclic digraphs D (Subsection 3.2).

3.1 *Upper bounds on* $\ell_{\min}(D)$

Las Vergnas [19] proved the following upper bound on $\ell_{\min}(D)$ for general digraphs.

Theorem 3.1 (Las Vergnas' theorem). *For a digraph D, we have $\ell_{\min}(D) \le \alpha(D)$.*

We will prove the following proposition which immediately implies the theorem.

Proposition 3.2 ([19]). *Let B be an out-branching of D with more than $\alpha(D)$ leaves. Then D contains an out-branching B' such that $L(B')$ is a proper subset of $L(B)$.*

Proof. We will prove this claim by induction on the number n of vertices in D. For $n \le 2$ the result holds; thus, we may assume that $n \ge 3$ and consider an out-branching B of D with $|L(B)| > \alpha(D)$. Clearly, D has an arc xy such that x, y are leaves of B. If the in-neighbor p of y in B is of out-degree at least 2, then $L(B') \subset L(B)$, where $B' = B + xy - py$. So, we may assume that $d^+_B(p) = 1$. Observe that $\alpha(D - y) \le \alpha(D) < |L(B)| = |L(B - y)|$. Hence by the induction hypothesis, $D - y$ has an out-branching B'' such that $L(B'') \subset L(B - y)$. Notice that $L(B - y) = L(B) \cup \{p\} \backslash \{y\}$. If $p \in L(B'')$, then observe that $L(B'' + py) \subset L(B)$. Otherwise, $L(B'' + xy) \subseteq L(B) \backslash \{x\} \subset L(B)$. □

The bound of Las Vergnas' theorem is tight as there are many digraphs D for which $\ell_{\min}(D) = \alpha(D)$, see, e.g., Theorem 3.5. It would be interesting to find other tight upper bounds on $\ell_{\min}(D)$.

While it is easy to show that the problem of checking whether a digraph has an out-branching with at most k leaves is $\mathcal{NP}$-hard for each fixed natural number k, Gutin, Razgon and Kim [15,16] proved that the problem of checking whether a digraph D of order n has an out-branching with at most $n - k$ leaves (or, equivalently, at least k non-leaves) is fixed-parameter tractable.

It is easy to show that Las Vergnas' theorem implies the well-known Gallai–Millgram theorem, Theorem 3.4. However, first we need the following simple result.

Lemma 3.3 ([16]). *Let $D = (V, A)$ be a digraph and let $\hat{D}$ be the digraph obtained from D by adding a new vertex s and all possible arcs from s to V. Then* $\mathrm{pc}(D) = \ell_{\min}(\hat{D})$.

Proof. Since a collection of p disjoint directed paths in D covering $V(D)$ corresponds to an out-branching of $\hat{D}$ with p leaves, we have $\mathrm{pc}(D) \ge \ell_{\min}(\hat{D})$. Let B be an out-branching of $\hat{D}$ with p leaves. We say that a vertex x of B is **branching** if $d^+_B(x) > 1$. Consider a maximal directed path Q of B not containing branching vertices. Observe that $B - V(Q)$ has $p - 1$ leaves. Thus, we can decompose the vertices of B into p disjoint directed paths. Deleting the vertex s from this collection of paths, we see that $\mathrm{pc}(D) \le \ell_{\min}(\hat{D})$. Thus, $\mathrm{pc}(D) = \ell_{\min}(\hat{D})$. □

Theorem 3.4 (Gallai–Milgram theorem [12]). *For every digraph D,* $\mathrm{pc}(D) \le \alpha(D)$.

Proof. Consider the digraph $\hat{D}$ defined in Lemma 3.3. By Lemma 3.3 and Las Vergnas' theorem, $\mathrm{pc}(D) = \ell_{\min}(\hat{D}) \le \alpha(\hat{D}) = \alpha(D)$. □

The bound of Las Vergnas' theorem is sharp as one can see from the following:

Theorem 3.5. *If D is a transitive acyclic digraph with a unique source s, then $\ell_{\min}(D) = \alpha(D)$.*

Proof. By Las Vergnas' theorem, D contains an out-branching B with $k \leq \alpha(D)$ leaves. Observe that B is rooted at s and the vertices of every path in B starting at s and terminating at a leaf induce a clique in $UG(D)$. Thus, the vertices of $UG(D)$ can be covered by k cliques and, hence, $\alpha(UG(D)) \leq k$. We conclude that $\ell_{\min}(D) = \alpha(D)$. □

The next theorem is equivalent to Theorem 3.5. Indeed, by Theorem 3.5 and Lemma 3.3, we have $\mathrm{pc}(D) = \ell_{\min}(\hat{D}) = \alpha(\hat{D}) = \alpha(D)$ for every transitive acyclic digraph D which implies Dilworth's theorem. Since $\mathrm{pc}(D) \leq \ell_{\min}(D) \leq \alpha(D)$ for each transitive acyclic digraph with a unique source, Dilworth theorem implies Theorem 3.5.

Theorem 3.6 (Dilworth's theorem [11]). *Every transitive acyclic digraph D has $\mathrm{pc}(D) = \alpha(D)$.*

Theorems 3.5 and 3.6 give rise to the following natural research problem.

Problem 3.7. *Find other non-trivial digraph classes for which the equalities of Theorem 3.5 and/or Theorem 3.6 hold.*

Las Vergnas proved another upper bound on $\ell_{\min}(D)$.

Theorem 3.8 ([19]). *Let D be a digraph on n vertices such that any two distinct non-adjacent vertices have degree sum at least $2n - 2h - 1$, where $1 \leq h \leq n - 1$. Then $\ell_{\min}(D) \leq h$.*

Settling a conjecture of Las Vergnas [19], Thomassé [21] proved the following:

Theorem 3.9. *If D is a strong, then $\ell_{\min}(D) \leq \max\{\alpha(D) - 1, 1\}$.*

3.2 *Acyclic digraphs*

Demers and Downing [10] suggested a heuristic approach for finding, in an acyclic digraph, an out-branching with minimum number of leaves. However, no argument or assertion has been made to provide the validity of their approach or to investigate its computational complexity. Using another approach, Gutin *et al.* [15] showed that a minimum leaf out-branching in an acyclic digraph can be found in polynomial time.

The following algorithm MINLEAF introduced by Gutin *et al.* [15] returns an out-branching with minimum number of leaves in an acyclic digraph. Observe that an acyclic digraph D has an out-branching if and only if it has exactly one source. It is not difficult to prove that MINLEAF is correct and of running time $O(n^{1.5}\sqrt{m})$, where n (m) is the order (size) of the input digraph.

MINLEAF
Input: An acyclic digraph D with vertex set V.
Output: A minimum leaf out-branching T of D if $\ell_{\min}(D) > 0$ and "NO", otherwise.

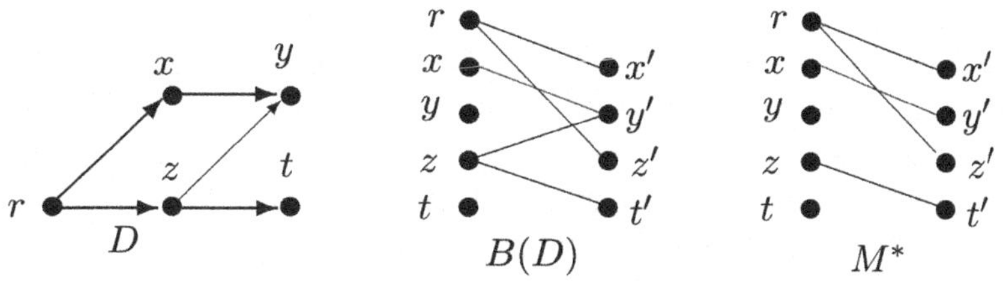

Figure 2. Illustration for MINLEAF

Step 1 Find a source r in D. If there is another source in D, return "no out-branching". Let $V' = \{v' : v \in V\}$.
Step 2 Construct a bipartite graph $B = B(D)$ of D with partite sets V, $V' - r'$ and edge xy' for each arc $xy \in A(D)$.
Step 3 Find a maximum matching M in B.
Step 4 $M^* := M$. For all $y' \in V'$ not covered by M, set $M^* := M^* \cup \{$an arbitrary edge incident with $y'\}$.
Step 5 $A(T) := \emptyset$. For all $xy' \in M^*$, set $A(T) := A(T) \cup \{xy\}$.
Step 6 Return T.

Figure 2 illustrates MINLEAF. There $M = \{rx', xy', zt'\}$ and $T = D - zy$.

The parameters directed tree-width, directed path-width and DAG-width of digraphs are analogs of tree-width of undirected graphs; for definitions, see [9]. Acyclic digraphs are of directed tree-width, directed path-width and DAG-width equal zero. It follows from one of the main results of the paper [17] by Johnson *et al.* that there is a polynomial-time algorithm for verifying whether a digraph of bounded directed tree-width (directed path-width, DAG-width, respectively) has a Hamilton directed path. In contrast, Dankelmann *et al.* [9] proved that the problem of finding the minimum number of leaves in an out-branching of a digraph of directed tree-width (directed path-width, DAG-width, respectively) equal one is NP-hard.

4. Maximum leaf out-branchings

Alon *et al.* [3] proved the following complexity result.

Proposition 4.1. *The problem of finding an out-branching of maximum number of leaves in an acyclic digraph is $\mathcal{NP}$-hard.*

Proof. Consider a bipartite graph G with bipartition X, Y and a vertex $s \notin V(G)$. To obtain an acyclic digraph D from G and s, orient the edges of G from X to Y and add all arcs sx, $x \in X$. Let B be an out-branching in D. Then the set of leaves of B is $Y \cup X'$, where $X' \subset X$, and for each $y \in Y$ there is a vertex $z \in Z = X \setminus X'$ such that $zy \in A(D)$. Observe that B has maximum number of leaves if and only if $Z \subseteq X$ is of minimum size among all sets $Z' \subseteq X$ such that $N_G(Z') = X$. However, the problem of finding Z' of minimum size such that $N_G(Z') = X$ is equivalent to the Set Cover problem ($\{N_G(y) : y \in Y\}$ is the family of sets to cover), which is $\mathcal{NP}$-hard. □

Bonsma and Dorn [6,7] showed that the problem of checking whether a digraph has an out-branching with at least k leaves is fixed-parameter tractable. In [7], they presented an algorithm for the problem of running time $2^{O(k \log k)} \cdot n^{O(1)}$. Kneis *et al.* [18] designed an algorithm of running time $4^k \cdot n^{O(1)}$; an algorithm of running time $3.72^k \cdot n^{O(1)}$ was consequently obtained by Daligault *et al.* [8].

Lower bounds on the maximum number of leaves in an out-branching of a digraph were investigated by Alon *et al.* [1,2] and Bonsma and Dorn [7].

For a digraph D let $\ell^t_{\max}(D)$, denote the maximum possible number of leaves in an out-tree of D. Notice that $\ell^t_{\max}(D) \geq \ell_{\max}(D)$ for every digraph D. Let $\mathcal{L}$ be the family of digraphs D for which either $\ell_{\max}(D) = 0$ or $\ell^t_{\max}(D) = \ell_{\max}(D)$. It is easy to see that $\mathcal{L}$ contains all strong and acyclic digraphs.

The following assertion from [1] shows that $\mathcal{L}$ includes a large number of digraphs including all strong digraphs, acyclic digraphs, semicomplete multipartite digraphs and quasi-transitive digraphs.

Proposition 4.2. *Suppose that a digraph D satisfies the following property: for every pair R and Q of distinct strong components of D, if there is an arc from R to Q then each vertex of Q has an in-neighbor in R. Then $D \in \mathcal{L}$.*

Let $P = u_1u_2\ldots u_q$ be a directed path in a digraph D. An arc u_iu_j of D is a **forward (backward) arc for** P if $i \leq j-2$ ($j < i$, respectively). Every backward arc of the type $v_{i+1}v_i$ is called **double**.

The following assertion is a slight refinement of a result by Alon *et al.* [1]. Better bounds were proved in [2] and [7] (the bound of Bonsma and Dorn [7] is optimal in a sense, see Remark 4.5), but our proof is significantly shorter than the proofs of the corresponding results in [1], [2] and [7].

Lemma 4.3. *Let D be an oriented graph of order n with every vertex of in-degree 2 and let D have an out-branching. If D has no out-tree with k leaves, then $n \leq k^5$.*

Proof. Assume that D has no out-tree with k leaves. Consider an out-branching T of D with p leaves so that this is the maximum number of leaves over all out-branchings in D. By the assumption $p < k$.

First observe that if $Q = v_1v_2\ldots v_s$ is an arbitrary path in T from the root to a leaf and v_iv_j is a forward arc, then, by the maximality of p, T must branch at v_{j-1}, that is, $d^+_T(v_{j-1}) \geq 2$. Since T has at most $k-1$ leaves and no two forward arcs end in the same vertex, this implies that Q has at most $(k-2)$ forward arcs.

Now fix a path $P = u_1u_2\ldots u_q$ from the root to a leaf in T which has $q \geq n/p$ vertices. When we delete all vertices of P from T we obtain a collection of out-trees covering $V(D) - V(P)$. It is easy to show by induction on the number of leaves that T can be decomposed into a collection $\mathcal{P} = \{P_1, P_2, \ldots, P_p\}$ of vertex-disjoint directed paths covering all vertices of D so that $P = P_1$.

Let $P' \in \mathcal{P} \setminus \{P\}$ be arbitrary. There are at most $k-1$ vertices on P with in-neighbors on P' since otherwise we could choose a set X of at least k vertices on P for which there were in-neighbors on P'. The vertices of X would be leaves of an out-tree formed by the vertices $V(P') \cup X$. Thus, there are

$m \le (k-1)(p-1) \le (k-1)(k-2)$ vertices of P with in-neighbors outside P and at least $q-(k-2)(k-1)$ vertices of P have both in-neighbors on P.

Let f be the number of forward arcs for P. By the argument above $f \le k-2$. Let uv be an arc of $A(D) \setminus A(P)$ such that $v \in V(P)$. There are three possibilities: (i) $u \notin V(P)$, (ii) $u \in V(P)$ and uv is forward for P, (iii) $u \in V(P)$ and uv is backward for P. By the inequalities above for m and f, we conclude that there are at most $k(k-2)$ vertices on P which are not terminal vertices (i.e., heads) of a backward arc. Consider a path $R = v_0 v_1 \dots v_r$ formed by backward arcs. Observe that the arcs $\{v_i v_{i+1} : 0 \le i \le r-1\} \cup \{v_j v_j^+ : 1 \le j \le r\}$ form an out-tree with r leaves, where v_j^+ is the successor of v_j on P. Thus, there is no path of backward arcs of length more than $k-1$.

If the in-degree of u_1 in the subdigraph of D induced by $V(P)$ is 2, remove one of the backward arcs terminating at u_1. Observe that now the backward arcs for P form a vertex-disjoint collection of out-trees with roots at vertices that are not terminal vertices of backward arcs. Therefore, the number of the out-trees in the collection is at most $k(k-2)$. Observe that each out-tree in the collection has at most $k-1$ leaves and thus its arcs can be decomposed into at most $k-1$ paths, each of length at most $k-1$. Hence, the original total number of backward arcs for P is at most $k(k-2)(k-1)^2+1$ (where the last one comes from the possible extra arc into u_1). On the other hand, it is at least $q-k(k-2)$ as every vertex on P is the head of an arc not in $A(P)$. Thus, $q-k(k-2) \le k(k-2)(k-1)^2+1$. Combining this inequality with $q \ge n/(k-1)$, we conclude that $n \le k^5$. □

Theorem 4.4 ([1]). *Let D be a digraph in $\mathcal{L}$ with $\ell_{\max}(D) > 0$.*

(a) *If D is an oriented graph with minimum in-degree at least 2, then $\ell_{\max}(D) \ge n^{1/5}-1$.*

(b) *If D is a digraph with minimum in-degree at least 3, then $\ell_{\max}(D) \ge n^{1/5}-1$.*

Proof.

(a) Let B^+ be an out-branching of D. Delete some arcs from $A(D) \setminus A(B^+)$, if needed, such that the in-degree of each vertex of D becomes 2. Now the inequality $\ell_{\max}(D) \ge n^{1/5}-1$ follows from Lemma 4.3 and the definition of $\mathcal{L}$.

(b) Let B^+ be an out-branching of D. Let P be the path formed in the proof of Lemma 4.3. (Note that $A(P) \subseteq A(B^+)$.) Delete every double arc of P, in case there are any, and delete some more arcs from $A(D) \setminus A(B^+)$, if needed, to ensure that the in-degree of each vertex of D becomes 2. It is not difficult to see that the proof of Lemma 4.3 remains valid for the new digraph D. Now the inequality $\ell_{\max}(D) \ge n^{1/5}-1$ follows from Lemma 4.3 and the definition of $\mathcal{L}$. □

It is not difficult to give examples showing that the restrictions on the minimum in-degrees in Theorem 4.4 are optimal. Indeed, any directed cycle C is a strong oriented graph with all in-degrees 1 for which $\ell_{\max}(C) = 1$ and any directed double cycle D is a strong digraph with in-degrees 2 for which $\ell_{\max}(D) = 2$ (a **directed double cycle** is a digraph obtained from an undirected cycle by replacing every edge xy with two arcs xy and yx).

Remark 4.5. Bonsma and Dorn [7] proved that one can replace $n^{1/5} - 1$ in the bounds of Theorem 4.4 by $\sqrt{n}/4$. Alon *et al.* [1] proved that if $D \in \mathcal{L}$, then $\ell_{\max}(D) \leq c \cdot \sqrt{n}$ for some constant c. Thus, we may conclude that if $D \in \mathcal{L}$ and $\ell_{\max}(D) > 0$, then $\ell_{\max}(D) = \Theta(\sqrt{n})$.

Remark 4.6. For some subfamilies of $\mathcal{L}$, one can obtain better bounds on $\ell_{\max}(D)$. One example is the class of **multipartite tournaments**, i.e., orientations of complete multipartite graphs. Gutin [14] and Petrovic and Thomassen [20] proved that every multipartite tournament M with at most one source has a vertex x such that the distance from x to any vertex of M is at most 4. Thus, a BFS tree of M has at least $\frac{n-1}{4}$ leaves and, hence, $\ell_{\max}(M) \geq \frac{n-1}{4}$. Another example is the class of quasi-transitive digraphs. A digraph Q is **quasi-transitive** if the existence of arcs xy and yz in Q implies the existence of an arc between x and z. Bang-Jensen and Huang [5] proved that every quasi-transitive digraph Q with $\ell_{\max}(Q) > 0$ has a vertex x such that each vertex of Q is at distance at most 3 from x. This implies that $\ell_{\max}(Q) \geq \frac{n-1}{3}$.

Acknowledgements

We are thankful to Eun Jung Kim for comments on the paper. Research of Gregory Gutin was supported in part by an EPSRC grant and the IST Programme of the European Community, under the PASCAL 2 Network of Excellence.

References

[1] N. Alon, F. Fomin, G. Gutin, M. Krivelevich and S. Saurabh, Parameterized Algorithms for Directed Maximum Leaf Problems, Proc. ICALP 2007, *Lect. Notes Comput. Sci.*, **4596** (2007) 352–362.

[2] N. Alon, F. Fomin, G. Gutin, M. Krivelevich and S. Saurabh, Better Algorithms and Bounds for Directed Maximum Leaf Problems, Proc. FSTTCS 2007, *Lect. Notes Comput. Sci.*, **4855** (2007) 316–327.

[3] N. Alon, F. Fomin, G. Gutin, M. Krivelevich and S. Saurabh, Spanning directed trees with many leaves, *SIAM J. Discrete Math.*, **23** (2009) 466–476.

[4] J. Bang-Jensen and G. Gutin, Digraphs: Theory, Algorithms and Applications, 2nd Ed., Springer (2009).

[5] J. Bang-Jensen and J. Huang, Kings in quasi-transitive digraphs, *Discrete Math.*, **185** (1998) 19–27.

[6] P. Bonsma and F. Dorn, An FPT Algorithm for Directed Spanning k-Leaf, Preprint arXiv:0711.4052, November (2007).

[7] P. S. Bonsma and F. Dorn, Tight Bounds and a Fast FPT Algorithm for Directed Max-Leaf Spanning Tree, Proc. ESA 2008, *Lect. Notes Comput. Sci.*, **5193** (2008) 222–233.

[8] J. Daligault, G. Gutin, E. J. Kim and A. Yeo, FPT Algorithms and Kernels for the Directed k-Leaf Problem, *J. Comput. Sys. Sci.*, **76** (2010) 144-152.

[9] P. Dankelmann, G. Gutin and E. J. Kim, On Complexity of Minimum Leaf Out-Branching Problem, *Discrete Applied Math.*, **157** (2009) 3000–3004.

[10] A. Demers and A. Downing, Minimum leaf spanning tree, US Patent no. 6,105,018, August (2000).

[11] R. P. Dilworth, A decomposition theorem for partially ordered sets, *Annals of Mathematics*, **51** (1950) 161–166.

[12] T. Gallai and A. N. Milgram, Verallgemeinerung eines graphentheoretischen Satzes von Rédei, *Acta Sci. Math. Szeged*, **21** (1960) 181–186.
[13] M. R. Garey and D. S. Johnson, Computers and intractability, W. H. Freeman and Co., San Francisco (1979).
[14] G. Gutin, The radii of n-partite tournaments, *Math. Notes*, **40** (1986) 743–744.
[15] G. Gutin, I. Razgon and E. J. Kim, Minimum Leaf Out-Branching Problems, Proc. AAIM'08, *Lect. Notes Comput. Sci.*, **5034** (2008) 235–246.
[16] G. Gutin, I. Razgon and E. J. Kim, Minimum Leaf Out-Branching and Other Problems, *Theoretical Computer Sci.*, **410** (2009) 4571–4579.
[17] T. Johnson, N. Robertson, P. D. Seymour and R. Thomas, Directed Tree-Width, *J. Combin. Theory Ser. B*, **82** (2001) 138–154.
[18] J. Kneis, A. Langer and P. Rossmanith, A new algorithm for finding trees with many leaves, Proc. ISAAC 2008, *Lect. Notes Comput. Sci.*, **5369** (2008) 270–281.
[19] M. Las Vergnas, Sur les arborescences dans un graphe orienté, *Discrete Math.*, **15** (1976) 27–39.
[20] V. Petrovic and C. Thomassen, Kings in k-partite tournaments, *Discrete Math.*, **98** (1991) 237–238.
[21] S. Thomassé, Covering a strong digraph by $\alpha - 1$ disjoint paths: A proof of Las Vergnas' conjecture, *J. Combin. Theory Ser. B*, **83** (2001) 331–333.

Proc. ICDM 2008, RMS-Lecture Notes Series No. 13, 2010, pp. 101–108.

Stability of Irregularity Strength on tK_4 and $K_1 \cup C_n$

K. M. Kathiresan[1] and K. Muthugurupackiam[2]

[1]Department of Mathematics, Ayya Nadar Janaki Ammal College, Sivakasi 626 124, India
[2]Department of Mathematics, Kalasalingam University, Krishnankoil 626 190, India
e-mail: kathir2esan@yahoo.com; gurupackiam@yahoo.com

Abstract. The problem of finding irregularity strengths of graphs was proposed by Chartrand *et al.*, [1] and has proven to be difficult in general. There are not many graphs for which irregularity strength is known. In this paper, we discuss how the addition of a new edge changes the irregularity strength in tK_4 and $K_1 \cup C_n$. In particular, we prove that the graph tK_4 is stable if $t \equiv 1 (\text{mod } 3)$ and is negative in all other cases. We also prove that the graph $K_1 \cup C_n$ is mixed if $n \equiv 0, 1 (\text{mod } 4)$ and is negative if $n \equiv 2, 3 (\text{mod } 4)$.

Keywords. Irregular degree, irregularity strength.

1. Introduction

In this paper, we consider simple undirected and disconnected graphs with at most one isolated vertex. Let $G = (V, E)$ be a simple graph. A *network* $G(f)$ consists of the graph G together with an assignment $f : E(G) \to Z^+$. The *strength of the network* $G(f)$ is defined by $s(G(f)) = \max\{f(e) : e \in E(G)\}$. In a network $G(f)$ for each vertex $v \in V(G)$, define the *weight* $w(v)$ by the sum of the assignments of the edges incident to v and call $G(f)$ *irregular* if $w(u) \neq w(v)$ for all distinct $u, v \in V(G)$. The *irregularity strength* of the graph G is defined to be

$$s(G) = \begin{cases} \min_f\{s(G(f)) : G(f) \text{ is irregular}\}, & \text{if } G(f) \text{ is irregular for} \\ & \text{at least one } f, \\ \infty & \text{otherwise} \end{cases}$$

Thus, the irregularity strength of a graph G is the smallest strength of all irregular weight assignments.

Many research papers have been published regarding irregularity strength of a graph. In this paper we discuss the changes in irregularity strength of a graph when we add a new edge from the complement of the graph to the given graph. It is interesting to see that addition of an edge from the complement of the graph to the graph G may increase or decrease the irregularity strength of the graph G or remains the same. Thus we call it as positive, negative and stable edge of G respectively. If all the edges of $\bar{G}$ are positive (negative, stable respectively) then we call the graph G as positive graph (negative graph, stable graph respectively). Otherwise we call it as mixed graph.

Definition 1.1. *Let G be any graph which is not complete. Let e be any edge of $\bar{G}$, then e is called a positive edge of G, if $s(G+e) > s(G)$, e is called a negative edge if $s(G+e) < s(G)$ and e is called a stable edge if $s(G+e) = s(G)$. If all the edges of $\bar{G}$ are positive, (negative, stable) edges of G, then G is called a positive (negative, stable) graph. Otherwise G is called a mixed graph.*

Example 1.2. P_3 is a positive graph.

The irregularity strength of P_3 is 2. But the addition of the edge from $\bar{P}_3$ to P_3 increases the irregularity strength of P_3 as shown below.

P_3 :

$P_3 + e$:

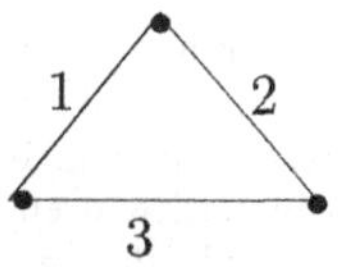

Star graphs $K_{1,n}$, are negative graphs for $n \geq 3$.

The graph $K_{1,3} + e$ is called *paw* [5]. Paw is a stable graph.

Consider P_4. It is easy to verify that $s(P_4) = 2$. Let v_1, v_2, v_3 and v_4 be the consecutive vertices of P_4. Clearly, v_2v_4 is a stable edge of P_4, v_1v_4 is a positive edge of P_4. Hence, P_4 is a mixed graph.

2. Stable, negative and mixed graphs

R. J. Faudree *et al.* [3], determined the irregularity strength of tK_4 as follows.

$$s(tK_4) = \left\lceil \frac{4t+2}{3} \right\rceil \quad \text{for } t \geq 2.$$

The following theorem proves that the disjoint union of t copies of K_4 is either a stable graph or a negative graph.

Theorem 2.1. *For $t \geq 2$, tK_4 is a stable graph if $t \equiv 1 (\text{mod } 3)$ and is a negative graph otherwise.*

Proof. Let us consider the disjoint union of t copies of K_4.

For $1 \leq i \leq t$, let v_{i1}, v_{i2}, v_{i3} and v_{i4} be the vertices of the ith copy of K_4. For any $e \in \overline{tK_4}$, $tK_4 + e$ is isomorphic to $tK_4 + v_{(t-1)3}v_{t3}$. We claim that $v_{(t-1)3}v_{t3}$ is a stable edge of tK_4 if $t \equiv 1 (\text{mod } 3)$ and is a negative edge of tK_4 if $t \equiv 1 (\text{mod } 3)$.

Define the edge labeling $f : E(tK_4 + v_{(t-1)3}v_{t3}) \to Z^+$ as follows:

For $1 \leq i \leq t-2$,

$$\text{(i)} \quad f(v_{i1}v_{i2}) = \begin{cases} 1 & \text{if } i = 1, \\ \frac{4i-3}{3} & \text{if } i \equiv 0 (\text{mod } 3), \\ \frac{4i-4}{3} & \text{if } i \equiv 1 (\text{mod } 3), i \neq 1, \\ \frac{4i-5}{3} & \text{if } i \equiv 2 (\text{mod } 3). \end{cases}$$

(ii) $$f(v_{i1}v_{i3}) = \begin{cases} 1 & \text{if } i = 1, \\ \frac{4i}{3} & \text{if } i \equiv 0 (\text{mod } 3), \\ \frac{4i+2}{3} & \text{if } i \equiv 1 (\text{mod } 3), i \neq 1, \\ \frac{4i+4}{3} & \text{if } i \equiv 2 (\text{mod } 3). \end{cases}$$

(iii) $$f(v_{i1}v_{i4}) = \begin{cases} \frac{4i}{3} & \text{if } i \equiv 0 (\text{mod } 3), \\ \frac{4i-1}{3} & \text{if } i \equiv 1 (\text{mod } 3), \\ \frac{4i-2}{3} & \text{if } i \equiv 2 (\text{mod } 3). \end{cases}$$

(iv) $$f(v_{i2}v_{i3}) = \begin{cases} \frac{4i+3}{3} & \text{if } i \equiv 0 (\text{mod } 3), \\ \frac{4i+2}{3} & \text{if } i \equiv 1 (\text{mod } 3), \\ \frac{4i+1}{3} & \text{if } i \equiv 2 (\text{mod } 3). \end{cases}$$

(v) $$f(v_{i3}v_{i4}) = \begin{cases} 3 & \text{if } i = 1, \\ \frac{4i+3}{3} & \text{if } i \equiv 0 (\text{mod } 3), \\ \frac{4i+2}{3} & \text{if } i \equiv 1 \ (\text{mod } 3), i \neq 1, \\ \frac{4i+1}{3} & \text{if } i \equiv 2 (\text{mod } 3). \end{cases}$$

(vi) $$f(v_{i4}v_{i2}) = \begin{cases} 1 & \text{if } i = 1, \\ \frac{4i}{3} & \text{if } i \equiv 0 (\text{mod } 3), \\ \frac{4i+2}{3} & \text{if } i \equiv 1 (\text{mod } 3), i \neq 1, \\ \frac{4i+4}{3} & \text{if } i \equiv 2 (\text{mod } 3). \end{cases}$$

(vii) $$f(v_{(t-1)1}v_{(t-1)2}) = \begin{cases} \frac{4t-9}{3} & \text{if } t \equiv 0 (\text{mod } 3), \\ \frac{4t-1}{3} & \text{if } t \equiv 1 (\text{mod } 3), \\ \frac{4t-5}{3} & \text{if } t \equiv 2 (\text{mod } 3). \end{cases}$$

(viii) $$f(v_{(t-1)1}v_{(t-1)3}) = \begin{cases} \frac{4t}{3} & \text{if } t \equiv 0 (\text{mod } 3), \\ \frac{4t-4}{3} & \text{if } t \equiv 1 (\text{mod } 3), \\ \frac{4t-5}{3} & \text{if } t \equiv 2 (\text{mod } 3). \end{cases}$$

(ix) $$f(v_{(t-1)1}v_{(t-1)4}) = \begin{cases} \frac{4t-3}{3} & \text{if } t \equiv 0 (\text{mod } 3), \\ \frac{4t-10}{3} & \text{if } t \equiv 1 (\text{mod } 3), \\ \frac{4t-2}{3} & \text{if } t \equiv 2 (\text{mod } 3). \end{cases}$$

(x) $$f(v_{(t-1)2}v_{(t-1)3}) = \begin{cases} \frac{4t-6}{3} & \text{if } t \equiv 0 (\text{mod } 3), \\ \frac{4t-1}{3} & \text{if } t \equiv 1 (\text{mod } 3), \\ \frac{4t-5}{3} & \text{if } t \equiv 2 (\text{mod } 3). \end{cases}$$

$$\text{(xi)}\quad f(v_{(t-1)3}v_{(t-1)4}) = \begin{cases} \frac{4t}{3} & \text{if } t \equiv 0 (\text{mod } 3), \\ \frac{4t-4}{3} & \text{if } t \equiv 1 (\text{mod } 3), \\ \frac{4t-2}{3} & \text{if } t \equiv 2 (\text{mod } 3). \end{cases}$$

$$\text{(xii)}\quad f(v_{(t-1)4}v_{(t-1)2}) = \begin{cases} \frac{4t}{3} & \text{if } t \equiv 0 (\text{mod } 3), \\ \frac{4t+2}{3} & \text{if } t \equiv 1 (\text{mod } 3), \\ \frac{4t-5}{3} & \text{if } t \equiv 2 (\text{mod } 3). \end{cases}$$

$$\text{(xiii)}\quad f(v_{(t-1)3}v_{t3}) = \begin{cases} 3 & \text{if } t \equiv 0 (\text{mod } 3), \\ 4 & \text{if } t \equiv 1 (\text{mod } 3), \\ 2 & \text{if } t \equiv 2 (\text{mod } 3). \end{cases}$$

$$\text{(xiv)}\quad f(v_{t1}v_{t2}) = \begin{cases} \frac{4t-6}{3} & \text{if } t \equiv 0 (\text{mod } 3), \\ \frac{4t-4}{3} & \text{if } t \equiv 1 (\text{mod } 3), \\ \frac{4t-2}{3} & \text{if } t \equiv 2 (\text{mod } 3). \end{cases}$$

$$\text{(xv)}\quad f(v_{t1}v_{t3}) = \begin{cases} \frac{4t}{3} & \text{if } t \equiv 0 (\text{mod } 3), \\ \frac{4t-1}{3} & \text{if } t \equiv 1 (\text{mod } 3), \\ \frac{4t-2}{3} & \text{if } t \equiv 2 (\text{mod } 3). \end{cases}$$

$$\text{(xvi)}\quad f(v_{t1}v_{t4}) = \begin{cases} \frac{4t}{3} & \text{if } t \equiv 0 (\text{mod } 3), \\ \frac{4t-1}{3} & \text{if } t \equiv 1 (\text{mod } 3), \\ \frac{4t+1}{3} & \text{if } t \equiv 2 (\text{mod } 3). \end{cases}$$

$$\text{(xvii)}\quad f(v_{t2}v_{t3}) = \begin{cases} \frac{4t-3}{3} & \text{if } t \equiv 0 (\text{mod } 3), \\ \frac{4t-4}{3} & \text{if } t \equiv 1 (\text{mod } 3), \\ \frac{4t+1}{3} & \text{if } t \equiv 2 (\text{mod } 3). \end{cases}$$

$$\text{(xviii)}\quad f(v_{t3}v_{t4}) = \begin{cases} \frac{4t}{3} & \text{if } t \equiv 0 (\text{mod } 3), \\ \frac{4t-1}{3} & \text{if } t \equiv 1 (\text{mod } 3), \\ \frac{4t+1}{3} & \text{if } t \equiv 2 (\text{mod } 3). \end{cases}$$

$$\text{(xix)}\quad f(v_{t4}v_{t2}) = \begin{cases} \frac{4t}{3} & \text{if } t \equiv 0 (\text{mod } 3), \\ \frac{4t-1}{3} & \text{if } t \equiv 1 (\text{mod } 3), \\ \frac{4t+1}{3} & \text{if } t \equiv 2 (\text{mod } 3). \end{cases}$$

Then the weights of the vertices of $tK_4 + v_{(t-1)3}v_{t3}$ are $3, 4, 5, \ldots, 4t + 2$. All are pairwise distinct.

If $t \equiv 0 (\text{mod } 3)$, then we have an irregular network with maximum label $\frac{4t}{3}$. Hence,

$$s(tK_4 + v_{(t-1)3}v_{t3}) \leq \frac{4t}{3} < s(tK_4).$$

Thus, $v_{(t-1)3}v_{t3}$ is a negative edge of tK_4.

If $t \equiv 1 (\text{mod } 3)$, then we have an irregular network with maximum label $\frac{4t+2}{3}$. Hence,

$$s(tK_4 + v_{(t-1)3}v_{t3}) \leq \frac{4t+2}{3} \tag{1}$$

In $tK_4 + v_{(t-1)3}v_{t3}$, there are $4t - 2$ vertices of degree 3 and its minimum possible weights are $3, 4, 5, \ldots, 4t$. So, it is not possible to obtain the weight $4t$ by using the labels which are fewer than $\lceil \frac{4t}{3} \rceil$. Hence, $s(tK_4 + v_{(t-1)3}v_{t3}) \geq \lceil \frac{4t}{3} \rceil$. If $t \equiv 1 (\text{mod } 3)$, then $\lceil \frac{4t}{3} \rceil = \frac{4t+2}{3}$. Thus,

$$s(tK_4 + v_{(t-1)3}v_{t3}) \geq \frac{4t+2}{3} \tag{2}$$

From (1) and (2) we get $s(tK_4 + v_{(t-1)3}v_{t3}) = \frac{4t+2}{3} = s(tK_4)$. Thus, $v_{(t-1)3}v_{t3}$ is a stable edge of tK_4.

If $t \equiv 2 (\text{mod } 3)$, then we have an irregular network with maximum label $\frac{4t+1}{3}$. Hence, $s(tK_4 + v_{(t-1)3}v_{t3}) \leq \frac{4t+1}{3} < s(tK_4)$. Thus, $v_{(t-1)3}v_{t3}$ is a negative edge of tK_4. □

We shall use the following functions to simplify the later notations.

$$\text{Let} \quad \alpha(n) = \begin{cases} 1 & \text{if } n \equiv 1 \pmod 2, \\ 0 & \text{otherwise.} \end{cases}$$

$$\beta(n) = \begin{cases} 1 & \text{if } n \equiv 1 \pmod 4, \\ 0 & \text{otherwise.} \end{cases}$$

$$\gamma(n) = \begin{cases} 1 & \text{if } n \equiv 2 \pmod 4, \\ 0 & \text{otherwise.} \end{cases}$$

In [2], Faudree *et al.*, proved that

$$s(C_n) = \begin{cases} \lceil \frac{n}{2} \rceil & \text{if } n \equiv 1 (\text{mod } 4), \\ \lceil \frac{n}{2} \rceil + 1 & \text{otherwise.} \end{cases}$$

The following theorem proves that the disjoint union K_1 and the cycle $C_n, n \geq 3$, is either a mixed graph or a negative graph.

Theorem 2.2. *Disjoint union of K_1 and the cycle $C_n, n \geq 3$ is*

(i) *a mixed graph if $n \equiv 0, 1 (\text{mod } 4)$,*
(ii) *a negative graph if $n \equiv 2, 3 (\text{mod } 4)$.*

Proof. Let u_0 be the isolated vertex of $K_1 \cup C_n$. Let $v_1, v_2, v_3, \ldots, v_n$ be the consecutive vertices of C_n. Let e be any edge of $\overline{K_1 \cup C_n}$, then either one vertex of e is u_0 and the another vertex of e is in C_n or both vertices of e are in C_n.

Case 1. Assume that both vertices of e are in $C_n, n \geq 4$. Since, the weight of the isolated vertex u_0 is zero, $s(C_n) = s(K_1 \cup C_n)$. KM. Kathiresan *et al.* [4], proved that $s(C_n + e) < s(C_n)$, for $n \geq 4$.

Hence, $s((K_1 \cup C_n) + e) < s(K_1 \cup C_n)$, for $n \geq 4$. Thus, e is a negative edge of $K_1 \cup C_n$, when both ends of e are in $C_n, n \geq 4$.

Case 2. Assume that one vertex of e is u_0 and the another vertex of e is in $C_n, n \geq 3$.

Subcase 2.1. Suppose $n \equiv 0 (\text{mod } 4)$. Let $n = 4k$, for some integer k. For $1 \leq i \leq n$, $(K_1 \cup C_n) + u_0v_i$, where $i \neq 2k+1$, is isomorphic to $(K_1 \cup C_n) + u_0v_{2k+1}$. We claim that u_0v_{2k+1} is a stable edge of $K_1 \cup C_n$.

Define the edge labeling $f : E((K_1 \cup C_n) + u_0v_{2k+1}) \to Z^+$ as follows

(i) $f(u_0v_{2k+1}) = f(v_1v_n) = 1$,
(ii) $f(v_iv_{i+1}) = i$ if $1 \leq i \leq 2k$,
(iii) $f(v_{n-i}v_{n+1-1}) = i + 2\alpha(i) + \alpha(i+1)$ if $1 \leq i \leq 2k-1$.

Therefore, the weights of the vertices of $K_1 \cup C_n + u_0v_{2k+1}$ are $1, 2, 3, \ldots, 4k, 4k+2$. All are pairwise distinct. Hence, we have an irregular network with maximum label $2k+1$. Therefore,

$$s((K_1 \cup C_n) + u_0v_{2k+1}) \leq 2k+1 \tag{3}$$

Since, the total sum of the weights of the vertices is twice the sum of the labels in the network, the minimum possible weights are $1, 2, 3, \ldots, 4k, 4k+2$. Since, u_0 is the only pendant vertex of $(K_1 \cup C_n) + u_0v_{2k+1}$, the edge u_0v_{2k+1} will receive the label 1 to obtain the minimum possible weights. Therefore, it is not possible to obtain the weight $4k+2$ by using the labels $2k$ or numbers smaller than $2k$. Hence,

$$s((K_1 \cup C_n) + u_0v_{2k+1}) \geq 2k+1 \tag{4}$$

From (3) and (4) $s((K_1 \cup C_n) + u_0v_{2k+1}) = 2k+1 = s(K_1 \cup C_n)$. Thus, u_0v_{2k+1} is a stable edge of $K_1 \cup C_n$.

Subcase 2.2. Suppose $n \equiv 1 (\text{mod } 4)$. Let $n = 4k+1$, for some integer k. For $1 \leq i \leq n$, $(K_1 \cup C_n) + u_0v_i$ where $i \neq 2k+2$ is isomorphic to $(K_1 \cup C_n) + u_0v_{2k+2}$. We claim that u_0v_{2k+2} is a stable edge of $K_1 \cup C_n$.

Define the edge labeling $f : E((K_1 \cup C_n) + u_0v_{2k+2}) \to Z^+$ as follows:

(i) $f(u_0v_{2k+2}) = f(v_1v_n) = 1$,
(ii) $f(v_iv_{i+1}) = i$ if $1 \leq i \leq 2k+1$,
(iii) $f(v_{n-i}v_{n+1-1}) = i + 2\alpha(i) + \alpha(i+1)$ if $1 \leq i \leq 2k-1$.

Therefore, the weights of the vertices of $K_1 \cup C_n + u_0v_{2k+2}$ are $1, 2, 3, \ldots, 4k+1$, $4k+3$. All are pairwise distinct. Hence, we have an irregular network with maximum label $2k+1$. Therefore,

$$s((K_1 \cup C_n) + u_0v_{2k+2}) \leq 2k+1 \tag{5}$$

Since, the total sum of the weights of the vertices is twice the sum of the labels in the network, the minimum possible weights are $1, 2, 3, \ldots, 4k+1, 4k+3$. Since, u_0 is the only pendant vertex of $(K_1 \cup C_n) + u_0v_{2k+2}$, the edge u_0v_{2k+2} will receive the label 1 to obtain the minimum possible weights. Therefore, it is not possible to obtain the weight $4k+3$ by using the labels $2k$ or numbers smaller than $2k$. Therefore,

$$s((K_1 \cup C_n) + u_0v_{2k+2}) \geq 2k+1. \tag{6}$$

From (5) and (6) $s((K_1 \cup C_n) + u_0v_{2k+2}) = 2k+1 = s(K_1 \cup C_n)$. Thus, u_0v_{2k+2} is a stable edge of $K_1 \cup C_n$.

Subcase 2.3. Suppose $n \equiv 2 (\text{mod } 4)$. Let $n = 4k+2$, for some integer k. For $1 \leq i \leq n$, $(K_1 \cup C_n) + u_0v_i$, where $i \neq 2k+2$, is isomorphic to $(K_1 \cup C_n) + u_0v_{2k+2}$. We claim that u_0v_{2k+2} is a negative edge of $K_1 \cup C_n$.

Define the edge labeling $f : E((K_1 \cup C_n) + u_0v_{2k+2}) \to Z^+$ as follows:

(i) $f(v_1v_n) = f(u_0v_{2k+2}) = 1$,
(ii) $f(v_iv_{i+1}) = i$ if $1 \leq i \leq 2k+1$,
(iii) $f(v_{n-i}v_{n+1-i}) = i + 2\alpha(i) + \alpha(i+1)$ if $1 \leq i \leq 2k$.

Therefore, the weights of the vertices of $K_1 \cup C_n + u_0v_{2k+2}$ are $1, 2, 3, \ldots, 4k+3$. All are pairwise distinct. Hence, we have an irregular network with maximum label $2k+1$. Therefore,

$$s((K_1 \cup C_n) + u_0v_{2k+2}) \leq 2k+1 < s(K_1 \cup C_n).$$

Hence, u_0v_{2k+2} is a negative edge of $K_1 \cup C_n$.

Subcase 2.4. Suppose $n \equiv 3 (\text{mod } 4)$. Let $n = 4k+3$, for some integer k. For $1 \leq i \leq n$, $K_1 \cup C_n + u_ov_i$, where $i \neq 2k+3$, is isomorphic to $K_1 \cup C_n + u_0v_{2k+3}$. We claim that u_0v_{2k+3} is a negative edge of $K_1 \cup C_n$.

Define the edge labeling $f : E((K_1 \cup C_n) + u_0v_{2k+3}) \to Z^+$ as follows:

(i) $f(v_1v_n) = f(u_0v_{2k+3}) = 1$,
(ii) $f(v_iv_{i+1}) = i$ if $1 \leq i \leq 2k+2$,
(iii) $f(v_{n-i}v_{n+1-i}) = i + 2\alpha(i) + \alpha(i+1)$ if $1 \leq i \leq 2k$.

Therefore, the weights of the vertices of $(K_1 \cup C_n) + u_0v_{2k+3}$ are $1, 2, 3, \ldots, 4k+4$. All are pairwise distinct. Hence, we have an irregular network with maximum label $2k+2$. Therefore,

$$s((K_1 \cup C_n) + u_0v_{2k+3}) \leq 2k+2 < s(K_1 \cup C_n).$$

Hence, u_0v_{2k+3} is a negative edge of $K_1 \cup C_n$.

Thus, $K_1 \cup C_n$ is a mixed graph if $n \equiv 0, 1 (\text{mod } 4)$ and $K_1 \cup C_n$ is a negative graph if $n \equiv 2, 3 (\text{mod } 4)$. □

Conclusion and scope

In this paper tK_4 are classified as stable graph and $K_1 \cup C_n$ is classified as a mixed graph if $n \equiv 0, 1 (\mathrm{mod}\ 4)$, negative graph if $n \equiv 2, 3 (\mathrm{mod}\ 4)$. The following are some interesting problems for further investigation.

(i) For $t \geq 2$, tK_n is a negative or stable graph for $n \geq 5$.
(ii) Characterize the mixed, stable and negative graphs.

Acknowledgment

The second author wishes to thank S. Arumugam Director (Research and Development), Kalasalingam University for many fruitful discussion. The authors wish to thank the referees for their comments.

References

[1] G. Chartrand, M. S. Jacobson, J. Lehel, O. R. Oellermann, S. Ruiz and F. Saba, Irregular networks, *Congr. Numer.*, **64** (1988) 187–192.
[2] R. J. Faudree, M. S. Jacobson, J. Lehel and R. H. Schelp, Irregular networks, regular graphs and integer matrices with distinct row and column sums, *Discrete Math.*, **76** (1988) 223–240.
[3] R. J. Faudree, M. S. Jacobson, L. Knich and J. Lehel, Irregularity strength of dense graphs, *Discrete Math.*, **91** (1991) 45–59.
[4] K M. Kathiresan and K. Muthugurupackiam, Change in irregularity strength by an edge, *J. Combin. Math. Combin. Comput.*, **64** (2008) 49–64.
[5] D. B. West, Introduction to Graph Theory, Prentice-Hall of India, New Delhi (2003).

Proc. ICDM 2008, RMS-Lecture Notes Series No. 13, 2010, pp. 109–117.

Some Inequalities for Imbalances in Oriented Graphs

Merajuddin[1] and Madhukar Sharma[2]

[1]Department Applied Mathematics, Z.H. College of Engg. & Tech., A.M.U., Aligarh 202 002 (U.P.), India
[2]Department Applied Mathematics, G.L. Bajaj Inst. of Tech. & Manag., Knowledge Park-III, Greater Noida 201 001 (U.P.), India
e-mail: meraj1957@rediffmail.com; madhukar13777@rediffmail.com

Abstract. In this paper we obtain some stronger inequalities for imbalances in oriented graphs that are individually stronger than the set of inequalities due to Dhruv *et al.*

Keywords. Imbalance, imbalance sequence, oriented graph.

Subject Classification: 05C20

1. Introduction

An oriented graph $D = (V, E)$ with vertex set $V = \{1, 2, \ldots, n\}$ is a digraph with no symmetric pair of directed arcs and with no loops. The imbalance $b(i)$ (or simply b_i) of a vertex i in a digraph is defined as $d^+(i) - d^-(i)$, where $d^+(i)$ and $d^-(i)$ are out-degree and in-degree of vertex i respectively. A sequence $B = (b_1, b_2, \ldots, b_n)$ with $b_1 \le b_2 \le \cdots \le b_n$ is called an imbalance sequence, if there exists a oriented graph D with imbalances $b_i, i = 1, 2, \ldots, n$. Dhruv *et al.* characterized the imbalance sequence of an oriented graph.

Theorem 1 ([5]). *A sequence of integers $B = (b_1, b_2, \ldots, b_n)$ with $b_1 \le b_2 \le \cdots \le b_n$ is an imbalance sequence of an oriented graph if and only if for every subset $I \subseteq [n] = \{1, 2, \ldots, n\}$,*

$$\sum_{i \in I} b_i \ge |I|(|I| - n), \tag{1}$$

with equality when $|I| = n$.

The inequalities (1) *are equivalent to*

$$\sum_{i=1}^{k} b_i \ge k(k - n), \quad 1 \le k \le n. \tag{2}$$

2. Main results

Theorems 2 and 5 are main results of the paper.

Some stronger inequalities for scores in tournaments are given by Brualdi and Shen [1]. Now we show that the imbalance sequence of an oriented graph satisfies the inequalities (3) below, which are individually stronger than the inequalities (1).

Theorem 2. *Asequence of integers $B = (b_1, b_2, \ldots, b_n)$ with $b_1 \leq b_2 \leq \cdots \leq b_n$ is an imbalance sequence of an oriented graph if and only if for every subset $I \subseteq [n]$,*

$$\sum_{i \in I} b_i \geq \sum_{i \in I} (i-1) + \frac{1}{2}|I|(|I| - 2n + 1) \tag{3}$$

with equality when $|I| = n$.

Proof. The sufficiency part of the Theorem follows from Theorem 1, since

$$\sum_{i \in I} (i-1) \geq \sum_{i=1}^{|I|} (i-1) = \frac{1}{2}|I|(|I| - 1)$$

Hence,

$$\sum_{i \in I} b_i \geq \frac{1}{2}|I|(|I| - 1) + \frac{1}{2}|I|(|I| - 2n + 1)$$

$$= |I|(|I| - n).$$

Thus inequalities (3) imply inequalities (1).

Now we prove that imbalance sequence of an oriented graph satisfies (3).

Suppose $B = (b_1, b_2, \ldots, b_n)$ is an imbalance sequence of an oriented graph such that $b_1 \leq b_2 \leq \cdots \leq b_n$. For any subset $I \subseteq [n]$, define

$$f(I) = \sum_{i \in I} b_i - \sum_{i \in I} (i-1) - \frac{1}{2}|I|(|I| - 2n + 1)$$

We choose I such that $f(I)$ and $|I|$ are minimum.

Claim. $I = \{i : 1 \leq i \leq |I|\}$.

Otherwise there exist $i \notin I$ and $j \in I$ such that $j = i + 1$. Since $b_i \leq b_j$,

$$f(I) - f(I \backslash \{j\})$$

$$= \sum_{i \in I} b_i - \sum_{i \in I} (i-1) - \frac{1}{2}|I|(|I| - 2n + 1)$$

$$- \left[\sum_{i \in I} b_i - b_j - \sum_{i \in I} (i-1) + (j-1) - \frac{1}{2}(|I| - 1)(|I| - 2n) \right]$$

$$= b_j - (j-1) + \frac{1}{2}(2n - 2|I|)$$

$$= b_j - (j - n + |I| - 1)$$

So,

$$f(I) - f(I \backslash \{j\}) = b_j - (j - n + |I| - 1) < 0$$

$$\Rightarrow b_j < j - n + |I| - 1 \tag{4}$$

Also,

$$f(I \cap \{i\}) - f(I)$$

$$= \sum_{m\in I} b_m + b_i - \left[\sum_{m\in I}(m-1) + (i-1)\right] - \frac{1}{2}(|I|+1)(|I|-2n+2)$$

$$- \sum_{m\in I} b_m + \sum_{m\in I}(m-1) + \frac{1}{2}|I|(|I|-2n+1)$$

$$= b_i - (i-1) + \frac{1}{2}(2n - 2|I| - 2)$$

$$= b_i - (i - n + |I|)$$

Thus

$$f(I \cap \{i\}) - f(I) = b_i - (i - n + |I|) \geq 0$$

$$\Rightarrow b_i \geq (i - n + |I|) \tag{5}$$

From conditions (4) and (5), we have

$$i - n + |I| \leq b_i \leq b_j < j - n + |I| - 1$$

$$\Rightarrow \quad i - n + |I| \leq b_i \leq b_j < i - n + |I|, \quad (\text{as } j = i + 1)$$

a contradiction. This proves the claim.

$$f(I) = \sum_{i=1}^{|I|} b_i - \sum_{i=1}^{|I|}(i-1) - \frac{1}{2}|I|(|I|-2n+1)$$

$$= \sum_{i=1}^{|I|} b_i - \frac{1}{2}|I|(|I|-1) - \frac{1}{2}|I|(|I|-2n+1)$$

$$= \sum_{i=1}^{|I|} b_i - |I|(|I|-n) \geq 0$$

where the inequality follows from Theorem 1. By the choice of subset I, Theorem 2 follows. □

The following example compares inequalities (3) with inequalities (1) for imbalance sequence $(-2, 1, 1)$.

I :	$\{1\}$	$\{2\}$	$\{3\}$	$\{1,2\}$	$\{1,3\}$	$\{2,3\}$
$\lvert I\rvert(\lvert I\rvert - n)$:	-2	-2	-2	-2	-2	-2
$\sum_{i\in I}(i-1) + \frac{1}{2}\lvert I\rvert(\lvert I\rvert - 2n + 1)$:	-2	-1	0	-2	-1	0

Clearly Theorem 2 gives a lower bound for $\sum_{i\in I} b_i$. It is natural to ask for an upper bound for $\sum_{i\in I} b_i$. The following Corollary 3 gives an upper bound for $\sum_{i\in I} b_i$.

Corollary 3. *A sequence of integers $B = (b_1, b_2, \ldots, b_n)$ with $b_1 \le b_2 \le \cdots \le b_n$ is an imbalance sequence of an oriented graph if and only if for every subset $I \subseteq [n]$,*

$$\sum_{i \in I} b_i \le \sum_{i \in I} (i-1) - \binom{|I|}{2} \tag{6}$$

with equality when $|I| = n$.

Proof. Let $J = [n] \backslash I$, then

$$\sum_{i \in I} b_i \le \sum_{i \in I} (i-1) - \binom{|I|}{2},$$

with equality when $|I| = n$, if and only if

$$\sum_{i \in J} b_i \ge \sum_{i \in J} (i-1) + \frac{1}{2}|J|(|J| - 2n + 1),$$

with equality when $|J| = n$, which implies that

$$\sum_{i \in [n]} b_i - \sum_{i \in I} b_i \ge \sum_{i \in [n]} (i-1) - \sum_{i \in I} (i-1) + \frac{1}{2}(n - |I|)(1 - |I| - n)$$

$$\Rightarrow \quad 0 - \sum_{i \in I} b_i \ge \frac{1}{2} n(n-1) - \sum_{i \in I} (i-1) + \frac{1}{2}(n - n^2 - |I| + |I|^2)$$

$$\Rightarrow \quad \sum_{i \in I} b_i \le \sum_{i \in I} (i-1) - \binom{|I|}{2},$$

with equality when $|I| = n$. □

Merajuddin and Madhukar [4] have obtained the bounds for imbalance b_i of a vertex i of an oriented graph. Now we are giving different proof of the Theorem, which is shorter than earlier proof.

Theorem 4 ([4]). *Suppose that a sequence of integers $B = (b_1, b_2, \ldots, b_n)$ with $b_1 \le b_2 \le \cdots \le b_n$ is an imbalance sequence of an oriented graph. Then for each i,*

$$i - n \le b_i \le i - 1.$$

Proof. Putting $I = \{i\}$ in Theorem 2 we get

$$\sum_{i \in \{i\}} b_i \ge \sum_{i \in \{i\}} (i-1) + \frac{1}{2}(1 - 2n + 1)$$

$$\Rightarrow \quad b_i \ge (i - n). \tag{7}$$

Again putting $I = \{i\}$ in Corollary 3 we have

$$b_i \le (i-1) - \binom{1}{2}$$

$$\Rightarrow \quad b_i \le (i-1). \tag{8}$$

The inequalities (7) and (8) yield the desired result. □

We also remark that the bounds obtained in Theorem 4 are best possible, because for any real number $\delta > 0$, the following inequalities

$$i - n + \delta \leq b_i \leq i - 1 - \delta$$

fail for some oriented graphs (for example, a transitive tournaments), as for transitive tournament of order n always $b_1 = 1 - n$ and $b_n = n - 1$.

Now we investigate existence of equality in inequalities (3). For any integers r and s with $r \leq s$, let $[r, s]$ denote the set of all integers between r and s.

We have following results.

Theorem 5. *Let $B = (b_1, b_2, \ldots, b_n)$ with $b_1 \leq b_2 \leq \cdots \leq b_n$ be an imbalance sequence of an oriented graph. If*

$$\sum_{i \in I} b_i = \sum_{i \in I} (i - 1) + \frac{1}{2}|I|(|I| - 2n + 1) \tag{9}$$

for some $I \subseteq [n]$, then one of the following holds

(i) $I = [1, |I|]$ *and*

$$\sum_{i=1}^{|I|} b_i = |I|(|I| - n)$$

(ii) $I = [t, t + |I| - 1]$ *for some* $t, 2 \leq t \leq n - |I| + 1$,

$$\sum_{i=1}^{t+|I|-1} b_i = (t + |I| - 1)(t + |I| - 1 - n)$$

and $b_i = t + |I| - 1 - n$, *for all* $i \leq t + |I| - 1$.

(iii) $I = [1, r] \cap [r + t, t + |I| - 1]$ *for some* r *and* t *such that* $1 \leq r \leq |I| - 1$ *and* $2 \leq t \leq n - |I| + 1$,

$$\sum_{i=1}^{r} b_i = r(r - n), \quad \sum_{i=1}^{t+|I|-1} b_i = (t + |I| - 1)(t + |I| - 1 - n) \text{ and}$$
$$b_i = (r + t + |I| - n - 1) \text{ for all } i, r + 1 \leq i \leq t + |I| - 1.$$

Proof. For any subset $J \subseteq [n]$, we define $f(J)$ as in Theorem 2, i.e.,

$$f(J) = \sum_{i \in J} b_i - \sum_{i \in J} (i - 1) - \frac{1}{2}|J|(|J| - 2n + 1)$$

So,

$$f(I) = \sum_{i \in I} b_i - \sum_{i \in I} (i - 1) - \frac{1}{2}|I|(|I| - 2n + 1).$$

By condition (9), we have

$$f(I) = 0 \text{ and } f(J) \geq 0 \text{ for all } J \subseteq [n].$$

Now we prove that I is one of the three types as given in the statement of Theorem 5. Suppose this is not true. Then there exist indices i, j, k, and l with $j = i + 1$ and $l = k + 1$ such that $\{i, k\} \cap I = \phi$ and $\{j, l\} \subseteq I$. Thus $b_i \leq b_j$ and $b_k \leq b_l$.

We have

$$\begin{aligned} &f(I) - f(I \backslash \{j, l\}) \\ &\quad = \sum_{i \in I} b_i - \sum_{i \in I} (i-1) - \frac{1}{2}|I|(|I| - 2n + 1) - \sum_{i \in I} b_i + b_j \\ &\quad\quad + b_l + \left\{ \sum_{i \in I} (i-1) - (j-1) - (l-1) \right\} + \frac{1}{2}(|I| - 2)(|I| - 2n - 1) \\ &\quad = b_j + b_l - (j + l + 2|I| - 2n - 3). \end{aligned} \tag{10}$$

And

$$\begin{aligned} &f(I \cap \{i, k\}) - f(I) \\ &\quad = \sum_{m \in I} b_m + b_i + b_k - \left\{ \sum_{m \in I} (m-1) + (i-1) + (k-1) \right\} \\ &\quad\quad - \frac{1}{2}(|I| + 2)(|I| - 2n + 3) - \sum_{m \in I} b_m + \sum_{m \in I} (m-1) + \frac{1}{2}|I|(|I| - 2n + 1) \\ &\quad = b_i + b_k - (i + k - 2n + 2|I| + 1). \end{aligned} \tag{11}$$

Since $f(I) - f(I \backslash \{j, l\}) \leq 0$ and $f(I \cap \{i, k\}) - f(I) \geq 0$, therefore from equations (10) and (11), we get

$$b_j + b_l - (j + l + 2|I| - 2n - 3) \leq 0 \text{ and}$$

$$b_i + b_k - (i + k - 2n + 2|I| + 1) \geq 0$$

$$\Rightarrow \quad b_j + b_l \leq (j + l + 2|I| - 2n - 3) \text{ and } b_i + b_k \geq (i + k - 2n + 2|I| + 1)$$

$$\Rightarrow \quad i + k - 2n + 2|I| + 1 \leq b_i + b_k \leq b_j + b_l \leq j + l + 2|I| - 2n - 3,$$

this gives,

$$i + k - 2n + 2|I| + 1 \leq i + k - 2n + 2|I| - 1 \text{ (as } j = i + 1 \text{ and } l = k + 1)$$

which imply $1 \leq -1$, a contradiction. This proves that I satisfies one of the conditions (i), (ii) or (iii).

Case 1. $I = [1, |I|]$, then

$$f(I) = \sum_{i=1}^{|I|} b_i - \sum_{i=1}^{|I|}(i-1) - \frac{1}{2}|I|(|I| - 2n + 1)$$

$$= \sum_{i=1}^{|I|} b_i - \frac{1}{2}|I|(|I| - 1) - \frac{1}{2}|I|(|I| - 2n + 1)$$

$$= \sum_{i=1}^{|I|} b_i - |I|(|I| - n),$$

and so, $\sum_{i=1}^{|I|} b_i = |I|(|I| - n)$, as $f(I) = 0$.

For other two cases we require the following claim.

Claim 1. If there exit indices i and j with $j = i + 1$ such that $i \notin I$ and $j \in I$, then

$$f(I \cap \{i\}) = f(I \backslash \{j\}) = 0.$$

Since, $f(I) - f(I\backslash\{j\}) = b_j - (j - n + |I| - 1) \leq 0$, and $f(I \cap \{i\}) - f(I) = b_i - (i - n + |I|) \geq 0$

$$\begin{aligned} &\Rightarrow \quad b_j \leq j - n + |I| - 1 \text{ and } b_i \geq i - n + |I| \\ &\Rightarrow \quad i - n + |I| \leq b_i \leq b_j \leq j - n + |I| - 1 \\ &\Rightarrow \quad i - n + |I| \leq b_i \leq b_j \leq i - n + |I|. \end{aligned}$$

Hence equality holds in all the above inequalities. Thus

$$f(I \cap \{i\}) = f(I) = f(I\backslash\{j\}) = 0.$$

Case 2. $I = [t, t + |I| - 1]$ for some $t, 2 \leq t \leq n - |I| + 1$.

By applying Claim 1 recursively, we have

$$f(\{I + |I| - 1\}) = 0$$

i.e.,

$$b_{t+|I|-1} - (t + |I| - 2) - \frac{1}{2}(2 - 2n) = 0$$

$$\Rightarrow \quad b_{t+|I|-1} = (t + |I| - 1 - n).$$

Since, $\sum_{i=1}^{t+|I|-1} b_i \geq (t + |I| - 1)(t + |I| - 1 - n)$ and $b_1 \leq b_2 \leq \cdots \leq b_{t+|I|-1}$, hence equality holds throughout all the above inequalities.

Case 3. $I = [1, r] \cap [r + t, t + |I| - 1]$ for some r and t such that $1 \leq r \leq |I| - 1$ and $2 \leq t \leq n - |I| + 1$. Again applying claim 1 recursively, we have

$$f([1, r]) = f([1, t + |I| - 1]) = f([1, r] \cap \{t + |I| - 1\}) = 0.$$

Therefore,

$$f([1,r]) = \sum_{i=1}^{r} b_i - r(r-n) = 0 \tag{12}$$

$$f([1,t+|I|-1]) = \sum_{i=1}^{t+|I|-1} b_i - (t+|I|-1)(t+|I|-1-n) = 0 \tag{13}$$

and

$$\begin{aligned} & f([1,r] \cap \{t+|I|-1\}) - f([1,\ r]) \\ & = \sum_{i=1}^{r} b_i + b_{t+|I|-1} - \left\{ \sum_{i=1}^{r} (i-1) + (t+|I|-2) \right\} \\ & \quad - \frac{1}{2}(r+1)(r-2n+2) - \sum_{i=1}^{r} b_i + \sum_{i=1}^{r} (i-1) - \frac{1}{2} r(r-2n+1) \\ & = b_{t+|I|-1} - (|I|+r+t-n). \end{aligned}$$

As, $f([1,r] \cap \{t+|I|-1\}) - f([1,r]) = 0$, we have $b_{t+|I|-1} = |I|+r+t-n$.

Since,

$$\begin{aligned} \sum_{i=r+1}^{t+|I|-1} b_i &= \sum_{i=1}^{t+|I|-1} b_i - \sum_{i=1}^{r} b_i \\ &= (t+|I|-1)(t+|I|-1-n) - r(r-n) \\ & \qquad \text{(using equations (12) and (13))} \\ &= (|I|+r+t-n-1)(|I|-r+t-1), \end{aligned}$$

and $b_{r+1} \le b_{r+2} \le \cdots \le b_{t+|I|-1}$, we have

$$b_i = r+t+|I|-n-1 \text{ for all } i,\ r+1 \le i \le t+|I|-1.$$

This completes the proof of the Theorem 5. □

Acknowledgment

We thank the anonymous referees for their valuable suggestions which improved the presentation of the paper.

References

[1] R. A. Brualdi and Jian Shen, Landau's inequalities for tournament scores and a short proof of a theorem on transitive sub-tournaments, *J. Graph Theory*, **38** no. (4) (2001), 244–254.

[2] F. Harary, R. Z. Norman and D. Cartwright, Structural Models: An Introduction to the Theory Directed Graphs, John Wiley and Sons, New York (1965).

[3] H. G. Landau, On dominance relations and the structure of animal societies: III. The condition for a score structure, *Bull. Math. Biophys.*, **15** (1953) 143–148.

[4] Merajuddin and Madhukar Sharma, On imbalance sequences of oriented graphs, to appear.

[5] Dhruv Mubayi, T. G. Will and Douglas B. West, Realizing degree imbalances in directed graphs, *Discrete Math.*, **239(173)** (2001) 147–153.

Proc. ICDM 2008, RMS-Lecture Notes Series No. 13, 2010, pp. 119–123.

A Note on Weak Chain-Completeness and Fixed Point Property for Pseudo Ordered Sets

S. Parameshwara Bhatta and Shiju George

Department of Mathematics, Mangalore University, Mangalagangothri, 574 199, Karnataka State, India
e-mail: s_p_bhatta@yahoo.co.in; shijugeorgem@rediffmail.com

Abstract. It is well known that the weak chain-completeness implies the least fixed point property for pseudo ordered sets. A strengthened form of this result is presented in this paper. A counter example is given to show that the converse is not true in general. Further, using some generalized notions of bounds in pseudo ordered sets, some well known results of Markowsky for posets are extended to pseudo ordered sets.

Keywords. Pseudo ordered sets, weak chain-completeness, fixed point property.

1. Introduction

A reflexive and antisymmetric binary relation $\trianglelefteq$ on a non-empty set A is called a *pseudo order*. The set A together with this pseudo order $\trianglelefteq$ is called a *pseudo ordered set* or a *psoset*. For a subset of A, the notions of a lower bound, an upper bound, the greatest lower bound (or meet), the least upper bound (or join), the minimum (or the least) element and the maximum (or the greatest) element are defined analogous to the corresponding notions in a poset. A *trellis* is a psoset, any two of whose elements have a join and a meet. An extensive investigation of the notions of psosets and related concepts can be found in H. L. Skala [3] and H. Skala [4].

A subset C of A including $C = \phi$ is called a *chain* in A if the restriction of $\trianglelefteq$ to C is a complete order (i.e. it is a partial order on C such that every pair of elements of C are comparable). A subset C^1 of a chain C is said to be cofinal in C, if for every element $c \in C$ there exists an element $c^1 \in C^1$ such that $c \trianglelefteq c^1$. A chain C in A is said to be a *well ordered chain* if every non-empty subset of C has a least element. A psoset A is said to be *chain-complete* if every chain in A has a join. A is said to be *weakly chain-complete* if every well ordered chain in A has a join. The above definitions are due to S. P. Bhatta [1]. A subset T of a psoset $\langle A, \trianglelefteq\rangle$ is called a *tournament* if for every pair of elements $a, b \in T$ either $a \trianglelefteq b$ or $b \trianglelefteq a$ holds. We denote $a \triangleleft b$ for $a \trianglelefteq b$ and $a \neq b$.

A map $f : A \to A$ is said to be *isotone* if $a \trianglelefteq b$ implies $f(a) \trianglelefteq f(b)$. An element $a \in A$ is said to be a *fixed point* for f if $f(a) = a$. If every isotone map of A into itself has a fixed point (least fixed point), then A is said to have the *fixed point property* (*least fixed point property*). Define $\overline{f(A)} = \{u$: *u is the supremum of a chain contained in* $f(A)\}$ [6].

Let B be a subset of the psoset $\langle A, \trianglelefteq\rangle$. Define a new relation $\sqsubseteq_B$ on B by setting $b \sqsubseteq_B b^1$ for two elements b and b^1 of B if there exists a finite sequence $(b_1, b_2, \ldots, b_n)$ of elements of B such that $b = b_1 \trianglelefteq b_2 \trianglelefteq \cdots \trianglelefteq b_n = b^1$. If for each pair of elements b and b^1 of B, both $b \sqsubseteq_B b^1$ and $b^1 \sqsubseteq_B b$ holds, then B will be called a *cycle*. One element cycles are called *trivial cycles*.

2. Weak chain-completeness and fixed point property

Characterization of certain class of posets by fixed points of functions has been a study of interest by several mathematicians. Markowsky [2] has proved that a poset P has the least fixed point property if and only if P is chain-complete. This result has been extended by S. P. Bhatta [1] in one direction by proving that '*every weakly chain-complete psoset has the least fixed point property*'. The theorem given below, further sharpens this result.

Theorem 2.1. *Let $\langle A, \trianglelefteq\rangle$ be a weakly chain-complete psoset. Then every map $f : A \to A$ such that the restriction of f to $\overline{f(A)}$ is isotone, has the least fixed point.*

Proof. Let $\langle A, \trianglelefteq\rangle$ be a weakly chain-complete psoset and $f : A \to A$ be a map such that the restriction of f to $\overline{f(A)}$ is isotone. Let S be the set of all elements x_α, defined by transfinite recursion as follows.

$x_0 = 0$, the least element of A.

$x_\alpha = \vee\{x_\beta : \beta < \alpha\}$, if α is a limit ordinal provided the supremum exists.

$x_\alpha = f(x_\beta)$, if α is the immediate successor of β and x_β exists.

In fact, $S = \{x_\alpha : \alpha$ is an ordinal number for which x_α is defined$\}$.

Claim. S is well ordered and $S \subseteq \overline{f(A)}$.

Assume the contrary. Then choose α to be the least ordinal for which either $\{x_\eta : \eta \leq \alpha\}$ is not well ordered or $x_\alpha \notin \overline{f(A)}$. Then $\{x_\eta : \eta < \alpha\}$ is well ordered.

If α is a limit ordinal, then $\{x_\eta : \eta \leq \alpha\}$ is well ordered. Define $S_\alpha^* = \{f(x_\eta) : \eta < \alpha\}$. Note that $S_\alpha^* \subseteq \{x_\eta : \eta \leq \alpha\}$ which is well ordered. Hence $\vee S_\alpha^*$ exists. Further, by transfinite induction it follows that $x_\alpha = \vee S_\alpha^* \in \overline{f(A)}$, a contradiction to the choice of α.

Thus α is the successor of an ordinal γ. Therefore $\alpha = \gamma + 1$ and hence by the definition we get $x_\alpha = f(x_\gamma) \in \overline{f(A)}$. Let $\beta = \min\{\eta : \eta < \alpha$ and $x_\eta \ntrianglelefteq x_\alpha\}$. Clearly $\beta \neq 0$. If β is a limit ordinal, then x_α is an upper bound for $\{x_\eta : \eta < \beta\}$ so that $\vee\{x_\eta : \eta < \beta\} = x_\beta \trianglelefteq x_\alpha$. If β is not a limit ordinal, then $\beta = \delta + 1$ for some ordinal number δ. Hence $x_\delta \trianglelefteq x_\gamma$ and therefore $x_\beta = f(x_\delta) \trianglelefteq f(x_\gamma) = x_\alpha$. Thus in both cases $x_\beta \trianglelefteq x_\alpha$ for every $\beta < \alpha$ so that $\{x_\beta : \beta \leq \alpha\}$ is well ordered, a contradiction to the choice of α.

Thus S is a well ordered chain contained in $\overline{f(A)}$.

Now $u = \vee S$ exists and $u \in S \subseteq \overline{f(A)}$. Also for every $x_\alpha \in S$ we have $x_\alpha \trianglelefteq u$ so that $f(x_\alpha) \trianglelefteq f(u)$. Further, if α is a limit ordinal, then by transfinite induction we

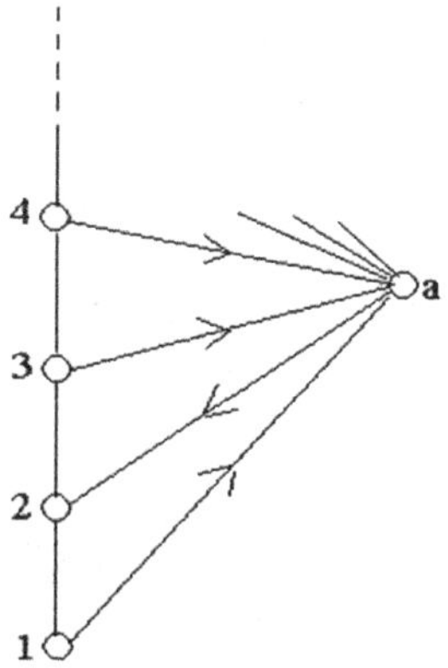

Figure 1.

get $x_\alpha \trianglelefteq f(u)$. Hence $f(u)$ is an upper bound for S so that $S \cup \{f(u)\} = S$. Thus $f(u) \in S$ and therefore $f(u) \trianglelefteq u$. Hence u is a fixed point of f.

If v is any other fixed point for f, then by transfinite induction it follows that v is an upper bound for S, so that $u \trianglelefteq v$. Hence u is the least fixed point for f.

Corollary 2.2 ([1]). *Every weakly chain-complete psoset has the least fixed point property.*

The converse of Corollary 2.2 was posed as an open problem in [1]. Even though Schröder in his review of [1] (See M. R. No. 2005m:06011) has asserted that the answer to this problem is affirmative, the counter example constructed below shows that this assertion is not true in general, even for a tournament. In fact a counter example is constructed for which even the converse of Theorem 2.1 is not true.

Counter example

Let $A = N \cup \{a\} = \{1, 2, 3, \dots\} \cup \{a\}$. We define a pseudo-order $\trianglelefteq$ on A as follows. The elements of N are ordered by the usual natural order of the reals. Further, $a \lhd 2$ and for any $k \in N, k \neq 2$ we have $k \lhd a$. This psoset is represented by the digraph in Figure 1.

Note that A is a tournament. Also A is not weakly chain-complete as the well ordered chain $C = \{1, 2, 3, \dots\}$ is not bounded above.

Let $f : A \to A$ be any map such that the restriction of f to $\overline{f(A)}$ is isotone. Suppose f does not have any fixed point. Note that 1 is the supremum of the empty chain. Hence $1 \in \overline{f(A)}$ so that $1 \lhd f(1) \lhd f^2(1) \lhd \dots$ is a chain in $\overline{f(A)}$. We also observe that $f^n(1) \neq a$ for any $n \in N$. For, if $f^n(1) = a$ for some $n \in N$, then as $f^n(1) \lhd f^{n+1}(1)$, we get $f^{n+1}(1) = 2$. We also have $f^{n+1}(1) \lhd f^{n+2}(1)$ so that $f^{n+2}(1) = k$ for some $k \in N$ where $k \geq 3$. But then $f^n(1) \lhd f^{n+2}(1)$ would imply that $a \lhd k$ for some $k \in N, k \geq 3$, a contradiction. Hence $f^n(1) \neq a$ so that $1 \lhd f(1) \lhd f^2(1) \lhd \dots$ is a chain contained in N.

As $1 \lhd f(1) \lhd f^2(1) \lhd \dots$, we have $f(1) \geq 2$ and $f^2(1) \geq 3$. Hence $\{f^2(1), f^3(1), \dots\}$ is a chain contained in $\{3, 4, \dots\}$. But a is an upper bound for

$\{f^2(1), f^3(1), \ldots\}$. Thus $f(a)$ is an upper bound for $\{f^3(1), f^4(1), \ldots\}$ in A. But the only upper bound of $\{f^3(1), f^4(1), \ldots\}$ is a, so that $f(a) = a$, a contradiction to the assumption that f has no fixed points. Hence f has a fixed point.

Now we shall show that f has the least fixed point.

If $f(1) = 1$, we have done. Hence we may assume that $f(1) \neq 1$. Suppose $f(2) = 2$. Then we cannot have $f(a) = a$. For, if $f(a) = a$, then as f is order preserving in $\overline{f(A)}$ we should have $f(1) = 1$, a contradiction. Thus when $f(2) = 2$, we have 2 as the least fixed point. Now we are left with the case when $f(1) \neq 1$ and $f(2) \neq 2$. If $f(k) = k$ for some $k \in N$, $k \geq 3$, then choose the least such k and that is the least fixed point. Finally, if $f(k) \neq k$ for any $k \in N$, then a is the only one fixed point of f. Thus f has the least fixed point.

Corollary 2.3. *The least fixed point property does not imply weak chain-completeness for trellises (in particular for psosets).*

3. Generalized bound notions in psosets

Definitions. *Let $\langle A, \trianglelefteq \rangle$ be a psoset with the pseudo order $\trianglelefteq$ and B be a subset of A. An element $a \in A$ is called a pseudo upper bound if $b \sqsubseteq_A a$ for every $b \in B$. A pseudo upper bound c of B is called the pseudo supremum if $c \trianglelefteq a$ for every pseudo upper bound a of B. Denote the pseudo supremum of B by $\sqcup B$.*

A map $f : A \to A$ is said to be pseudo isotone if for every $x, y \in A$, $x \sqsubseteq_A y$ implies $f(x) \sqsubseteq_A f(y)$. A psoset is said to be pseudo chain-complete if for every chain C in A, the pseudo supremum $\sqcup C$ exists. If every well ordered chain C in A has the pseudo supremum, then A is called pseudo weakly chain-complete.

Note 1. Let B be any subset of a psoset $\langle A, \trianglelefteq \rangle$. If both $\vee B$ and $\sqcup B$ exist, then $\sqcup B \trianglelefteq \vee B$

Note 2. Let B be a non-trivial cycle in a psoset $\langle A, \trianglelefteq \rangle$. Then the pseudo supremum of B doesn't exist.

Theorem 3.1. *A psoset A is pseudo weakly chain-complete if and only if it is pseudo chain-complete.*

Proof. Clearly it suffices to show the converse part. Hence suppose that A is pseudo weakly chain-complete. Let C be any chain in A. As C is a completely ordered set, we can find a cofinal well ordered chain C^1 of C [5]. By the hypothesis $a = \sqcup C^1$ exists in A. We shall show that $\sqcup C = a$. Clearly a is a pseudo upper bound of C. In fact, for any $x \in C$ there is $y \in C^1$ such that $x \trianglelefteq y$. Also $y \sqsubseteq_A a$ so that $x \sqsubseteq_A a$. Further, if b is any pseudo upper bound of C, then b is also a pseudo upper bound for C^1 so that $a \trianglelefteq b$. Therefore $\sqcup C = a$ exists in A. Thus A is pseudo chain-complete.

Theorem 3.2. *Let $\langle A, \trianglelefteq \rangle$ be a psoset such that every pseudo isotone map $f : A \to A$ has the least fixed point, then A is pseudo chain-complete.*

Proof. Suppose that A is not pseudo chain-complete. Then by the previous theorem we can find a well ordered chain C in A that does not have the pseudo supremum.

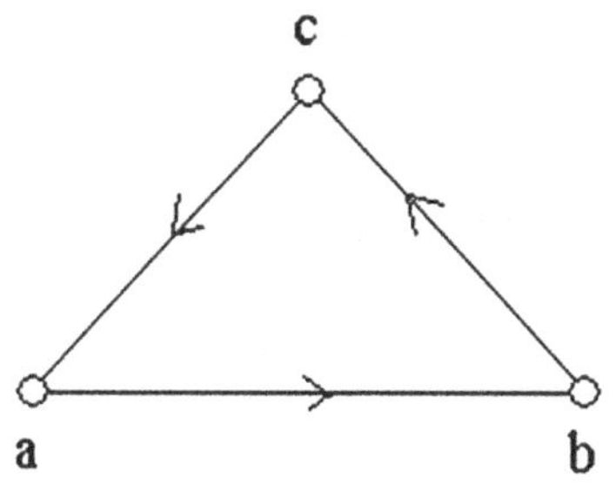

Figure 2.

We shall now construct a pseudo isotone map $f : A \to A$ that does not have the least fixed point. Let U be the set of all pseudo upper bounds of C.

For every $x \in A$, define $C(x) = \{c \in C : c \not\trianglelefteq_A x\}$. Now define the map f as follows: $f(x) = x$ if $x \in U$, and $f(x) = \min c(x)$ if $x \notin U$.

Clearly f is well defined and pseudo isotone. Also the set of all fixed points of f equals U and therefore U must have the least element by our hypothesis. However, the least element of U turns out to be the pseudo supremum of C, a contradiction to our choice of C. This proves the theorem.

Remark 3.3. However the converse of this theorem is not true in general. For example, the psoset represented by the digraph in Figure 2 is pseudo chain-complete, but the map given by $f(a) = b$, $f(b) = c$ and $f(c) = a$, is pseudo isotone but does not have the least fixed point.

Acknowledgment

The second author gratefully acknowledges the financial assistance by CSIR, New Delhi, INDIA, under the Research Fellowship Scheme. We also acknowledge some useful comments of unknown referees.

References

[1] S. P. Bhatta, Weak chain-completeness and fixed point property for pseudo-ordered sets, *Czechoslovak Mathematical Journal*, **55(130)** (2005) 365–369.

[2] G. Markowsky, Chain-complete posets and directed sets with applications, *Algebra Universalis*, **6** (1976) 53–68.

[3] H. L. Skala, Trellis Theory, *Algebra Universalis*, **1** (1971) 218–233.

[4] H. Skala, Trellis Theory, *Mem. Amer. Math. Soc.*, **1** Providence (1972).

[5] L. A. Skornjakov, Elements of lattice theory, Hindustan Publishing Corporation (India) (1977).

[6] M. R. Taskovic, Characterizations of inductive posets with applications, *Proc. Amer. Math. Soc.*, **104** (1988) 650–659.

Proc. ICDM 2008, RMS-Lecture Notes Series No. 13, 2010, pp. 125–132.

Connectivity of Random Addable Graphs

Paul Balister[1], Béla Bollobás[2] and Stefanie Gerke[3]

[1]Department of Mathematical Sciences, University of Memphis, TN 38152-3240, USA
[2]Trinity College Cambridge, CB2 1TQ, UK, and University of Memphis, TN 38152-3240, USA
[3]Royal Holloway College, University of London, TW20 0EX, UK
e-mail: pbalistr@memphis.edu; B.Bollobas@dpmms.cam.ac.uk; stefanie.gerke@rhul.ac.uk

Abstract. A non-empty class $\mathcal{A}$ of labelled graphs is weakly addable if for each graph $G \in \mathcal{A}$ and any two distinct components of G, any graph that can be obtain by adding an edge between the two components is also in $\mathcal{A}$. For a weakly addable graph class $\mathcal{A}$, we consider a random element R_n chosen uniformly from the set of all graph in $\mathcal{A}$ on the vertex set $\{1, \ldots, n\}$. McDiarmid, Steger and Welsh conjecture [5] that the probability that R_n is connected is at least $e^{-1/2} + o(1)$ as $n \to \infty$, and showed that it is at least e^{-1} for all n. Balister, Bollobás and Gerke improved the result by showing that this probability is at least $e^{-0.7983}$ for sufficiently large n. In this paper results on the connectivity of random addable graphs are surveyed and some extensions of the conjecture are discussed.

1. Introduction

Motivated by [5] and following [2] we call a non-empty class $\mathcal{A}$ of labelled graphs *weakly addable*, if for each graph G in $\mathcal{A}$, whenever u and v are vertices in distinct components of G the graph obtained from G by adding an edge joining u and v is also in $\mathcal{A}$. In [5] a weakly addable graph class is defined so that it is also closed under isomorphism, but we do not need this additional requirement. Examples of weakly addable graph classes include forests, planar graphs, and triangle-free graphs, or more generally any H-free or H-minor-free class of graphs for any 2-edge-connected graph H.

For a class $\mathcal{A}$ of labelled graphs, we let $\mathcal{A}_n$ denote the set of graphs in $\mathcal{A}$ on the vertex set $[n] = \{1, \ldots, n\}$. We want to consider a random element R_n of $\mathcal{A}$, that is, we draw R_n uniformly at random from $\mathcal{A}_n$. C. McDiarmid, A. Steger and D. Welsh [5] showed several properties that hold asymptotically almost surely for R_n. (We say that a property P holds asymptotically almost surely for a random addable graph if the probability that $R_n \in P$ tends to one as n tends to infinity.) Many results in their paper relied on the fact that a constant proportion of graphs in any nonempty weakly addable graph class is connected. More precisely, C. McDiarmid, A. Steger and D. Welsh showed [4] that $\mathbb{P}[R_n \text{ is connected }] \geq 1/e$ for an element R_n drawn uniformly at random from $\mathcal{A}_n \neq \emptyset$ of an addable class $\mathcal{A}$ for any n. They conjectured that this result can be strengthened in the following way.

Conjecture 1.1 ([5]). Let $\mathcal{A}$ be any weakly addable class of graphs. Suppose that $\mathcal{A}_n$ is non-empty for all sufficiently large n, and let R_n be drawn uniformly at random from $\mathcal{A}_n$. Then

$$\liminf_{n\to\infty} \mathbb{P}[R_n \text{ is connected}] \geq \frac{1}{\sqrt{e}}.$$

Since an element F_n chosen uniformly at random from the set $\mathcal{F}_n$ of forests with n vertices satisfies $\lim_{n\to\infty} \mathbb{P}[F_n \text{ is connected }] = 1/\sqrt{e}$ [7], the lower bound in Conjecture 1.1 cannot be strengthened. Actually, the class of forests may be in some sense the least connected class of addable graphs, which is made precise in the following conjecture.

Conjecture 1.2. Let $\mathcal{A}$ be weakly addable class of graphs, and let $\mathcal{A}_n$ be non-empty. Let R_n be drawn uniformly at random from $\mathcal{A}_n$, and let F_n be drawn uniformly at random form the class $\mathcal{F}_n$ of forests with n vertices. Then

$$\mathbb{P}[F_n \text{ is connected}] \leq \mathbb{P}[R_n \text{ is connected}].$$

At the moment the best bound on the connectivity in random addable graphs is due to P. Balister, B. Bollobás and S. Gerke and is given in the following theorem.

Theorem 1.3 ([2]). *Let $\mathcal{A}$ be any weakly addable class of graphs. Suppose that $\mathcal{A}_n$ is non-empty, and let R_n be drawn uniformly at random from $\mathcal{A}_n$. Then for sufficiently large n*

$$\mathbb{P}[R_n \textit{ is connected}] \geq e^{-0.7983}.$$

There have been several extension to the notion of addability. The first extension is 2-addable graphs. We say that a non-empty class $\mathcal{B}$ of labelled graphs is 2-addable, if for each graph G in $\mathcal{B}$ and for any pair (C_1, C_2) of distinct components of G, any graph obtained from G by adding at most 2 edges between C_1 and C_2 also lies in $\mathcal{B}$. Note that a 2-addable graph class is weakly addable and thus for large n the probability that an element R_n chosen uniformly at random from a 2-addable class of graph is connected is at least $e^{-0.7983}$. It is known that in fact this probability tends to 1 as n tends to infinity as stated in the next theorem.

Theorem 1.4 ([2]). *Let $\mathcal{B}$ be any 2-addable class of graphs. Suppose that $\mathcal{B}_n$ is non-empty for all sufficiently large n, and let R_n be drawn uniformly at random from $\mathcal{B}_n$. Then*

$$\lim_{n\to\infty} \mathbb{P}[R_n \textit{ is connected}] = 1.$$

It is natural to ask about that probability that a random graph of a 2-addable class is 2-edge-connected. If asked in this way, the answer is that the probability may be zero as the class consisting of a single connected graph with a bridge is 2-addable but not 2-edge-connected. When the definition of 2-addability is slightly strengthened to 2*-addability where for every graph G in a 2*-addable classe $\mathcal{E}$, any graph that can be obtained by adding an edge between two vertices of G that do not have at least two internally edge-disjoint path between them in G is also in $\mathcal{E}$. For 2*-addable graph classes it is unknown whether a constant fraction of graph in such a class are 2-edge-connected.

Conjecture 1.5. There exist a constant $c > 0$ such that for every 2*-addable class $\mathcal{E}$ with $\mathcal{E}_n \neq \emptyset$ for all sufficiently large n an element R_n drawn uniformly at random from $\mathcal{E}_n$ satisfies

$$\liminf_{n\to\infty} \mathbb{P}[R_n \text{ is 2-edge-connected}] \geq c.$$

Another extension allows you to also delete bridges. Following [1] we call a class $\mathcal{D}$ downwards addable if it is weakly addable and if for each graph $G \in \mathcal{D}$ any graph G' that can be obtained from G by deleting an edge of G that increases the number of components is also in $\mathcal{D}$. All the examples given for weakly addable graphs are in fact downwards addable. In particular, forests are downwards addable and hence the best general lower bound on the probability that a random element of $\mathcal{D}_n$ is connected for a downwards addable graph class $\mathcal{D}$ cannot exceed $1/\sqrt{e} + o(1)$. It is known that this lower bound is essentially correct [6].

In the remainder of the paper we shall sketch the proof of Theorem 1.3.

2. Forests

We have already seen that the class of forests plays an important role as it is conjectured that it is the least connected graph class. The following Lemma from [2] shows that in order to prove Conjecture 1.1 it suffices to consider graph classes in which all connected components are trees.

Lemma 2.1 ([2]). *If for every weakly addable graph class consisting of forests only, an element R_n drawn uniformly at random from all elements of this class on $\{1, \ldots, n\}$ satisfies $\mathbb{P}[R_n \text{is connected}] \geq x$ for some $0 \leq x \leq 1$, then an element R'_n drawn uniformly at random from all graphs $\{1, \ldots, n\}$ of any weakly addable graph class satisfies $\mathbb{P}[R'_n$ is connected$] \geq x$.*

Proof. Let $\mathcal{A}$ be a weakly addable graph class. We say that two graphs G, G' in $\mathcal{A}_n$ are equivalent if the graphs obtained from G and G' by removing all bridges are identical; see Figure 1. In other words, G and G' are equivalent, if they have the same 2-edge-connected blocks of size at least 3. Consider a fixed equivalence class $\mathcal{E}_n$, and the collection of 2-edge-connected blocks of size at least 3 obtained by removing all bridges from a graph $G \in \mathcal{E}_n$. Note that it does not matter which graph $G \in \mathcal{E}_n$ is chosen as all graphs in $\mathcal{E}_n$ have the same 2-edge-connected blocks of size at least 3. For each such block, we fix a tree on the same set of vertices. For each graph in the equivalence class $\mathcal{E}_n$, we replace each 2-edge-connected block by its assigned tree. Note that this yields a weakly addable class $\mathcal{C}_n$ such that all of its elements are forests. Moreover, there is a bijection between $\mathcal{E}_n$ and $\mathcal{C}_n$ such that each graph $G \in \mathcal{E}_n$ has the

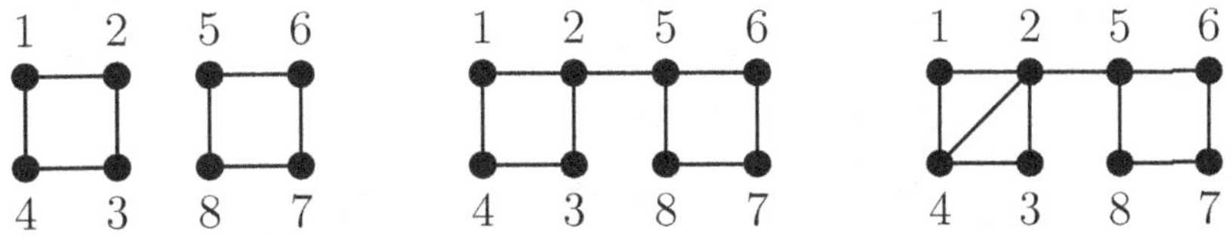

Figure 1. The first two graphs are equivalent but the third is not equivalent to the first two.

same number of components as its image. Hence if at least an x fraction of the graphs in $\mathcal{C}_n$ is connected, then the same is true for the equivalence class $\mathcal{E}_n$ and similarly for all other equivalence classes. This in turn implies the result. □

Note that if we are interested in downwards addable graphs Lemma 2.1 implies that one only has to consider classes of addable graphs whose minimal elements are graphs consisting of a collection of say cliques. (Here, we mean that an element is minimal if it is minimal in the natural partial order on graphs where a graph H is smaller than G when H is a subgraph of G.)

3. Proof of theorem 1.3

Because of Lemma 2.1, it remains to prove Theorem 1.3 if $\mathcal{A}$ consists of forests only. Let $\mathcal{A}_n^i \subseteq \mathcal{A}_n$ be the set of forests of $\mathcal{A}$ on $\{1, \dots, n\}$ with i components. Thus $\mathcal{A}_n^1$ consists of trees on n vertices. Assume that there exists an $0 \le x \le 1$ such that

$$i|\mathcal{A}_n^{i+1}| \le x|\mathcal{A}_n^i| \qquad \text{for all } i = 1, \dots, \lfloor \log n \rfloor, \tag{1}$$

and

$$i|\mathcal{A}_n^{i+1}| \le |\mathcal{A}_n^i| \qquad \text{for all } i = \lfloor \log n \rfloor + 1, \dots, n-1. \tag{2}$$

Then either $|\mathcal{A}_n^i| = 0$ or

$$|\mathcal{A}_n^i| = \frac{|\mathcal{A}_n^i|}{|\mathcal{A}_n^{i-1}|}\frac{|\mathcal{A}_n^{i-1}|}{|\mathcal{A}_n^{i-2}|}\cdots\frac{|\mathcal{A}_n^2|}{|\mathcal{A}_n^1|}|\mathcal{A}_n^1|$$

and hence for n sufficiently large

$$\frac{\sum_{i=1}^n |\mathcal{A}_n^i|}{|\mathcal{A}_n^1|} \le \sum_{i=1}^{\lfloor \log n \rfloor} \frac{x^{i-1}}{(i-1)!} + \sum_{i=\lfloor \log n \rfloor+1}^{n} \frac{x^{\lfloor \log n \rfloor - 1}}{(i-1)!} \le \sum_{i=0}^{\infty} \frac{x^i}{i!} + \sum_{i=\lfloor \log n \rfloor}^{\infty} \frac{1}{i!} = e^x + o(1).$$

Thus

$$\mathbb{P}[R_n \text{ is connected}] = \frac{|\mathcal{A}_n^1|}{\sum_{i=1}^n |\mathcal{A}_n^i|} \ge e^{-x} + o(1).$$

To prove (2), we consider the bipartite graph $B = (\mathcal{A}_n^i \cup \mathcal{A}_n^{i+1}, E)$ with an edge in E between a forest $F \in \mathcal{A}_n^i$ and a forest $F' \in \mathcal{A}_n^{i+1}$ if F' can be obtained from F by removing an edge, see Figure 2. Since any forest in $\mathcal{A}_n^i$ has $n - i$ edges, such a

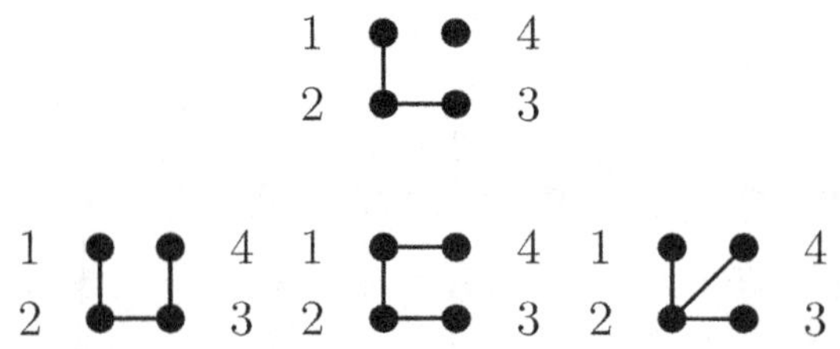

Figure 2. The graph on the top is adjacent to all the graphs on the bottom.

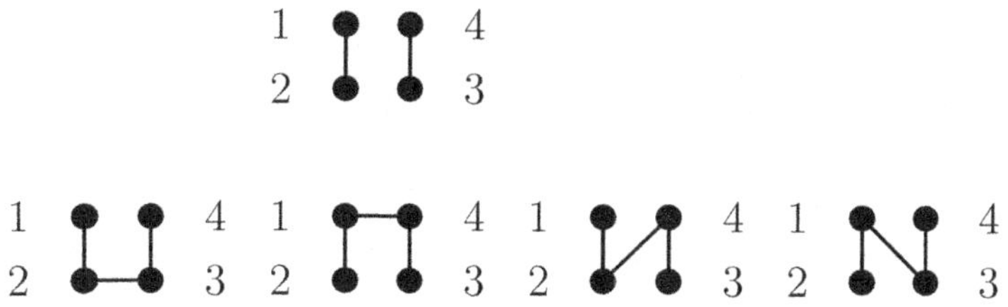

Figure 3. The graph on top is adjacent to all graphs on the bottom.

forest is adjacent to at most $n-i$ forests in $\mathcal{A}_n^{i+1}$. In addition, as the class $\mathcal{A}$ is weakly addable, each forest in $F' \in \mathcal{A}_n^{i+1}$ with components of size $k_1, \ldots, k_{i+1}$ is adjacent to $\sum_{j=1}^{i+1} \sum_{l=j+1}^{i+1} k_j k_l$ forests in $\mathcal{A}_n^i$. In particular, not all vertices in our bipartite graph have the same degree, see Figure 2 and Figure 3.

As $k_j \geq 1$, it follows that

$$\begin{aligned}\sum_{j=1}^{i+1} \sum_{l=j+1}^{i+1} k_j k_l &= \sum_{j=1}^{i+1} \sum_{l=j+1}^{i+1} ((k_j-1)(k_l-1)+k_j+k_l-1)\\ &\geq \sum_{j=1}^{i+1} \sum_{l=j+1}^{i+1} (k_j+k_l-1)\\ &= i\sum_{j=1}^{i+1} k_j - \frac{i(i+1)}{2} = i(n-i) + \binom{i}{2} \geq i(n-i).\end{aligned}$$

thus each forest in $\mathcal{A}_n^{i+1}$ is adjacent to at least $i(n-i)$ forests in $\mathcal{A}_n^i$ (and in fact any forest consisting of i isolated vertices and one component of size $n-i$ has minimal degree in B). Counting the edges of B in two different ways yields

$$|\mathcal{A}_n^i|(n-i) \geq |\mathcal{A}_n^{i+1}| i(n-i)$$

and (2) follows.

To prove (1) we again consider the graph B but with a weighting on the edges. Ideally, we would like to find a weighting such that the weighted degree of each vertex is the same. Such an ideal weighting would yield the optimal result immediately. However, it is conceivable that such a weighting does not exist even though the conjecture is true, which gives a further indication why the proof of the conjecture does not seem to be straightforward.

We use the following weighting to obtain our bound. We assign to the edge $\{F, F'\}$ a weight depending on the degrees of the endvertices of the edge $\{u, v\}$ that we remove from $F \in \mathcal{A}_n^i$ to obtain $F' \in \mathcal{A}_n^{i+1}$. More precisely, for some fixed α, $0 < \alpha \leq 1$, that we shall determine later, we assign the weight $1/(d(u)d(v))^\alpha$ to $\{F, F'\}$ where $d(u), d(v)$ are the degrees of u and v in F.

Consider a forest F' consisting of trees $T_1, \ldots, T_{i+1}$. Because the class $\mathcal{A}_n$ is weakly addable, every forest that is obtained by adding an edge between two trees T_i, T_j is in $\mathcal{A}_n^i$. Hence the sum of the weight over all edges incident to F' equals

$$\sum_{\substack{i<j \\ u\in T_i,\, v\in T_j}} \frac{1}{(d(u)+1)^\alpha}\frac{1}{(d(v)+1)^\alpha} = \sum_{i<j}\left(\sum_{u\in T_i}\frac{1}{(d(u)+1)^\alpha}\cdot\sum_{v\in T_j}\frac{1}{(d(v)+1)^\alpha}\right). \tag{3}$$

We want to give a lower bound on this sum of weights. To do so we consider

$$\min_{T,|V(T)|=n}\sum_{u\in T}\frac{1}{(d(u)+1)^\alpha}$$

where the minimum is taken over all trees with n vertices. Note that $\sum_{u\in T} d(u) = 2n-2$. In addition $(x+1)^{-\alpha}$ is a convex function and hence it is minimised if $d(u) \le d(v)+1$ for all $u, v \in T$. Thus the minimum is attained if T is a path and

$$\min_{T,|V(T)|=n}\sum_{u\in T}\frac{1}{(d(u)+1)^\alpha} = \frac{n-2}{3^\alpha}+\frac{2}{2^\alpha}.$$

Thus if the forest $F' \in \mathcal{A}_n^{i+1}$ consists of components of size $k_1 \ge \cdots \ge k_{i+1}$, then the graph consisting of paths $P_1, \ldots, P_{i+1}$ of length $k_1, \ldots, k_{i+1}$ has smaller weighted degree in our bipartite graph. Moreover if $k_2 \ge 3$ then the graph consisting of paths $P_1', \ldots, P_{i+1}'$ of length $k_1+1, k_2-1, k_3, \ldots, k_{i+1}$ has a weighted degree as most as large as one can see as follows. First

$$\begin{aligned}\left(\frac{k_1-2}{3^\alpha}+\frac{2}{2^\alpha}\right)\left(\frac{k_2-2}{3^\alpha}+\frac{2}{2^\alpha}\right) &= \frac{k_1k_2-2k_1-2k_2+4}{3^{2\alpha}}+\frac{2k_1+2k_2-8}{2^\alpha 3^\alpha}+\frac{4}{2^{2\alpha}}\\ &\ge \frac{k_1k_2-3k_1-k_2+3}{3^{2\alpha}}+\frac{2k_1+2k_2-8}{2^\alpha 3^\alpha}+\frac{4}{2^{2\alpha}}\\ &= \left(\frac{k_1-1}{3^\alpha}+\frac{2}{2^\alpha}\right)\left(\frac{k_2-3}{3^\alpha}+\frac{2}{2^\alpha}\right)\end{aligned}$$

and thus the summand in (3) with $i = 1$, $j = 2$ is at least as large when considering P_1 and P_2 instead of P_1' and P_2'. All summands that do not contain P_1, P_2 and P_1', P_2' respectively, do not change, and finally for all $j = 3, \ldots, i+1$, we have

$$\begin{aligned}&\left(\sum_{u\in P_1}\frac{1}{(d(u)+1)^\alpha}\cdot\sum_{v\in P_j}\frac{1}{(d(v)+1)^\alpha}\right)+\left(\sum_{u\in P_2}\frac{1}{(d(u)+1)^\alpha}\cdot\sum_{v\in P_j}\frac{1}{(d(v)+1)^\alpha}\right)\\ &= \left(\frac{(k_1+2)+(k_1+2)}{3^\alpha}+\frac{4}{2^\alpha}\right)\sum_{v\in P_j}\frac{1}{(d(v)+1)^\alpha}\\ &= \left(\frac{(k_1+3)+(k_1+1)}{3^\alpha}+\frac{4}{2^\alpha}\right)\sum_{v\in P_j}\frac{1}{(d(v)+1)^\alpha}\\ &= \left(\sum_{u\in P_1'}\frac{1}{(d(u)+1)^\alpha}\cdot\sum_{v\in T_j}\frac{1}{(d(v)+1)^\alpha}\right)+\left(\sum_{u\in P_2'}\frac{1}{(d(u)+1)^\alpha}\cdot\sum_{v\in T_j}\frac{1}{(d(v)+1)^\alpha}\right).\end{aligned}$$

Thus the weighted degree of a graph in $\mathcal{A}_{i+1}$ attains it minimum at a graph that contains one long path of size at least $n-2i$ and the remaining i components consist of one or two vertices. Note that an isolated vertex v satisfies $1/(d(v)+1)^{\alpha}=1$ and a component with two vertices u, v satisfies $1/(d(u)+1)^{\alpha}+1/(d(v)+1)^{\alpha}=2/2^{\alpha}\geq 1$. By only considering the summands involving the large component of such a minimum weight graph one obtains for any forest F' consisting of trees $T_1, \ldots, T_{i+1}$, that the sum of the weight over all edges incident to F' satisfies

$$\sum_{\substack{k<j\\ u\in T_k,\, v\in T_j}} \frac{1}{(d(u)+1)^{\alpha}}\frac{1}{(d(v)+1)^{\alpha}} = \sum_{k<j}\left(\sum_{u\in T_k}\frac{1}{(d(u)+1)^{\alpha}}\cdot\sum_{v\in T_j}\frac{1}{(d(v)+1)^{\alpha}}\right)$$

$$\geq \sum_{j=2}^{i+1}\left(\frac{n-2i-2}{3^{\alpha}}+\frac{2}{2^{\alpha}}\right)\geq\frac{i(n-2i)}{3^{\alpha}}. \qquad (4)$$

To obtain an upper bound on the sum of weights over all edges incident to a forest $F\in\mathcal{A}_n^i$, we consider a tree T and $R_{-\alpha}(T)=\sum_{\{v,u\}\in E(T)}(d(u)d(v))^{-\alpha}$. The value $R_{-\alpha}(T)$ is called the Randić index. It is known [3] that there exists a computable constant $\beta_0(\alpha)$ such that for each tree on at least 3 vertices $R_{-\alpha}(T)\leq\beta_0(\alpha)$ $(n+1)$, see Figure 4. Hence in general, $R_{-\alpha}(T)\leq\beta_0(\alpha)(n+C)$ for some constant C. It follows that for all $F\in\mathcal{A}_n^i$ the sum of weights over all edges incident to F is at most $\beta_0(\alpha)(n+Ci)$.

Thus by counting the edge weights of the bipartite graph in two different ways by (4) we obtain

$$|\mathcal{A}_n^i|\ \beta_0(\alpha)(n+Ci)\geq|\mathcal{A}_n^{i+1}|\frac{i(n-2i)}{3^{\alpha}},$$

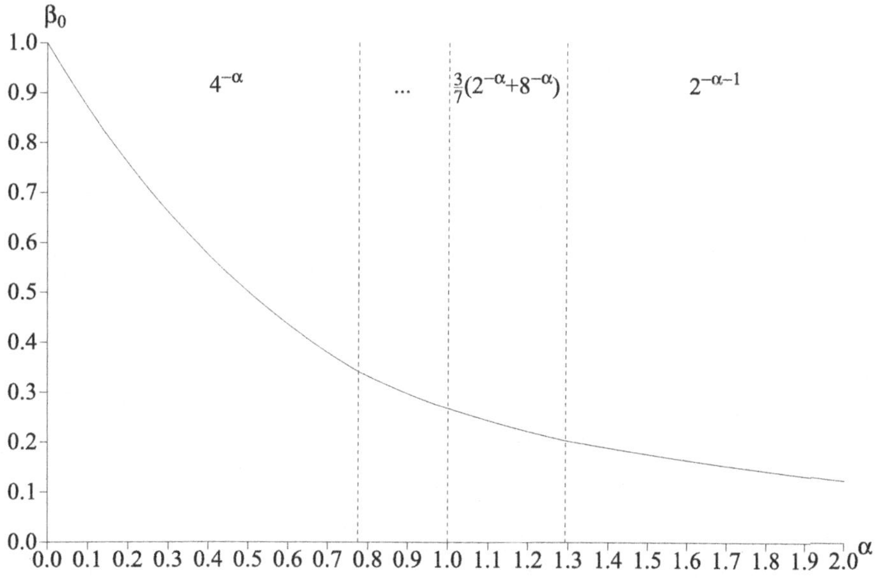

Figure 4. The function $\beta_c(\alpha)$.

and thus

$$\frac{|\mathcal{A}_n^{i+1}|}{|\mathcal{A}_n^i|} \le \frac{\beta_0(\alpha)(n+Ci)}{i(n-2i)/3^\alpha} \le \frac{\beta_0(\alpha)3^\alpha}{i}\left(1+O\left(\frac{i}{n-i}\right)\right).$$

As $i \le \log n$, it remains to find α such that $\beta_0(\alpha)3^\alpha$ is as small as possible. Using the algorithm described in [3] we computed $\beta_0(\alpha)$ for various values of α to estimate the optimal value of α. Setting $\alpha = 0.868$ yields $\beta_0(\alpha) \le 0.30762$ and $3^\alpha\beta_0(\alpha) < 0.7983$, and the claimed result follows.

References

[1] L. Addario-Berry, C. McDiarmid and B. Reed, private communication.

[2] P. Balister, B. Bollobás and S. Gerke, Connectivity of addable graph classes, *Journal of Combinatorial Theory, Series B*, **98** (2008) no. 3, 577–584.

[3] P. Balister, B. Bollobás and S. Gerke, The generalized Randić index of trees, *Journal of Graph Theory*, **56** (2007) no. 4, 270–286.

[4] C. McDiarmid, A. Steger and D. J. A. Welsh, Random planar graphs, *J. Combin. Theory Ser. B.*, **93** (2005) pp. 187–205.

[5] C. McDiarmid, A. Steger and D. J. A. Welsh, Random graphs from planar and other addable classes in Topics in Discrete Mathematics, Dedicated to Jarik Nešetril on the Occasion of its 60th birthday, M. Klazar, J. Kratochvil, M. Loebl, J. Matousek, R. Thomas and P. Valtr, eds., Algorithms and Combinatorics **26**, Springer-Verlag, Berlin (2006) pp. 231–246.

[6] K. Panagiotou, private communication.

[7] A. Rényi, Some remarks on the theory of trees, *Magyar Tud. Akad. Mat. Kutató Int. Közl*, **4** (1959) pp. 73–85.

Proc. ICDM 2008, RMS-Lecture Notes Series No. 13, 2010, pp. 133–152.

Variations of Graph Coloring, Domination, and Combinations of Both: A Brief Survey

Renu Laskar, Jobby Jacob and Jeremy Lyle

Clemson University, Clemson, SC 29631, USA
e-mail: rclsk@clemson.edu

Abstract. A k-coloring of a graph $G = (V, E)$ in general is a partitioning of $V(G)$ into vertex sets $V_1, V_2, \ldots, V_k$, such that each class V_i has a certain prescribed property $\mathcal{P}$. Thus, graph coloring falls into the category of partitioning a set of objects into classes according to certain rules, which is a fundamental process in mathematics. In this brief survey, we will discuss varieties of graph coloring including Grundy coloring, partial Grundy coloring, order chromatic coloring, b-coloring, fall coloring, $L(2, 1)$-coloring, rankings, harmonious coloring, line-distinguishing coloring, t-partitions, etc. We will also look at the color classes where the property $\mathcal{P}$ is related to domination concepts. We conclude with some open problems.

Keywords. Graph coloring, domination, Grundy coloring, partial Grundy coloring, order chromatic coloring, b-coloring, fall coloring, ranking, $L(2, 1)$-coloring, harmonious coloring, line-distinguishing coloring, t-partitions, domatic partitions, dominator partitions.

1. Introduction: Definitions

1.1 *Coloring*

1.1.1 *Chromatic and achromatic*

In general, a *k-coloring* of a graph $G = (V, E)$ is a partition $\Pi = \{V_1, V_2, \ldots, V_k\}$ of the vertices of G into sets V_i, called color classes, having some property $\mathcal{P}$.

The most commonly used property $\mathcal{P}$ is independence, that is, no two vertices in the same color class are adjacent, and the coloring is called *proper*. A graph is proper *k-colorable* if it has a proper k-coloring.

A k-coloring is *complete* if for every $1 \leq i < j \leq k$, there is a vertex $u \in V_i$ and a vertex $v \in V_j$ such that $(u, v) \in E(G)$. The *chromatic number* of G, $\chi(G)$, is defined as the minimum value k such that G has a proper k-coloring, and the *achromatic number* of G, $\psi(G)$, is defined as the largest value k for which G has a complete, proper k-coloring. Notice that every proper coloring of G with $\chi(G)$ colors must be a complete coloring, and $\chi(G) \leq \psi(G)$. We will use loosely k-coloring to mean a proper k-coloring unless stated otherwise.

1.1.2 *Grundy and partial grundy*

A *Grundy k-coloring* of a graph G is a k-coloring $\Pi = \{V_1, V_2, \ldots, V_k\}$ such that for each color class V_i, $1 < i \leq k$, every vertex $u \in V_i$ is adjacent to at least one

Figure 1. Two proper colorings of P_5, to demonstrate that $\chi(G) = 2$ and $\psi(G) = 3$.

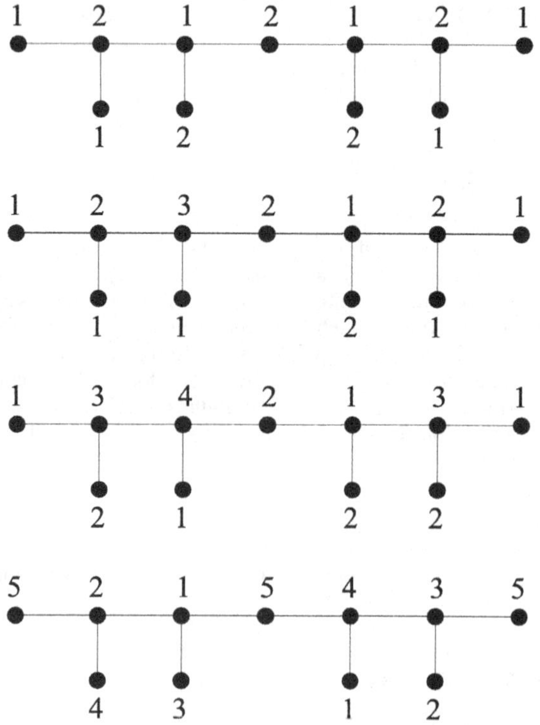

Figure 2. A graph for which $\chi(G) = 2$, $\Gamma(G) = 3$, $\partial\Gamma(G) = 4$, and $\psi(G) = 5$.

vertex in color class V_j for every $j < i$. The *Grundy number*, $\Gamma(G)$, is defined as $\Gamma(G) = \max\{k : G \text{ has a Grundy } k\text{-coloring}\}$. Notice that every Grundy coloring of a graph G is a complete coloring. One can also observe that the minimum integer k for which G has a Grundy k-coloring is always equal to the chromatic number, but there exist graphs G for which $\Gamma(G) < \psi(G)$. Thus for any graph G, $\chi(G) \leq \Gamma(G) \leq \psi(G)$.

In a k-coloring Π, a vertex $v \in V_i$ is called a *Grundy vertex* if v is adjacent to at least one vertex in each color class V_j, for every $j < i$. Note that every vertex $v \in V_1$ vacuously satisfies the definition of being a Grundy vertex and further, observe that every vertex in a Grundy coloring is a Grundy vertex.

A *partial Grundy coloring* is a k-coloring Π such that every color class V_i, $1 \leq i \leq k$, contains at least one Grundy vertex. The *partial Grundy number*, $\partial\Gamma(G)$, is the maximum integer k such that G has a partial Grundy k-coloring. Note that for every graph, $\Gamma(G) \leq \partial\Gamma(G)$. This implies the following inequality chain:

$$\chi(G) \leq \Gamma(G) \leq \partial\Gamma(G) \leq \psi(G) \tag{1}$$

There are graphs for which all of these parameters are different; such an example is shown in Figure 2.

Figure 3. Two proper 2-colorings of P_3 with different chromatic sum.

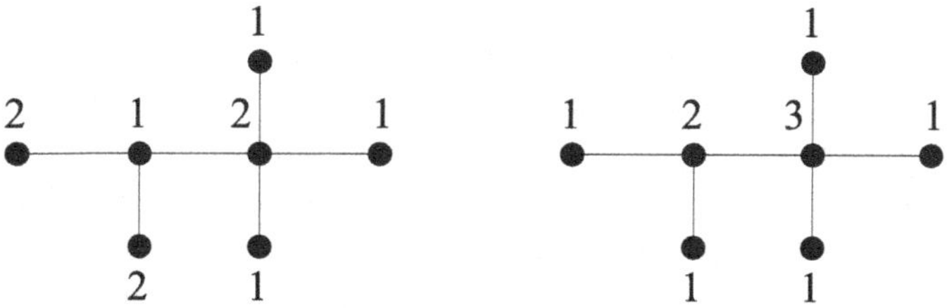

Figure 4. An example where the chromatic sum is 10, accomplished with a proper 2-coloring and with a proper 3-coloring.

1.1.3 *Chromatic sum: Smallest and largest best coloring*

Grundy colorings arise in the study of the problem of determining the *chromatic sum* of a graph G, that is, the minimum sum of colors used in a proper coloring of G.

It is easy to see that every proper coloring of a graph which uses a minimum sum of colors must be a Grundy coloring. Such a minimum sum coloring is called a *best coloring*. This in turn gives rise to two interesting coloring invariants: $\eta^-(G)$, the minimum number of colors used in a best coloring of G, and $\eta^+(G)$, the maximum number of colors used in a best coloring of G.

Thus, for any graph G:

$$\chi(G) \leq \eta^-(G) \leq \eta^+(G) \leq \Gamma(G) \leq \psi(G). \tag{2}$$

1.1.4 *Parsimonious proper coloring (PPC): Ochromatic and upper ochromatic*

Another two types of proper colorings are variants on *parsimonious proper colorings (ppc)*, similar to Grundy and partial Grundy colorings, and are defined as follows: Let $\phi = v_1, v_2, \ldots, v_n$ be an arbitrary ordering of $V(G)$. Consider coloring the vertices of G in the following manner: the vertices are colored in the given order ϕ; when a vertex v_i is to be colored, it must be assigned one of the colors that has been used to color the vertices before it, $v_1, v_2, \ldots, v_{i-1}$, provided that a proper coloring will result. If v_i is adjacent to vertices which have been colored with all currently used colors, then a new color must be assigned. If v_i can be assigned more than one used color, one must choose a color that results in the least number of colors being used to color G. The minimum number of colors used to color G in this way, for the given ordering ϕ, is called the *parsimonious proper ϕ-coloring number*, and is denoted $\chi_\phi(G)$. The *ochromatic number*, $\chi^o(G)$, is defined as the maximum parsimonious proper ϕ-coloring number over all orderings ϕ.

The *upper ochromatic number*, $\psi^o(G)$, is analogous to the ochromatic number $\chi^o(G)$, except that in the ppc-coloring, if a vertex v_i can be assigned more than one used color, you must choose a color that results in the largest number of colors being used to color $V(G)$. The maximum number of colors used to color G in this way, for the given ordering ϕ, is denoted $\psi_\phi(G)$. The *upper ochromatic number*, $\psi^o(G)$, is defined as the maximum value for $\psi_\phi(G)$ over all orderings ϕ. Note that for any graph G, $\chi^o(G) \le \psi^o(G)$.

1.1.5 *b-coloring and fall coloring*

Recall that a vertex $v \in V_i$ is a Grundy vertex in a coloring $\Pi = \{V_1, V_2, \ldots, V_k\}$ if v is adjacent to at least one vertex in each color class V_j, for every $j < i$. A vertex $v \in V_i$ will be called a *colorful vertex* if it is adjacent to at least one vertex in each color class V_j, for $j \neq i$. A color class V_i is called a *colorful color class* if it contains at least one colorful vertex, and a coloring Π is called a *colorful coloring* if every color class is colorful. The *b-chromatic number*, $\varphi(G)$, is the maximum value k for which the graph G has a colorful k-coloring. It is easy to see that every colorful coloring is a complete coloring, and so

$$\chi(G) \le \varphi(G) \le \psi(G) \tag{3}$$

It is also worth noticing that there is no inequality relation between $\varphi(G)$ and $\Gamma(G)$.

A colorful k-coloring $\Pi = \{V_1, V_2, \ldots, V_k\}$ is a *fall k-coloring* if every vertex $v \in V(G)$ is colorful with respect to the coloring. The *fall chromatic number*, $\chi_f(G)$, is the minimum value k such that G has a fall k-coloring. The *fall achromatic number*, $\psi_f(G)$, is the maximum value k such that G has a fall k-coloring. Not every graph has a fall k-coloring for some k. For example, $\chi(C_5) = 3$, but there is no fall k-coloring of C_5 for any value of k. Considering all the complete colorings, we note that if G has a fall coloring, then:

$$\chi(G) \le \chi_f(G) \le \psi_f(G) \le \begin{Bmatrix} \varphi(G) \\ \Gamma(G) \end{Bmatrix} \le \partial\Gamma(G) \le \psi(G). \tag{4}$$

1.1.6 *Line-distinguishing and Harmonious coloring*

The colorings defined in the preceding sections are all examples of complete colorings, that is, between any two color classes V_i and V_j, there exists an edge $(u, v) \in E(G)$, where $u \in V_i$ and $v \in V_j$. If the condition that *at least* one edge exists between any two color classes is replaced by the condition that *at most* one edge exists between any two color classes, the scenario changes drastically.

A k-coloring (not necessarily proper) partition $\Pi = \{V_1, V_2, \ldots, V_k\}$ is *line-distinguishing* if each V_i contains at most one edge and between any two color classes there is at most one edge. The *line-distinguishing number*, $h'(G)$, is the minimum value k for which G has a k-coloring which is line-distinguishing. If a line-distinguishing coloring is also a proper coloring, then it is called a *harmonious coloring*. The *harmonious coloring number*, $h(G)$, is the minimum value k such that G has a harmonious k-coloring. A line-distinguishing coloring $\Pi = \{V_1, V_2, \ldots, V_k\}$ is minimal if any coloring Π' obtained from Π by uniting any two color classes V_i and V_j of Π as one class in Π' is no longer line-distinguishing. A minimal harmonious coloring is similarly defined. The *upper line-distinguishing coloring number*, $H'(G)$, is the maximum value k for which G has a minimal k-line-distinguishing coloring.

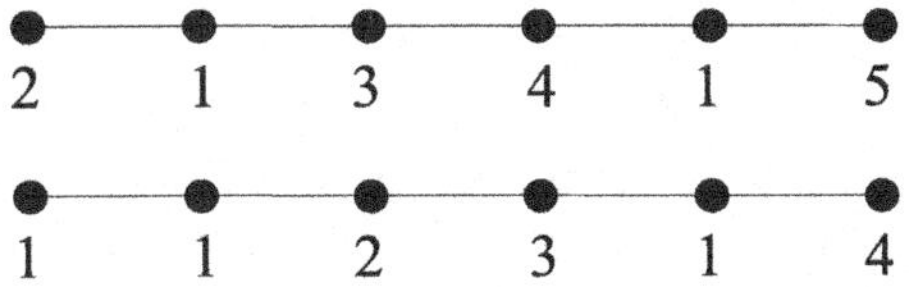

Figure 5. The minimal harmonious coloring of P_6, and a minimal coloring of P_6 which is line-distinguishing.

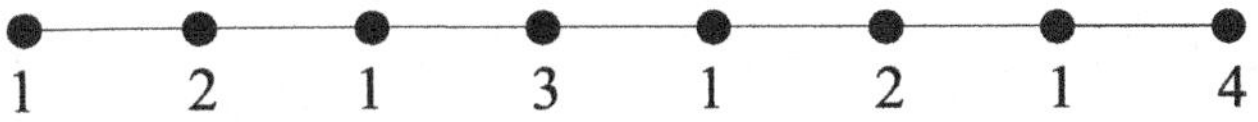

Figure 6. An example of a ranking of P_8.

The *upper harmonious coloring number*, $H(G)$, is the maximum value k for which G has a minimal harmonious k-coloring.

1.1.7 *t-partitions and s-complete t-partitions*

These two types of colorings are generalizations of line-distinguishing and harmonious colorings. Let $\Pi = \{V_1, V_2, \ldots, V_k\}$ be a partition of $V(G)$. Let $e(V_i)$ denote the number of edges in $\langle V_i \rangle$, the induced subgraph of V_i, and $e(V_i, V_j)$ denote the number of edges of G having one end vertex in V_i and one in V_j, in other words, $e(V_i, V_j)$ is the number of edges between V_i and V_j.

The partition Π is a *t-partition* of G if $e(V_i) \leq t$ for $1 \leq i \leq k$ where t is a non-negative integer. The *t-chromatic number*, $\chi_t(G)$, is the minimum value k for which G has a t-partition of order k. Since a 0-partition is a partition into independent sets, $\chi_0(G) = \chi(G)$. A t-partition $\{V_1, V_2, \ldots, V_k\}$ is *s-complete* if $e(V_i, V_j) \geq s$ for $1 \leq i < j \leq k$. The *s-complete t-chromatic number*, $\chi_t^s(G)$ is the minimum value k for which G has a s-complete t-partition of order k.

Since every t-partition of G is a t'-partition for all $t' \geq t$, the t-chromatic numbers of G with e edges form an inequality chain.

$$\chi(G) = \chi_0(G) \geq \chi_1(G) \geq \cdots \geq \chi_{e-1}(G) \geq \chi_e = 1 \tag{5}$$

The idea of t-partitions is not new, but the problem has been presented until now under the following different form. Given a partition $\Pi = \{V_1, \ldots, V_p\}$ of $V(G)$ into p classes, let $\gamma_p(\Pi) = \max\{e(V_i); 1 \leq p \leq |V(G)|\}$. Furthermore, let $\gamma_p(G)$ be the minimum value of $\gamma_p(\Pi)$ over all partitions Π of $V(G)$ into p classes. The problem is to find a good upper bound on $\gamma_p(G)$. For $p = 2$, Erdös conjectured in 1988 that $\gamma_2(G) \leq e/4 + O(\sqrt{e})$ for every graph G with e edges [18]. Later, a sharp bound for any integer p was established by Bollobás and Scott [7].

1.1.8 *Ranking and $L(2, 1)$-coloring*

We include these two proper colorings in the listing because of their interesting aspects and applicability in some other areas. A *k-ranking* of G is a mapping $f : V(G) \to \{1, 2, \ldots, k\}$ such that if $u, v \in V(G)$, where $u \neq v$ and $f(u) = f(v)$, then every $u - v$ path contains a vertex $w \in V(G)$ such that $f(w) > f(u)$.

The *rank number* of a graph, $\chi_r(G)$, is the minimum value k for which G has a k-ranking. Note that a ranking is a proper coloring. A k-ranking, $f : V(G) \to \{1, 2, \ldots, k\}$, is minimal if for any vertex $v \in V(G)$ such that $f(v) = i > 1$ and for $g : V(G) \to \{1, 2, \ldots, k\}$ such that $g(u) = f(u)$ for $u \neq v$ and $g(v) < i$, then g is not a ranking. The *arank number*, $\psi_r(G)$, is the maximum value k for which G has a minimal k-ranking.

The problem of finding rank numbers of a graph has received a lot of attention recently because of the growing number of applications, such as Cholesky factorizations of matrices in parallel, VLSI layout, and scheduling problems of assembly steps in manufacturing systems (see [6], [15] [49], [48], [55], [3]). Determining the rank and arank number of various classes of graphs is an interesting problem. The rank number of a path is determined in [5]. Determining the arank number of a general graph is very difficult. However, the arank numbers of paths and cycles are determined in [43] and [42].

The $L(2,1)$*-coloring* is modeled for channel assignment problems. The channel assignment problem is to assign frequencies, or channels, to transmitters at various locations in a way that permits interference-free communication. Frequencies assigned to neighboring transmitters are required to be substantially different, and channels given to pairs of transmitters which are slightly farther away are required to be different so as to prevent interference.

The channel assignment problem can be modeled as a graph coloring problem. A vertex is assigned to each transmitter. Based on the proximity of the transmitters and the power of the transmitters, edges are placed between vertices to represent possible interference. The frequencies are denoted by nonnegative integers $0, 1, 2, \ldots, \lambda$. The channel assignment problem is then the problem of prescribing integer labels to vertices so that neighboring vertices receive labels that differ by at least two while vertices with a common neighbor have different labels. Such a coloring is called an $L(2,1)$*-coloring* of G. A coloring $f : V(G) \to \{0, 1, \ldots, \lambda\}$ is an $L(2,1)$-coloring if:

(1) $|f(u) - f(v)| \geq 2$, if $(u, v) \in E(G)$ and
(2) $|f(u) - f(v)| \geq 1$, if $d(u, v) = 2$, where $d(u, v)$ is the distance between u and v.

The *span* of G is denoted $\lambda(G)$ and represents the minimum value λ such that G has a $L(2,1)$-coloring with colors from the set $\{0, 1, \ldots, \lambda\}$.

An $L(2,1)$-coloring $f : V(G) \to \{0, 1, \ldots, k\}$ is an *irreducible* coloring if no color of any vertex can be reduced without violating the distance constraints. An $L(2,1)$-coloring $f : V(G) \to \{0, 1, \ldots, k\}$ is a *no-hole* coloring if every color is used. A $L(2,1)$-coloring which is both irreducible and no-hole is called an *inh-coloring*. See Figure 7 for an example.

It was conjectured for any tree $T \neq K_{1,n}$, that $\lambda(T) = \lambda_{inh}(T)$, where $\lambda_{inh}(G)$ denotes the *inh-span* of T, the minimum value k such that G has an inh-coloring with k colors. Recently, the conjecture has been shown to be true [47].

These are only a few of the variations of colorings in the literature. Many more coloring parameters are studied, including the general chromatic number by E. Sampathkumar [65], the k-size chromatic number by E. Sampathkumar and G. D. Kamath [66], the semi-strong chromatic number of a graph by E. Sampathkumar

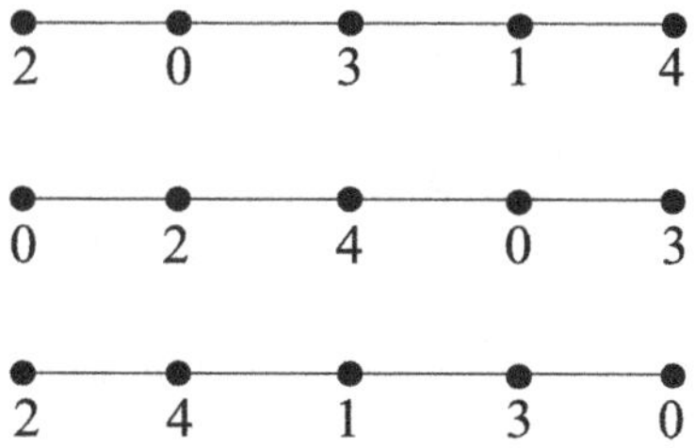

Figure 7. An example of a $L(2, 1)$-coloring which is: (1) an inh-coloing, (2) reducible, with a hole, (3) no-hole, but reducible.

and L. Pushpalatha [67] and the independent saturation number by S. Arumugam and M. Subramanian [69].

1.2 *Domination*

1.2.1 *The domination chain*

A subset of the vertices, $S \subseteq V(G)$, is called a *dominating set* if it satisfies the property that for all vertices $v \in V(G)$, $N[v] \cap S \neq \emptyset$. Note that $S = V(G)$ is trivially a dominating set. A dominating set S is a *minimal dominating set* if the removal of any vertex from S leaves a set which is not a dominating set. One observation to make is that any vertex contained in a minimal dominating set (in a graph with no isolates) has at least one neighbor which is not in the set S. This observation can be used to get the following result due to Ore [53].

Theorem 1.1 ([53]). *If G has no isolates and $S \subseteq V(G)$ is a minimal dominating set of G, then $V(G) - S$ is a dominating set.*

This gives a bound on the cardinality of the smallest dominating set of a graph. The *domination number* of a graph, $\gamma(G)$, is the cardinality of the smallest dominating set of a graph G, and from Theorem 1.1, one can easily see that $\gamma(G) \leq n/2$. The *upper domination number* of a graph, $\Gamma(G)$, is the largest cardinality of a minimal dominating set. There are many variations on the concept of a dominating set, and we discuss only a few of them.

An *independent dominating set* is a dominating set S with the further property that the set S is independent. Since S is independent, each independent dominating set is also a minimal dominating set. Note that if a set is a *maximal independent set*, that is, no vertex can be added to this set such that the resulting set is independent, then it is an independent dominating set. In fact, a set is an independent dominating set if and only it is a maximal independent set. The *independent domination number* of a graph, $i(G)$, is the cardinality of the smallest independent dominating set of a graph G. The *independence number* of a graph, $\beta(G)$, is the cardinality of the largest independent set of a graph.

A subset of the vertices, $S \subseteq V(G)$, is called an *irredundant set* if it satisfies the property that for all vertices $u \in S$, the set $N[u] - N[S - u]$ is not empty. A *maximal irredundant set* S is an irredundant set such that if any vertex is added, the resulting

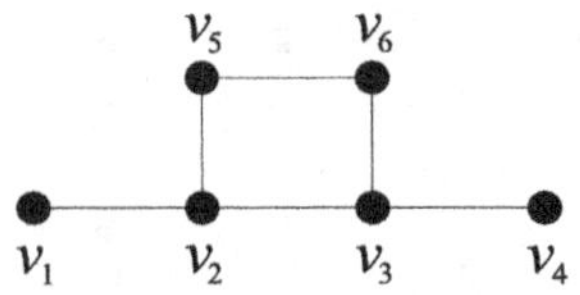

Figure 8. The set $\{v_2, v_3\}$ form a minimal dominating set which is necessarily a maximal irredundant set. The set $\{v_5, v_6\}$, however, forms an maximal irredundant set which is not a dominating set.

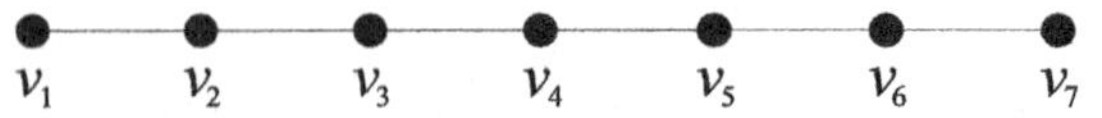

Figure 9. Total and connected dominating sets of P_8. Note that the set $\{v_2, v_3, v_5, v_6\}$ is a total dominating set, but the smallest connected dominating set is $\{v_2, v_3, v_4, v_5, v_6\}$.

set is not an irredundant set. The *upper irredundance number* of a graph, $IR(G)$, is the cardinality of the largest irredundant set. The *irredundance number* of a graph, $ir(G)$, is the minimum cardinality of a maximal irredundant set.

The definitions of the parameters described above lead to the following inequality chain:

$$ir(G) \le \gamma(G) \le i(G) \le \beta(G) \le \Gamma(G) \le IR(G). \tag{6}$$

This interesting chain of inequalities of has produced so many research papers that they are impossible to cite. The interested reader is referred to the following references [31,32].

1.2.2 *Total domination, connected domination, and k-tuple domination*

A subset of the vertices, $S \subseteq V(G)$, is called a *total dominating set* if it satisfies the property that for all vertices $v \in V(G)$, $N(v) \cap S \neq \emptyset$. Note that a total dominating set is a dominating set with the additional property that each vertex in the set S has at least one other neighbor in S. The *total domination number* of a graph, $\gamma_t(G)$, is the cardinality of a smallest total dominating set. The property that each vertex in $V(G)$ has at least one neighbor in the set S can easily be generalized to say that each vertex must have k neighbors in the set S. If every vertex in a dominating set S has at least k neighbors in S, then S is a *k-tuple dominating set.* The *k-tuple domination number* of G, denoted $\gamma_{\times k}(G)$, is the cardinality of a smallest k-tuple dominating set.

Another way to extend this concept is to consider a *connected dominating set.* A connected dominating set is a dominating set S such that the induced subgraph $\langle S \rangle$ is connected. The *connected domination number* of a graph, $\gamma_c(G)$, is the cardinality of a smallest connected dominating set.

As in the case of the discussion of colorings, these are only a few examples of the domination parameters that are studied in the literature. There are many other domination-related parameters, including, among others, the dom-saturation number introduced by B. D. Acharya [68].

2. Background and known results

Graph coloring has a central position in discrete mathematics. It appears in many places without any apparent connection to coloring. It essentially deals with partitioning a set of objects into subsets satisfying some conditions. Scheduling problems, timetabling, in many variations are essentially graph coloring problems. It is not known who first started graph coloring, but certainly the four color conjuncture, which is now a four color theorem inspired many mathematicians for more than a century. The readers are referred to the book *Graph Coloring Problems* by Jensen and Toft [40]. There are so many variations of graph coloring and consequently so many problems are coming up, it is almost impossible to note those down. In this brief survey we deal only a handful of coloring concepts. The following table gives a quick view of the origin of the concepts we are dealing with.

Coloring Variation	Parameters	Origins
Proper Complete Coloring	Chromatic Number, $\chi(G)$	*Kempe, Heawood (1879) [33]*
	Achromatic Number, $\psi(G)$	*Harary, S.T. Hedetniemi (1970) [29]*
Grundy Coloring	Grundy Function Grundy Number, $\Gamma(G)$	*Grundy (1939) [28]* *Christen and Selkow (1979) [11]*
Partial Grundy Coloring	Partial Grundy Number, $\partial\Gamma(G)$	*Erdös, Hedetniemi, Laskar, Prins (2003) [19]*
Chromatic Sum	Best Coloring Minimum Best Coloring, $\eta^-(G)$ Maximum Best Coloring, $\eta^+(G)$	*Kubicka, Schwenk (1989) [46]* *Erdös, Hedetniemi, Laskar, Prins (2003) [19]* *Erdös, Hedetniemi, Laskar, Prins (2003) [19]*
Parsimonious Proper Coloring (PPC)	Ochromatic Number, $\chi^o(G)$ Upper Ochromatic Number, $\psi^o(G)$	*Simmons (1983) [57], (1985) [59], (1985) [58]* *Simmons (1983) [57], (1985) [59], (1985) [58]*
b-Coloring	b-Chromatic Number, $\phi(G)$	*Irving, Manlove, (1999) [37]*
Fall Coloring	Fall-Chromatic Number, $\chi_f(G)$	*Dunbar, Hedetniemi, Hedetniemi, Jacobs, Knisely, Laskar, Rall, (2000) [16]*
	Fall-Achromatic Number, $\psi_f(G)$	*Dunbar, Hedetniemi, Hedetniemi, Jacobs, Knisely, Laskar, Rall, (2000) [16]*
Line-Distinguishing	Line-Distinguishing Number, $h'(G)$	*Frank, Harary, Plantholt (1982) [23], Hopcroft, Krishnamoorty (1983) [36]*
	Upper Line-Distinguishing Number, $H'(G)$	*Chen, Domke, Hattingh, Laskar (1999) [10]*

Coloring variation	Parameters	Origins
Harmonious Coloring	Harmonious Chromatic Number, $h(G)$ Upper Harmonious Chromatic Number, $H(G)$	*Miller, Pritikin, (1991) [51]* *Chen, Domke, Hattingh, Laskar, (1999) [10]*
t-Partitions	t-Chromatic Number, $\chi_t(G)$	*Favaron, Laskar, Rautenbach, (2006) [21]*
s-Complete t-Partitions	s-Complete t-Chromatic Number, $\chi_t^s(G)$	*Favaron, Laskar, Rautenbach, (2006) [21]*
Ranking	Ranking Number, $\chi_r(G)$ Aranking Number, $\psi_r(G)$	*Bodlaender, Deogun, Jansen, Kloks, Kratsch, Müller, and Tuza (1994) [5]* *Ghoshal, Laskar, Pillone, (1996) [26]*
$L(2,1)$-**coloring**	Span, $\lambda(G)$	*Roberts (1988) [54] Griggs, Yeh (1992) [27]*

In addition to presenting a table of relevant references, we also consider the complexity of determining many of these parameters, in the case of a general graph G, as well as common subclasses.

	Arbitrary	Bipartite	Chordal	Tree	Interval
$\chi(G)$	NPC *Karp (1972) [41]*	Trivial	Polynomial *Gavril (1972) [25]*	Trivial	Polynomial *Gavril (1972) [25]*
$\Gamma(G)$	NPC *McRae (1994) [50]*	NPC *McRae (1994) [50]*	NPC *McRae (1994) [50]*	Linear *Hedetniemi, Hedetneimi, Beyer (1982) [35]*	
$\partial\Gamma(G)$	NPC *McRae (1994) [50] Chaudhary Vishwanathan (1997) [9]*			Linear *Shi, Goddard, Hedetniemi, Kennedy, Laskar, McRae (2005) [56]*	
$\psi(G)$	NPC *Yannakakis, Gavril (1980) [61]*	NPC *Cairnie, Edwards (1997) [8]*	NPC *Cairnie, Edwards (1997) [8]*	NPC *Cairnie, Edwards (1997) [8]*	NPC *Bodlaender (1989) [4]*
$h(G)$	NPC *Hopcroft, Krishnamurthy (1983) [36] Garey, Johnson (1979) [24]*	NPC *Edwards, McDiarmid (1995) [17]*	NPC *Asdre, Ioannidou, Nikolopoulos (2007) [1]*	NPC *Edwards, McDiarmid (1995) [17]*	NPC *Asdre, Ioannidou, Nikolopoulos (2007) [1]*

	Arbitrary	Bipartite	Chordal	Tree	Interval
$h'(G)$	NPC *Hoproft, Krishnamurthy (1983) [36]*				
$H(G)$	NPC *Chen, Domke, Hattingh, Laskar (1999) [10]*				
$H'(G)$	NPC *Chen, Domke, Hattingh, Laskar (1999) [10]*				
$\psi_t(G)$ $t = 1$	NPC *Favaron, Laskar, Rantenbach (2006) [21]*				
$\pi_d(G)$	NPC *Hedetniemi, Hedetniemi, Laskar, McRae, Wallis, [34]*	NPC *Hedetniemi, Hedetniemi, Laskar, McRae, Wallis, [34]*	NPC *Hedetniemi, Hedetniemi, Laskar, McRae, Wallis, [34]*		

3. Combinations of coloring and domination

3.1 *Domatic partitions*

In Section 1, we discussed varieties of vertex colorings; that is partitioning $V(G)$ into classes where each class has a property $\mathcal{P}$. In most of these cases the property $\mathcal{P}$ was some restriction on vertex independence. Now we choose $\mathcal{P}$ to be some domination-related concept.

The partition $\Pi = \{V_1, V_2, \ldots, V_k\}$ of $V(G)$ is called a *k-domatic partition* if each V_i is a dominating set [1]. The *domatic number* of a graph, dom(G), is the maximum value k for which G has a k-domatic partition. This concept was introduced by Cockayne and Hedetniemi in [13]. The authors observed that for a graph with no isolates, the domatic number is at least 2 (Theorem 1.1) and is bounded above by $\delta(G) + 1$. In fact, if the maximum degree of a graph is not too large, the authors of [22] were able to show that one can get very close to this bound.

Theorem 3.1 ([22]). *For every graph G,*

$$dom(G) \geq (1 - o(1))\frac{\delta + 1}{\ln(\Delta)}.$$

[1] The term "domatic" is used as a combination of "domination" and "chromatic".

In the case where the minimum degree of the graph is very large, Zelinka [62] gave an alternative bound.

Theorem 3.2 ([62]).

$$dom(G) \geq \left\lfloor \frac{n}{n - \delta(G)} \right\rfloor$$

In a domatic partition, if each V_i is additionally a total dominating set, then Π is a *total domatic partition*, introduced in [12]. The *total domatic number* of a graph G, $\mathrm{dom}_t(G)$ is the maximum value k for which a graph has a k-total domatic partition. Note that $\mathrm{dom}_t(G)$ is not guaranteed to be larger than 1 (consider the five cycle C_5), and now the upper bound on $\mathrm{dom}_t(G)$ is $\delta(G)$.

An *idomatic partition* has a similar definition, as it is a domatic partition where each V_i is additionally independent. The *idomatic number* of G, $i\mathrm{dom}(G)$, is the maximum value k for which G has a k-idomatic partition. It is interesting to observe that an idomatic partition is the same as a fall coloring of G [16]. Domatic partitions are well studied, and the interested reader is referred to a survey article by Zelinka in [64].

One can see that we can define and study various other types of domatic partitions and the corresponding domatic numbers of graphs relative to varieties of domination concepts. Some of these parameters may not exist for all graphs, such as idomatic partitions. However it would be interesting to determine which graphs have such partitions.

Another interesting partition of $V(G)$ is called an *antidomatic partition*, introduced by Zelinka [63], is defined as a partition of $V(G)$ such that none of its classes are dominating sets in G. The minimum value k for which G has a k-antidomatic partition is called the *antidomatic number* of a graph G, denoted $\overline{\mathrm{dom}}(G)$.

3.2 *Dominator partitions*

A *dominator partition* is a partition of the vertices of G, $\Pi = \{V_1, V_2, \ldots, V_k\}$, such that for each $v \in V$ there exists a V_i such that v dominates V_i (that is, v is adjacent to all vertices in V_i). This concept was introduced by Hedetniemi et al in [34]. Note that for any graph G on n vertices $\{\{v_1\}, \{v_2\}, \ldots, \{v_n\}\}$ is a dominator partition of G.

A dominator partition $\Pi = \{V_1, V_2, \ldots, V_k\}$ is *minimal* if for any partition Π' obtained from Π by forming union of any two blocks $V_i \cup V_j$, $i \neq j$ as one block

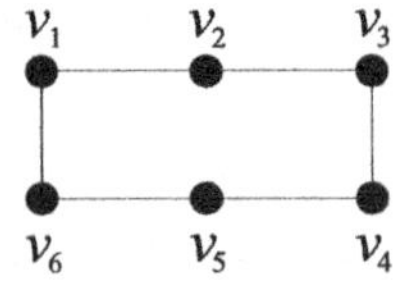

Figure 10. Consider the vertex partitions $\Pi_1 = \{\{v_1, v_2\}, \{v_3, v_4\}, \{v_5, v_6\}\}$, $\Pi_2 = \{\{v_1\}, \{v_4\}, \{v_2, v_3\}, \{v_5, v_6\}\}$, $\Pi_3 = \{\{v_1\}, \{v_4\}, \{v_2, v_3, v_5, v_6\}\}$ and $\Pi_4 = \{\{v_1\}, \{v_3\}, \{v_5\}, \{v_2, v_4, v_6\}\}$. The partitions $\Pi_1, \Pi_2, \Pi_3, \Pi_4$ are all dominator partitions.

of Π' is no longer a dominator partition. The *dominator partition number* of a graph, $\pi_d(G)$, is the minimum value of k for which the graph G has a dominator partition of size k, and similarly, the *upper dominator partition number* of a graph G is the maximal value of k for which the graph G has a minimal dominator partition of size k. Note that in Figure 10, $\pi_d(G) = 3$ and $\Pi_d(G) = 4$.

Let $\Pi = \{V_1, V_2, \ldots, V_k\}$ be a dominator partition of a graph G. A class $V_i \in \Pi$ is said to be a *private dominator class*, abbreviated as PDC, if there exists $v \in V(G)$ such that v is a private neighbor of V_i; *i.e*, v dominates V_i and v does not dominate any other class of Π.

Note that in the example from Figure 10, all the classes of Π_1 are PDCs. In Π_2 the classes $\{v_2, v_3\}$ and $\{v_5, v_6\}$ are not PDCs.

Theorem 3.3 ([39]). *Let Π be a dominator partition of a graph G. Π is a minimal dominator partition if and only if there is at most one class in Π which is not a PDC.*

In the example from Figure 10, Π_2 is not a minimal dominator partition because the classes $\{v_2, v_3\}$ and $\{v_5, v_6\}$ are not PDCs.

The following results about dominator partitions have been proved.

Theorem 3.4 ([34]). *For any graph G of order n,*

1. $1 \le \pi_d(G) \le \Pi_d(G) \le n$,
2. $\pi_d(G) = 1 = \Pi_d(G) \Leftrightarrow G = K_n$,
3. $\pi_d(G) = n = \Pi_d(G) \Leftrightarrow G = \overline{K_n}$,
4. $\pi_d(C_4) = \Pi_d(C_4) = 2$.

Theorem 3.5 ([34]). *For a graph $G \ne K_n$, $\pi_d(G) = 2$ iff $\bar{G}$ is disconnected, or bipartite.*

The next result is quite interesting and gives a tight bound for $\pi_d(G)$.

Theorem 3.6 ([34]). *For any graph G, $\gamma(G) \le \pi_d(G) \le \gamma(G) + 1$.*

So a natural question arises is: what are the graphs for which $\gamma(G) = \pi_d(G)$ and $\pi_d(G) = \gamma(G) + 1$? A graph G is a *dominator graph* if $\pi_d(G) = \gamma(G) + 1$.

Theorem 3.7 ([34]). *A tree T of order $n \ge 4$ is a dominator tree if and only if the leaves of T form a γ-set of T, or equivalently, if and only if T is a corona.*

Theorem 3.8 ([34]). *Let $\sigma = \{C_1, C_2, \ldots, C_l\}$ be the set of distinct simplicial subgraphs induced by the simplicial vertices of a chordal graph G. Then, $\pi_d(G) = \gamma(G)$ if and only if σ is a partition of $V(G)$.*

The bounds of $\Pi_d(G)$ have also been established.

Theorem 3.9 ([34]). *For any graph G of order n,*

$$\frac{n}{1 + \Delta(G)} \le \pi_d(G) \le \Pi_d(G) \le n - \delta(G)$$

where $\Delta(G)$ and $\delta(G)$ are the maximum and minimum degrees of G respectively.

When G is a star on n vertices, one can easily see that $\Pi_d(G) = n - \delta(G) = n - 1$. However, for a general graph, especially when G is a tree, this upper bound may not be a tight bound.

Theorem 3.10 ([39]). $\Pi_d(C_n) = \Pi_d(P_n) = \left\lfloor \frac{2n+1}{3} \right\rfloor$ *for* $n \geq 5$.

A generalization of the dominator partition is defined as the k-dominator partition. A *k dominator partition* of G is a partition of $V(G)$ into classes such that every vertex in $V(G)$ dominates at least k classes. The parameters $\pi_{kd}(G)$ and $\Pi_{kd}(G)$ represent the size of the minimum and maximum minimal k-dominator partitions.

Theorem 3.11 ([38]). *For any graph G which admits a k-dominator partiton,* $\gamma_{\times k}(G) \leq \pi_{kd}(G) \leq \gamma_{\times k}(G) + 1$

Theorem 3.12 ([38]). *Let G be a graph with* $\delta(G) \geq k$. *Then* $\Pi_{kd}(G) \leq n - \delta(G) + k - 1$.

4. Open problems

In this section, we present many open problems relating to the concepts discussed in previous sections.

4.1 *Nordhaus-gaddum-type theorems*

For any graph G, let $\bar{G}$ denote the complementary graph where $V(G) = V(\bar{G})$ and $(u, v) \in E(\bar{G})$ if and only if $(u, v) \notin E(G)$.

In 1956, Nordhaus and Gaddum [52] proved the following theorem on chromatic numbers and complementary graphs. Since then, this type of theorem for any property dealing with complementary graphs has become very popular and known in general as Nordhaus-Gaddum type theorems.

Theorem 4.1 (Nordhaus and Gaddum [52]). *If G is a graph of order n, then*

(1) $2\sqrt{n} \leq \chi(G) + \chi(\bar{G}) \leq n + 1$,
(2) $n \leq \chi(G) \cdot \chi(\bar{G}) \leq \left(\frac{n+1}{2}\right)^2$.

The only other such theorems on coloring parameters are the following:

Theorem 4.2 (Cockayne and Thomason [14]). *If G is a graph of order n, then*

(1) *For* $n \geq 10$, $2\sqrt{n} \leq \Gamma(G) + \Gamma(\bar{G}) \leq \left\lceil \frac{5n+2}{4} \right\rceil$,
(2) *For* $n > 50$, $n \leq \Gamma(G) \cdot \Gamma(\bar{G}) \leq \left\lfloor \frac{m^2}{n} \right\rfloor$, *where* $m = \frac{5n+1}{4}$.

Theorem 4.3 (Kouider, Mahéo, [44]). *If G is a graph of order n, then*

(1) $2\sqrt{n} \leq \varphi(G) + \varphi(\bar{G}) \leq n + 1$,
(2) $n \leq \varphi(G) \cdot \varphi(\bar{G}) \leq \left(\frac{n+1}{2}\right)^2$.

Theorem 4.4 (Erdös, Hedetniemi, Laskar, Prins, [19]). *If G is a graph of order n, then*

(1) $2\sqrt{n} \leq \partial\Gamma(G) + \partial\Gamma(\bar{G}) \leq n + 1 + \Delta(G) - \delta(G)$,
(2) $n \leq \partial\Gamma(G) \cdot \partial\Gamma(\bar{G}) \leq \left\lfloor (n + 1 + \Delta(G) - \frac{\delta(G)}{2} \right\rfloor^2$.

So the natural question arises: what other Nordhaus-Gaddum type theorems can be found for the other coloring parameters discussed in this survey, such as $h(G)$, $H(G)$, $\chi_f(G)$, $\psi_f(G)$, etc.?

4.2 *Interpolation theorems*

In 1967, Harary, Hedetniemi, and Prins [30] published the following interpolation theorem for complete colorings.

Theorem 4.5 (Harary, Hedetniemi, Prins [30]). *If G has two complete colorings of order m and n respectively, then for every integer k where $m \leq k \leq n$, G has a complete k-coloring.*

A similar interpolation theorem for Grundy colorings was proved by Erdös and Prins (unpublished). Alternatively, *b*-colorings do not interpolate for all graphs, and have been the object of study for several papers ([45], [2], [20]). So the natural questions arise: Does there exist interpolation theorems for the other colorings defined? If there is no interpolation theorem, can characterizations be found for graphs which do interpolate?

4.3 *$L(2,1)$-coloring*

(1) Characterize all unicyclic graphs with an inh-span of $\Delta(G) + 1$.
(2) (Conjecture) For any unicyclic graph G where $\Delta(G) < n + 1$ and $G \neq C_4$, $\lambda(G) = \lambda_{inh}(G)$.
(3) Develop an algorithm that gives an inh-span coloring of a hex graph.
(4) What other lattices are inh-colorable?
(5) Characterize all inh-colorable graphs.
(6) (Conjecture) For any graph G, $\lambda(G) \leq \Delta^2$. [27]
(7) Characterize all trees with span $\Delta + 1$. [60]
(8) For what graphs does $\lambda(G) = \Lambda(G)$, where $\Lambda(G)$ denotes the maximum no-hole span? [60]
(9) Is it true that for all $r \in \mathbb{Z}^+$ there exists an infinite family of graphs $\mathcal{F}$ such that $\Lambda(G) - \lambda(G) = r$ if $G \in \mathcal{F}$? [60]
(10) Characterize all trees with an inh-span $\Delta + 1$. [60]
(11) For what inh-colorable graphs does $\lambda_{inh}(G) = \Lambda_{inh}(G)$? [60]
(12) Is it true that for all $r \in \mathbb{Z}^+$ there exists an infinite family of graphs $\mathcal{F}$ such that $\Lambda_{inh}(G) - \lambda_{inh}(G) = r$ if $G \in \mathcal{F}$? [60]
(13) Does there exist a tree such that $\Lambda_{inh}(G) = \Delta^2 + 2\Delta$? [60]
(14) Characterize all unicyclic graphs with an inh-span of $\Delta + 1$. [60]

(15) (Conjecture) For any unicyclic graph G, $\lambda(G) = \lambda_{inh}(G)$. [60]
(16) Is is true that for all $r \in \mathbb{Z}^+$ there exists an infinite family of graphs $\mathcal{C}$ such that $\lambda_{inh}(G) = \lambda(G) + r$ for all $G \in \mathcal{C}$? [60]
(17) For any grid graph $G = P_r \times P_c$, find bounds $l < 9$ and $b < 22$ such that $l \leq \Lambda_{inh}(G) \leq b$. [60]
(18) Develop an algorithm that gives an inh-span coloring of a hex graph. [60]
(19) (Conjecture) For $n \geq 3$ the n-dimensional hypercube is inh-colorable. [60]
(20) Is the complexity of determining the inh-span of the hypercube polynomial? [60]
(21) (Conjecture) All bipartite graphs other than complete bipartite graphs are inh-colorable. [60]
(22) Are any of the following classes of graphs inh-colorable: planar, chordal, split? [60]
(23) Prove that almost all graphs are inh-colorable. [60]
(24) If G is inh-colorable, how many inh-colorings exist on G? [60]
(25) Detemine the graphs for which every span coloring contains the same number of holes. [60]
(26) Is it true that for all $r \in \mathbb{Z}$ there exists an infinite family of graphs $\mathcal{F}$ such that the the difference between the minimum amount of holes in a span coloring and the maximum number of holes in a span coloring is r if $G \in \mathcal{F}$? [60]
(27) Determine the values of h and k for which the path, cycle and trees are inh(h, k)-colorable. [60]
(28) Determine $\lambda(Q_n)$, $\lambda_{inh}(Q_n)$, and $\Lambda_{inh}(Q_n)$.
(29) Determine good bounds for $\lambda(h, k)$ and $\lambda(G(L, w))$.
(30) What is $\lambda(h, k)$ for trees? No hole colorings in each for certain classes of graphs.
(31) Complexity issues
(32) Is there a linear algorithm to determine $\lambda_{inh}(G)$ for trees? How about for other classes of graphs?

4.4 *Ranking*

(1) Is there an upper bound on $\psi_r(G)$ such that when the maximum size of an induced matching is 1, $\psi_r(G) = p - i(G) + 1$?
(2) We know that for any graph G, $\chi_r(G) \geq \chi(G)$. If G is connected, is $\psi_r(G) \geq \psi(G)$?
(3) Any algorithmic results for $\psi_r(G)$, especially trees and chordal graphs.
(4) How can we construct a minimal ranking of G?
(5) Can we determine $\chi_r(G)$ for any G? Can we determine $\psi_r(G)$ for any G?

4.5 *Dominator partitions*

(1) Characterize classes of dominator graphs other than dominator trees or dominator chordal graphs. [34]
(2) When is the complement $\bar{G}$ of a dominator graph G a dominator graph? [34]
(3) If G has a unique π_d-partition, is it a dominator partition? [34]

(4) For which graphs is $\pi_d(G) = \Pi_d(G)$? The equality holds for complete graphs, the complement of complete graphs and the 4-cycle, for example. [34]

(5) What can one say about dominator partitions in which every vertex must be a dominator of at least one class other than its own class. This is called *total dominator partition* of G. The *total dominator partition number* $\pi_{td}(G)$, equals the minimum order of a total dominator partition of G. It is easy to see that $\gamma_t(G) \leq \pi_{td}(G) \leq \gamma_t(G) + 1$, where $\gamma_t(G)$ is the total domination number. [34]

4.6 *Other coloring parameters*

(1) (Harmonius Colorings) Determine the upper line-distinguishing and upper harmonious chromatic numbers of cycles and classes of trees.

(2) (Chromatic Sum) [19] What can you determine about the chromatic sum invariants $\eta^-(G)$ and $\eta^+(G)$? In particular:

 (a) Can these two values be computed for any tree in polynomial time?
 (b) For what classes of trees $\chi(T) = \eta^-(T) = 2$?
 (c) For what classes of trees is $\eta^-(T) = \eta^+(T)$?
 (d) For what classes of trees is $\eta^+(T) = \Gamma(T)$?

(3) (Partial Grundy) Determine the algorithmic complexity of computing the value of $\partial\Gamma(G)$ for various classes of graphs. [19] (Note: For trees this is determined and a linear time algorithm is constructed [56]).

(4) (Fall Coloring) What is the smallest graph with $\delta = k$ which does not have a fall coloring? For $\delta = 1$ the smallest graph with no fall coloring is the graph K_3 with a pendant vertex. For $\delta = 2$ the smallest graph with no fall coloring is C_5. For $\delta = 3$ the smallest graphs with no fall colorings are the wheel of order 6 and the graph which is the complement of the graph $K_3 \cup P_3$. Finally, it can be shown that the complement of the cycle C_7 is the unique smallest graph with $\delta = 4$ having no fall coloring. [16]

(5) (Fall Coloring) Under what conditions does the categorical product of a set of complete graphs K_r, for every integer r in some specified set S, have a fall k-coloring for some integer k not in S? [16]

References

[1] Katerina Asdre, Kyriaki Ioannidou, and Stavros D. Nikolopoulos, The harmonious coloring problem is NP-complete for interval and permutation graphs, *Discrete Appl. Math.*, **155(17)** (2007) 2377–2382.

[2] Dominique Barth, Johanne Cohen, and Taoufik Faik, On the b-continuity property of graphs, *Discrete Appl. Math.*, **155(13)** (2007) 1761–1768.

[3] H. L. Bodlaender, J. S. Deogun, K. Jansen, T. Kloks, D. Kratsch, H. Müller and Zs. Tuza, Rankings of graphs, In: Graph-theoretic concepts in computer science (Herrsching, 1994), volume 903 of Lecture Notes in Comput. Sci., pages 292–304. Springer, Berlin (1995).

[4] Hans L. Bodlaender. Achromatic number is NP-complete for cographs and interval graphs, *Inform. Process. Lett.*, **31(3)** (1989) 135–138.

[5] Hans L. Bodlaender, Jitender S. Deogun, Klaus Jansen, Ton Kloks, Dieter Kratsch, Haiko Müller and Zsolt Tuza, Rankings of graphs, *SIAM J. Discrete Math.*, **11(1)** (1998) 168–181 (electronic).

[6] Hans L. Bodlaender, John R. Gilbert, Hjálmtýr Hafsteinsson and Ton Kloks, Approximating treewidth, pathwidth, and minimum elimination tree height. In: Graph-theoretic concepts in computer science (Fischbachau, 1991), volume 570 of Lecture Notes in Comput. Sci., pages 1–12. Springer, Berlin, (1992).

[7] B. Bollobás and A. D. Scott, Exact bounds for judicious partitions of graphs, *Combinatorica*, **19(4)** (1999) 473–486.

[8] Niall Cairnie and Keith Edwards, Some results on the achromatic number, *J. Graph Theory*, **26(3)** (1997) 129–136.

[9] Amitabh Chaudhary and Sundar Vishwanathan, Approximation algorithms for the achromatic number, In: Proceedings of the Eighth Annual ACM-SIAM Symposium on Discrete Algorithms (New Orleans, LA, 1997), pages 558–563, New York, (1997). ACM.

[10] Guantao Chen, Gayla S. Domke, Johannes H. Hattingh and Renu C. Laskar, On the upper line-distinguishing and upper harmonious chromatic numbers of a graph, *J. Combin. Math. Combin. Comput.*, **31** (1999) 227–239, Papers in honour of Stephen T. Hedetniemi.

[11] Claude A. Christen and Stanley M. Selkow, Some perfect coloring properties of graphs, *J. Combin. Theory Ser. B*, **27(1)** (1979) 49–59.

[12] E. J. Cockayne, R. M. Dawes and S. T. Hedetniemi, Total domination in graphs, *Networks*, **10(3)** (1980) 211–219.

[13] E. J. Cockayne and S. T. Hedetniemi, Towards a theory of domination in graphs, *Networks*, **7(3)** (1977) 247–261.

[14] E. J. Cockayne and A. G. Thomason, Ordered colourings of graphs, *J. Combin. Theory Ser. B*, **32(3)** (1982) 286–292.

[15] I. S. Duff and J. K. Reid, The multifrontal solution of indefinite sparse symmetric linear equations, *ACM Trans. Math. Software*, **9(3)** (1983) 302–325.

[16] J. E. Dunbar, S. M. Hedetniemi, S. T. Hedetniemi, D. P. Jacobs, J. Knisely, R. C. Laskar and D. F. Rall, Fall colorings of graphs, *J. Combin. Math. Combin. Comput.*, **33** (2000) 257–273, Papers in honour of Ernest J. Cockayne.

[17] Keith Edwards and Colin McDiarmid, The complexity of harmonious colouring for trees, *Discrete Appl. Math.*, **57(2–3)** (1995) 133–144, Combinatorial optimization 1992 (CO92) (Oxford).

[18] Paul Erdös, Sixth International Conference on the Theory and Applications of Graphs, Kalamazoo, MI, (1988), Invited talk.

[19] Paul Erdös, Stephen T. Hedetniemi, Renu C. Laskar and Geert C. E. Prins, On the equality of the partial Grundy and upper ochromatic numbers of graphs, *Discrete Math.*, **272(1)** (2003) 53–64, In honor of Frank Harary.

[20] Taoufik Faik, About the *b*-continuity of graphs (extended abstract), In: Workshop on Graphs and Combinatorial Optimization, volume 17 of *Electron. Notes Discrete Math.*, pages 151–156 (electronic), Elsevier, Amsterdam, (2004).

[21] Odile Favaron, Renu Laskar and Dieter Rautenbach, t-partitions and s-complete t-partitions of a graph, *Australas. J. Combin.*, **36** (2006) 295–302.

[22] Uriel Feige, Magnús M. Halldórsson, Guy Kortsarz and Aravind Srinivasan, Approximating the domatic number, *SIAM J. Comput.*, **32(1)** (2002/03) 172–195 (electronic).

[23] Ove Frank, Frank Harary and Michael Plantholt, The line-distinguishing chromatic number of a graph, *Ars Combin.*, **14** (1982) 241–252.

[24] Michael R. Garey and David S. Johnson. Computers and intractability. W. H. Freeman and Co., San Francisco, Calif., 1979. A guide to the theory of NP-completeness, A Series of Books in the Mathematical Sciences.

[25] Fănică Gavril, Algorithms for minimum coloring, maximum clique, minimum covering by cliques, and maximum independent set of a chordal graph, *SIAM J. Comput.*, **1(2)** (1972) 180–187.

[26] J. Ghoshal, R. Laskar and D. Pillone, Minimal rankings, *Networks*, **28(1)** (1996) 45–53.
[27] Jerrold R. Griggs and Roger K. Yeh, Labelling graphs with a condition at distance 2, *SIAM J. Discrete Math.*, **5(4)** (1992) 586–595.
[28] P. M. Grundy, Mathematics and games, *Eureka*, **2** (1939) 6–8.
[29] Frank Harary and Stephen Hedetniemi, The achromatic number of a graph, *J. Combinatorial Theory*, **8** (1970) 154–161.
[30] Frank Harary, Stephen Hedetniemi and Geert Prins, An interpolation theorem for graphical homomorphisms, *Portugal. Math.*, **26** (1967) 453–462.
[31] T. W. Haynes, S. T. Hedetniemi and P. J. Slater, Domination in graphs, Advanced topics, Marcel Dekker (1998).
[32] T. W. Haynes, S. T. Hedetniemi and P. J. Slater, Fundamentals of Domination in graphs, Marcel Dekker (1998).
[33] P. J. Heawood, Map-colour theorem, *Proc. London Math. Soc. (2)*, **51** (1949) 161–175.
[34] S. M. Hedetniemi, S. T. Hedetniemi, R. C. Laskar, A. A. McRae and C. K. Wallis, Dominator partitions of graphs, To Appear.
[35] Sandra Mitchell Hedetniemi, Stephen Hedetniemi and Terry Beyer, A linear algorithm for the Grundy (coloring) number of a tree, In: Proceedings of the thirteenth Southeastern conference on combinatorics, graph theory and computing (Boca Raton, Fla., 1982), **36** (1982) 351–363.
[36] J. E. Hopcroft and M. S. Krishnamoorthy, On the harmonious coloring of graphs, *SIAM J. Algebraic Discrete Methods*, **4(3)** (1983) 306–311.
[37] Robert W. Irving and David F. Manlove, The b-chromatic number of a graph, *Discrete Appl. Math.*, **91(1–3)** (1999) 127–141.
[38] Jobby Jacob and Wayne Goddard, Generalized dominator partitions of graphs, Presented in the Thirty-Nineth Southeastern International Conference on Combinatorics, Graph Theory and Computing (2008).
[39] Jobby Jacob, Renu Laskar and Wayne Goddard, The upper dominator partition number of some simple graphs, In: Proceedings of the Thirty-Seventh Southeastern International Conference on Combinatorics, Graph Theory and Computing, **182** (2006) 65–77.
[40] Tommy R. Jensen and Bjarne Toft, Graph coloring problems, Wiley-Interscience Series in Discrete Mathematics and Optimization. John Wiley & Sons Inc., New York, (1995), A Wiley-Interscience Publication.
[41] Richard M. Karp, Reducibility among combinatorial problems, In: Complexity of computer computations (Proc. Sympos., IBM Thomas J. Watson Res. Center, Yorktown Heights, N.Y., 1972), pages 85–103, Plenum, New York (1972).
[42] V. Kostyuk and D. Narayan, Maximum minimal k-rankings of cycles. to appear.
[43] V. Kostyuk, D. Narayan and V. Williams, Minimal rankings and the arank number of a path, *Discrete Math.*, **306** (2006) 1991–1996.
[44] Mekkia Kouider and Maryvonne Mahéo, Some bounds for the b-chromatic number of a graph, *Discrete Math.*, **256(1–2)** (2002) 267–277.
[45] Jan Kratochvíl, Zsolt Tuza, and Margit Voigt, On the b-chromatic number of graphs, In: Graph-theoretic concepts in computer science, volume 2573 of *Lecture Notes in Comput. Sci.*, (2002) 310–320. Springer, Berlin.
[46] E. Kubicka and A. Schwenk, Introduction to chromatic sums, Proceedings of the 17th Annual ACM Computer Science Conference (1989) 39–45.
[47] Renu Laskar, Beth Novick and Gretchen Matthews, On irreducible, no-hole $L(2,1)$-coloring of Trees, *Networks*, to appear.

[48] Charles E. Leiserson, 21st Annual Symposium on Foundations of Computer Science, Institute of Electrical and Electronics Engineers Inc. (IEEE), New York (1980).

[49] Joseph W. H. Liu, The role of elimination trees in sparse factorization, *SIAM J. Matrix Anal. Appl.*, **11(1)** (1990) 134–172.

[50] A. McRae, NP-completeness Proof for Grundy Coloring, unpublished manuscript, (1994).

[51] Z. Miller and D. Pritikin, The harmonious coloring number of a graph, *Discrete Math.*, **93(2–3)** (1991) 211–228.

[52] E. A. Nordhaus and J. W. Gaddum, On complementary graphs, *Amer. Math. Monthly*, **63** (1956) 175–177.

[53] Oystein Ore, Theory of graphs. American Mathematical Society Colloquium Publications, Vol. XXXVIII. American Mathematical Society, Providence, R.I. (1962).

[54] Fred Roberts, Private Communication (1988).

[55] Arunabha Sen, Haiyong Deng and Sumanta Guha, On a graph partition problem with application to VLSI layout, *Inform. Process. Lett.*, **43(2)** (1992) 87–94.

[56] Zhengnan Shi, Wayne Goddard, Stephen T. Hedetniemi, Ken Kennedy, Renu Laskar and Alice McRae, An algorithm for partial Grundy number on trees, *Discrete Math.*, **304(1–3)** (2005) 108–116.

[57] Gustavus J. Simmons, On the ochromatic number of a graph, In: *Proceedings of the fourteenth Southeastern conference on combinatorics, graph theory and computing (Boca Raton, Fla., 1983)*, **40** (1983) 339–366.

[58] Gustavus J. Simmons, A graph-theoretical exorcism, In: Proceedings of the sixteenth Southeastern international conference on combinatorics, graph theory and computing (Boca Raton, Fla., 1985), **49** (1985) 205–214.

[59] Gustavus J. Simmons, The ochromatic number of planar graphs, In: Graphs and applications (Boulder, Colo., 1982), Wiley-Intersci. Publ. (1985) 295–316, Wiley, New York.

[60] John Villalpando, Graph parameters: channel assignment as related to $L(2, 1)$-coloring, PhD thesis, Clemson University, (2002).

[61] M. Yannakakis and F. Gavril, Edge dominating sets in graphs, *SIAM J. Appl. Math.*, **38(3)** (1980) 364–372.

[62] Bohdan Zelinka, Domat ic number and degrees of vertices of a graph, *Math. Slovaca*, **33(2)** (1983) 145–147.

[63] Bohdan Zelinka, Antidomatic number of a graph, *Arch. Math. (Brno)*, **33(3)** (1997) 191–195.

[64] Bohdan Zelinka, Domatic numbers of graphs and their variants: A survey. In: Domination in graphs, volume 209 of *Monogr. Textbooks Pure Appl. Math.*, (1998) 351–377. Dekker, New York.

[65] E. Sampathkumar Generalizations of independence and chromatic numbers of a graph, In: *Discrete Math.*, **115** (1993) 245–251.

[66] E. Sampathkumar and G. D. Kamath, k-size chromatic number of a graph, In: *Sankhyā. The Indian Journal of Statistics*, **54** (1992) 393–397.

[67] E. Sampathkumar and L. Pushpalatha, Semi-strong chromatic number of a graph, In: *Indian Journal of Pure and Applied Mathematics*, **26** (1995) 35–40.

[68] B. D. Acharya, The strong domination number of a graph and related concepts, In: *Journal of Mathematical and Physical Sciences*, **14** (1980) 471–475.

[69] S. Arumugam and M. Subramanian, Independence saturation and extended domination chain in graphs, In: *AKCE International Journal of Graphs and Combinatorics*, **4** (2007) 171–181.

Proc. ICDM 2008, RMS-Lecture Notes Series No. 13, 2010, pp. 153–160.

Directionally n-Signed Graphs

E. Sampathkumar[1], P. Siva Kota Reddy[2] and M. S. Subramanya[1]

[1]Department of Studies in Mathematics, University of Mysore, Manasagangotri, Mysore 570 006, India
[2]Department of Mathematics, Acharya Institute of Technology, Soldevanahalli, Bangalore 560 090, India
e-mail: esampathkumar@eth.net; reddy_math@yahoo.com; subramanya_ms@rediffmail.com

Abstract. Let $G = (V, E)$ be a graph. By *directional labeling (or d-labeling)* of an edge $x = uv$ of G by an ordered n-tuple $(a_1, a_2, \ldots, a_n)$, we mean a labeling of the edge x such that we consider the label on uv as $(a_1, a_2, \ldots, a_n)$ in the direction from u to v, and the label on x as $(a_n, a_{n-1}, \ldots, a_1)$ in the direction from v to u. In this paper, we study graphs, called *(n, d)-sigraphs*, in which every edge is d-labeled by an n-tuple $(a_1, a_2, \ldots, a_n)$, where $a_k \in \{+, -\}$, for $1 \leq k \leq n$. Giving some motivation to study such graphs, we obtain some results by introducing some notions of balance and special types of complementations.

Keywords. Sigraphs, directional labeling, complementation, balance.

2000 Subject Classification: 05C 22

1. Introduction

For graph theory terminology and notation in this paper we follow the book [2]. All graphs considered here are finite and simple.

In this paper, we investigate the notion of complementation and generalize the notion of balance in sigraphs (a graph whose edges have signs) to the more general context of graphs with multiple signs on their edges. We look at two kinds of complementation: complementing some or all of the signs, and reversing the order of the signs on each edge. Also, we generalize the notion of balance in sigraphs to such graphs in two different ways which coincides, when restricted to sigraphs and obtain characterization of such graphs.

There are two ways of labeling the edges of a graph by an ordered n-tuple $(a_1, a_2, \ldots, a_n)$.

1. *Undirected labeling* or *labeling*. This is a labeling of each edge uv of G by an ordered n-tuple $(a_1, a_2, \ldots, a_n)$ such that we consider the label on uv as $(a_1, a_2, \ldots, a_n)$ irrespective of the direction from u to v or v to u.
2. *Directional labeling* or *d-labeling*. This is a labeling of each edge uv of G by an ordered n-tuple $(a_1, a_2, \ldots, a_n)$ such that we consider the label on uv as $(a_1, a_2, \ldots, a_n)$ in the direction from u to v, and $(a_n, a_{n-1}, \ldots, a_1)$ in the direction from v to u.

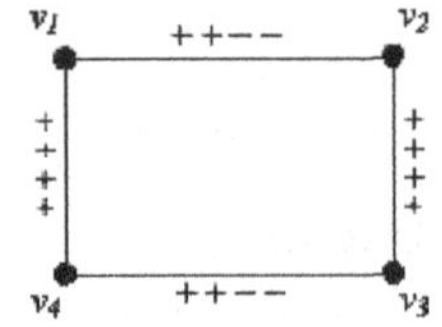

Figure 1.

Note that the d-labeling of edges of G by ordered n-tuples is equivalent to labeling the symmetric digraph $\overrightarrow{G} = (V, \overrightarrow{E})$, where uv is a symmetric arc in $\overrightarrow{G}$ if, and only if, uv is an edge in G, so that if $(a_1, a_2, \ldots, a_n)$ is the d-label on uv in G, then the labels on the arcs $\overrightarrow{uv}$ and $\overrightarrow{vu}$ are $(a_1, a_2, \ldots, a_n)$ and $(a_n, a_{n-1}, \ldots, a_1)$ respectively.

Let H_n be the *n-fold sign group*, $H_n = \{+, -\}^n = \{(a_1, a_2, \ldots, a_n) : a_1, a_2, \ldots, a_n \in \{+, -\}\}$ with co-ordinate-wise multiplication. Thus, writing $a = (a_1, a_2, \ldots, a_n)$ and $t = (t_1, t_2, \ldots, t_n)$ then $at := (a_1t_1, a_2t_2, \ldots, a_nt_n)$. For any $t \in H_n$, the *action* of t on H_n is $a^t = at$, the co-ordinate-wise product.

Let $n \geq 1$ be a positive integer. An *n-sigraph (n-sidigraph)* is a graph $G = (V, E)$ in which each edge (arc) is labeled by an ordered n-tuple of signs, i.e., an element of H_n. A *sigraph* $G = (V, E)$ is a graph in which each edge is labeled by $+$ or $-$. Thus a 1-sigraph is a sigraph. Sigraphs are well studied in literature (See for example [1,3,4,8] and [9]). In this paper, we study graphs in which each edge is labeled by an ordered n-tuple $a = (a_1, a_2, \ldots, a_n)$ of signs (i.e, an element of H_n) in one direction but in the other direction its label is the reverse: $a^r = (a_n, a_{n-1}, \ldots, a_1)$, called *directionally labeled n-signed graphs* (or *(n, d)-sigraphs*).

Note that an n-sigraph $G = (V, E)$ can be considered as a symmetric digraph $\overrightarrow{G} = (V, \overrightarrow{E})$, where both $\overrightarrow{uv}$ and $\overrightarrow{vu}$ are arcs if, and only if, uv is an edge in G. Further, if an edge uv in G is labeled by the n-tuple $(a_1, a_2, \ldots, a_n)$, then in $\overrightarrow{G}$ both the arcs $\overrightarrow{uv}$ and $\overrightarrow{vu}$ are labeled by the n-tuple $(a_1, a_2, \ldots, a_n)$.

In [1], the authors study voltage graph defined as follows: A *voltage graph* is an ordered triple $\overrightarrow{G} = (V, \overrightarrow{E}, M)$, where V and $\overrightarrow{E}$ are the vertex set and arc set respectively and M is a group. Further, each arc is labeled by an element of the group M so that if an arc $\overrightarrow{uv}$ is labeled by an element $a \in M$, then the arc $\overrightarrow{vu}$ is labeled by its inverse, a^{-1}.

Since each n-tuple $(a_1, a_2, \ldots, a_n)$ is its own inverse in the group H_n, we can regard an n-sigraph $G = (V, E)$ as a voltage graph $\overrightarrow{G} = (V, \overrightarrow{E}, H_n)$ as defined above. Note that the d-labeling of edges in an (n, d)-sigraph considering the edges as symmetric directed arcs is different from the above labeling. For example, consider a $(4, d)$-sigraph in Figure 1. As mentioned above, this can also be represented by a symmetric 4-sidigraph. Note that this is not a voltage graph as defined in [1], since for example; the label on $\overrightarrow{v_2v_1}$ is not the (group) inverse of the label on $\overrightarrow{v_1v_2}$.

2. Motivation

Some motivation to study (n, d)-sigraphs comes from the following examples. Certain relations among human beings can be represented by (n, d)-sigraphs, where $n = 3$ as follows:

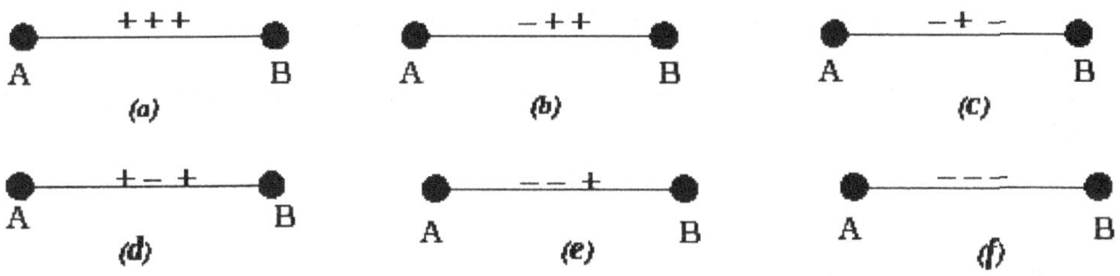

Figure 2.

Let $G = (V, E)$ be a graph whose vertices represents individuals, where for any two vertices u and v, uv is an edge if, and only if, u and v know each other. Note that the vertices of G also represent certain parts of a machine which connected together, chemicals which kept together, or plants which are planted together etc. Two persons who know each other/two parts of a machine,chemicals/plants may be related by a relation R as follows:

- A and B are business partners/life partners/living together/ working together in the same place.
- A and B may be two parts of a machine connected to each other
- A and B may be chemicals kept/mixed together
- A and B may be plants planted side by side
- A may be a person having a disease B
- A may be a teacher and B may be group of students taught by A.

In each of the above cases we say that A and B are *R-adjacent*. If not, A and B are *R-nonadjacent*. Further, the R-adjacency can be classified into two types, *wanted* and *unwanted R-adjacency*. Similarly, any R-nonadjacency between A and B can be classified into two types, *wanted and unwanted R-nonadjacency*.

The above situations can be conveniently represented by a $(3, d)$-sigraph.

Suppose A and B are two persons/objects and some relation R exists between them as mentioned above. Then the following cases may arise:

Case 1. The relation is good from the viewpoint of both A and B we can represent this by an edge AB labeled by the triple $+++$ as in Figure 2(a). Note that the middle sign $+$ indicates that the relation R exists between A and B, the first $+$ and the third $+$ sign indicate that the relation R is liked by both A and B.

Case 2. A does not like this relationship while B likes it. This can be represented by an edge AB labeled by the triple $-++$ as in Figure 2(b).

Case 3. Both A and B do not like this relation. This can be represented by an edge AB labeled by the triple $-+-$ as in Figure 2(c).

Suppose A and B are not related by the relation R. If this is liked by both A and B, it can be represented by an edge AB labeled by the triple $+-+$ as in Figure 2(d).

Note that in Figure 2(d), the middle sign $-$ indicates that the relation R does not exist between A and B. Further the first $+$ and the third $+$ sign indicate that this R-nonadjacency is liked by both A and B. Similarly, the edge AB labeled by the triple $--+$ as in Figure 2(e) indicates that R-nonadjacency between A and B is disliked by A where as liked by B, and the edge AB labeled by the triple $---$ as in Figure 1(f) indicates that this R-nonadjacency between A and B is not liked by both A and B.

3. Complementation

For any $t \in H_n$, the *t-complement* of $a = (a_1, a_2, \ldots, a_n)$ is: $a^t = at$. The *reversal* of $a = (a_1, a_2, \ldots, a_n)$ is: $a^r = (a_n, a_{n-1}, \ldots, a_1)$. For any $T \subseteq H_n$, and $t \in H_n$, the *t-complement* of T is $T^t = \{a^t : a \in T\}$.

For any $t \in H_n$, the *t-complement* of an (n, d)-sigraph $G = (V, E)$, written G^t, is the same graph but with each edge label $a = (a_1, a_2, \ldots, a_n)$ replaced by a^t. The *reversal* G^r is the same graph but with each edge label $a = (a_1, a_2, \ldots, a_n)$ replaced by a^r.

Let $G = (V, E)$ and $G' = (V', E')$ be two (n, d)-sigraphs. Then G is said to be *isomorphic* to G' and we write $G \cong G'$, if there exists a bijection $\phi : V \to V'$ such that if uv is an edge in G which is d-labeled by $a = (a_1, a_2, \ldots, a_n)$, then $\phi(u)\phi(v)$ is an edge in G' which is d-labeled by a, and conversely.

For each $t \in H_n$, an (n, d)-sigraph $G = (V, E)$ is *t-self complementary*, if $G \cong G^t$. Further, G is *self reverse*, if $G \cong G^r$.

Proposition 1. *For all $t \in H_n$, an (n, d)-sigraph $G = (V, E)$ is t-self complementary if, and only if, G^a is t-self complementary, for any $a \in H_n$.*

Proof. Suppose G is t-self complementary. Then, $G \cong G^t$. This implies $G^a \cong G^{at}$.

Conversely, suppose that G^a is t-self complementary. Then, $G^a \cong (G^a)^t$. Since $(G^a)^a = G$. Hence $G \cong (G^{at})^a = G^t$. □

3.1 *Product of the n-tuples on a cycle in an (n, d)-sigraph $G = (V, E)$*

Let $v_1, v_2, \ldots, v_m$ be a cycle C in G and $(a_{k1}, a_{k2}, \ldots, a_{kn})$ be the n-tuple on the edge $v_k v_{k+1}$, $1 \le k \le m-1$, and $(a_{m1}, a_{m2}, \ldots, a_{mn})$ be the n-tuple on the edge $v_m v_1$.

For any cycle C in G, let $P(\overrightarrow{C})$ denotes the product of the n-tuples on C given by $(a_{11}, a_{12}, \ldots, a_{1n})(a_{21}, a_{22}, \ldots, a_{2n}) \ldots (a_{m1}, a_{m2}, \ldots, a_{mn})$ and $P(\overleftarrow{C}) = (a_{mn}, a_{m(n-1)}, \ldots, a_{m1})(a_{(m-1)n}, a_{(m-1)(n-1)}, \ldots, a_{(m-1)1}) \ldots (a_{1n}, a_{1(n-1)}, \ldots, a_{11})$.

An n-tuple $(a_1, a_2, \ldots, a_n)$ is *identity n-tuple*, if each $a_k = +$, for $1 \le k \le n$, otherwise it is a *non-identity n-tuple*. Further an n-tuple $a = (a_1, a_2, \ldots, a_n)$ is *symmetric*, if $a^r = a$, otherwise it is a *non-symmetric n-tuple*. In (n, d)-sigraph $G = (V, E)$ an edge labeled with the identity n-tuple is called an *identity edge*, otherwise it is a *non-identity edge*.

Note that the above products $P(\overrightarrow{C})$ as well as $P(\overleftarrow{C})$ are n-tuples. In general, these two products need not be equal. However, the following holds.

Proposition 2. *For any cycle C of an (n, d)-sigraph $G = (V, E)$, $P(\overleftarrow{C}) = P(\overrightarrow{C})^r$.*

Proof. By the definition, we have,

$$
\begin{aligned}
P(\overleftarrow{C})^r &= ((a_{mn}, a_{m(n-1)}, \ldots, a_{m1})(a_{(m-1)n}, a_{(m-1)(n-1)}, \ldots, a_{(m-1)1}) \ldots \\
&\quad \times (a_{1n}, a_{1(n-1)}, \ldots, a_{11}))^r \\
&= ((a_{m1}, a_{m2}, \ldots, a_{mn})^r (a_{(m-1)1}, a_{(m-1)2}, \ldots, a_{(m-1)n})^r \ldots \\
&\quad \times (a_{11}, a_{12}, \ldots, a_{1n})^r)^r \\
&= (a_{m1}, a_{m2}, \ldots, a_{mn})(a_{(m-1)1}, a_{(m-1)2}, \ldots, a_{(m-1)n}) \ldots (a_{11}, a_{12}, \ldots, a_{1n}) \\
&= P(\overrightarrow{C})
\end{aligned}
$$

□

Corollary 3. *For any cycle C, $P(\overleftarrow{C}) = P(\overrightarrow{C})$ if, and only if, $P(\overrightarrow{C})$ is a symmetric n-tuple. Furthermore, $P(\overrightarrow{C})$ is the identity n-tuple if, and only if, $P(\overleftarrow{C})$ is.*

We make use of the above facts in defining two notions of balance in an (n, d)-sigraph.

4. Balance in an (n, d)-sigraph

A sigraph $G = (V, E)$ is *balanced*, if every cycle in G has an even number of negative edges (see[3]). Equivalently, a sigraph $G = (V, E)$ is balanced, if the product of signs on the edges of each cycle in G is positive. The following result is well known (see [3]).

Proposition 4 (Harary [3]). *A sigraph $G = (V, E)$ is balanced, if, and only if, its vertex set V can be partitioned into two sets V_1 and V_2 such that every negative edge joins a vertex in V_1 and a vertex in V_2, and every positive edge joins two vertices in V_1 or in V_2.*

4.1 *Two kinds of balance*

Motivated by the definition of balance in sigraphs, we define two notions of balance in an (n, d)-sigraph $G = (V, E)$ as follows:

Definition. *Let $G = (V, E)$ be an (n, d)-sigraph. Then,*

(i) *G is* identity balanced *(or* i-balanced*), if $P(\overrightarrow{C})$ on each cycle of G is the identity n-tuple, and*
(ii) *G is* balanced, *if every cycle contains an even number of non-identity edges.*

Note. An i-balanced (n, d)-sigraph need not be balanced and conversely. For example, consider the $(4, d)$-sigraphs in Figure 3. In Figure 3(a) G is an i-balanced but not balanced, and in Figure 3(b) G is balanced but not i-balanced.

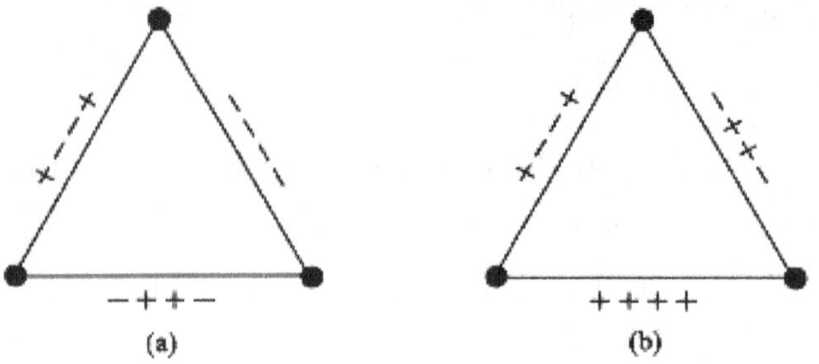

Figure 3.

4.2 *Comparison with group-valued graphs*

An (n, d)-sigraph is defined as a graph whose edges are d-labeled with the elements of the abelian group H_n. In [7], the authors studied group valued graphs defined as follows:

A *group valued graph* is an ordered triple (G, M, ϕ), where G is a graph, M is an finite abelian group and ϕ is a function from the set $X(G)$ of edges of G into M. For an edge $y \in X(G)$, $\phi(y)$ is *value* of y. The notion of balance introduced in [7] is as follows:

A cycle with n-edges in G having values $x_1, x_2, \ldots, x_n$ is said to be *balanced* in G if $x_1^{s_1} x_2^{s_2} \ldots x_n^{s_n} = e$, the identity element of M, for some $s_k \in \{1, -1\}, k = 1, 2, \ldots, n$. A group valued graph G is *balanced*, if every cycle in G is balanced.

Though the notion of i-balance in (n, d)-sigraphs is similar to the notion of balance defined in [7], they are different mainly because the labeling of the edges in (n, d)-sigraph is directional where as it is not so in [7]. Because of d-labelings of edges of an (n, d)-sigraph, we cannot consider an (n, d)-sigraph as a group valued graph even though the n-tuples on the edges are elements of the group H_n.

4.3 *Criteria for balance*

An (n, d)-sigraph $G = (V, E)$ is i-balanced if each non-identity n-tuple appears an even number of times in $P(\overrightarrow{C})$ on any cycle of G.

However, the converse is not true. For example see Figure 4(a). In Figure 4(b), the number of non-identity 4-tuples is even and hence it is balanced. But it is not i-balanced, since the 4-tuple $(++--)$ (as well as $(--++)$) does not appear an even number of times in $P(\overrightarrow{C})$ of 4-tuples.

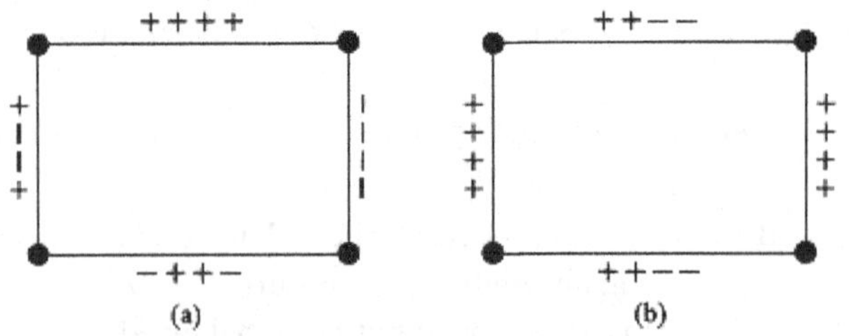

Figure 4.

Proposition 5 (Characterization of balanced (n, d)-sigraphs). *An (n, d)-sigraph $G = (V, E)$ is balanced if, and only if, there exists a partition $V_1 \cup V_2$ of V such that each identity edge joins two vertices in V_1 or V_2, and each non-identity edge joins a vertex of V_1 and a vertex of V_2.*

Proof. We associate a sigraph G' with G on the same vertex set V and the edge set E of G as follows: An edge ab in G' is lableled $+$ or $-$ according as ab is an identity edge or non identity edge in G. Clearly, the (n, d)-sigraph G is balanced if, and only if, the sigraph G' is balanced, and the result follows from Proposition-4. □

Let $P(C)$ denote the product of the n-tuples in $P(\vec{C})$ on any cycle C in an (n, d)-sigraph $G = (V, E)$.

Proposition 6 (Characterization of an i-balanced (n, d)-sigraph). *An (n, d)-sigraph $G = (V, E)$ is i-balanced if, and only if, for each k, $1 \leq k \leq n$, the number of n-tuples in $P(C)$ whose kth co-ordinate is $-$ is even.*

Proof. Suppose G is i-balanced and let C be a cycle in G with number of n-tuples in $P(C)$ whose kth co-ordinate is $-$ is odd, for some k, $1 \leq k \leq n$. Then, the kth co-ordinate in $P(C)$ of n-tuples on the edges of the cycle C is $-$ and C is i-unbalanced cycle in G. Hence G is i-unbalanced a contradiction.

Converse part follows from the definition of i-balanced (n, d)-sigraph. □

In H_n, let S_1 denote the set of non-identity symmetric n-tuples and S_2 denote the set of non-symmetric n-tuples. The product of all n-tuples in each S_k, $1 \leq k \leq 2$ is the identity n-tuple.

As above, let $P(C)$ denote the product of the n-tuples in $P(\vec{C})$ on any cycle C in an (n, d)-sigraph $G = (V, E)$.

Proposition 7. *An (n, d)-sigraph $G = (V, E)$ is i-balanced, if both of the following hold:*

(i) *In $P(C)$, each n-tuple in S_1 occurs an even number of times, or each n-tuple in S_1 occurs odd number of times (the same parity, or equal mod 2).*
(ii) *In $P(C)$, each n-tuple in S_2 occurs an even number of times, or each n-tuple in S_2 occurs an odd number of times.*

Proof. For any n-tuple, $a = (a_1, a_2, \ldots, a_n)$ in H_n, a^2 is the identity n-tuple. The product of any n-tuple an even number of times is the identity n-tuple. As a consequence, the product of any n-tuple $a = (a_1, a_2, \ldots, a_n)$ an odd number of times is a itself. Also, since the product of all n-tuples in S_k, $1 \leq k \leq 2$, is the identity n-tuple, it follows that in any product of n-tuples belonging to S_k, if each n-tuple of S_k occurs odd number of times, then this product is the identity n-tuple. □

Note.

1. Let $G = (V, E)$ be an (n, d)-sigraph and G^r be the reversal of G. Then, G is i-balanced (balanced) if, and only if, G^r is i-balanced (balanced).

2. If an (n, d)-sigraph $G = (V, E)$ is i-balanced (balanced), then its t-complement G^t, need not be i-balanced (balanced), where t can be any element of H_n. For example, let $G = (V, E)$ be the cycle of length 13. Label the 12 edges of G by the 12 non-symmetric 4-tuples of H_4, and label the remaining edge by the identity 4-tuple $(++++)$. We find that $(4, d)$-sigraph thus obtained is both i-balanced and balanced. But its t-complement G^t of G is neither i-balanced nor balanced, for any non-identity 4-tuple, $t \in H_4$. However, the following holds:

Proposition 8. *Let $G = (V, E)$ be an (n, d)-sigraph. Suppose the underlying graph of G is bipartite. Then, for any $t \in H_n$, G is i-balanced if, and only if, its t-complement G^t is i-balanced.*

Proof. Since G is i-balanced, by Proposition-6, for each k, $1 \leq k \leq n$, the number of n-tuples on any cycle C in G whose kth co-ordinate is $-$ is even. Also, since G is bipartite, for each k, $1 \leq k \leq n$, number of n-tuples on C whose kth co-ordinate is $+$ is also even. This implies that the same thing is true in any t-complement of G, where t can be any element of H_n. Hence G^t is i-balanced. Similarly, the converse follows, since for each $t \in H_n$, the underlying graph of G^t is also bipartite. □

For a more general notion of balance see [5] and [6]. We investigate these notions of balance in a separate paper.

Acknowledgement

The authors thank T. Zasalvasky and the referee for their valuable comments and suggestions. Also, the authors are thankful to Department of Science and Technology, Government of India, New Delhi for the financial support under the grant SR/S4/MS:275/05.

References

[1] B. D. Acharya and M. Acharya, New algebraic models of a Social System, *Indian J. of Pure and Appl. Math.*, **17(2)** (1986) 152–168.
[2] F. Harary, Graph Theory, Addison-Wesley Publishing Co. (1969).
[3] F. Harary, On the notion of balance of a signed graph, *Michigan Math. J.*, **2** (1953) 143–146.
[4] F. Harary, R. Norman and D. Cartwright, Structural models: An introduction to the theory of directed graphs, J. Wiley, New York (1965).
[5] E. Sampathkumar, k-Dimensional Graphs and Semigraphs, DST Project Report No. DST/MS/131/2K (2005).
[6] E. Sampathkumar, Generalized Graph Structures, *Bulletin of Kerala Math. Association*, **3(2)** (2006, December) 67–123.
[7] E. Sampathkumar and V. N. Bhave, Group Valued graphs, *The Karnatak University Journal*, XVIII (1973) 325–328.
[8] T. Zaslavsky, Signed Graphs, *Discrete Appl. Math.*, **4(1)** (1982) 47–74.
[9] T. Zaslavsky, A Mathematical bibliography of Signed and Gain graphs and its allied areas, *Electronic J. Combinatorics*, **8(1)** (1998), Dynamic Surveys No. 8, pp. 124 (1998a).

Proc. ICDM 2008, RMS-Lecture Notes Series No. 13, 2010, pp. 161–170.

3-Consecutive Vertex Coloring of a Graph

E. Sampathkumar, M. S. Subramanya and Charles Dominic

Department of Studies in Mathematics, Manasagangotri, University of Mysore, Mysore 570 006, India
e-mail: esampathkumar@eth.net; subramanya_ms@rediffmail.com; charlesdominicpu@gmail.com

Abstract. The *3-consecutive achromatic number* of a graph $G = (V, E)$, denoted $\psi_{3c}(G)$, equals the maximum number of colors that can be used in a coloring of the vertices of G such that if uv and vw are any two consecutive edges in G, then either u or w is colored the same as v. In this paper we show that for any connected graph G of order $n = |V| \geq 3$, with minimum degree δ and vertex covering number $\alpha_0(G)$, (i) $\psi_{3c}(G) \leq \lfloor n/2 \rfloor + 1$, (ii) $\psi_{3c}(G) \leq \lfloor n/\delta \rfloor$, and (iii) $\psi_{3c}(G) \leq \alpha_0(G) + 1$. Furthermore, if G is a unicyclic graph, then $\alpha_0(G) - 1 \leq \psi_{3c}(G)$.

Keywords. 3-consecutive coloring, 3-consecutive coloring number.

Subject Classification: 05C15

1. Introduction

Let $G = (V, E)$ be a graph of order $n = |V|$ and size $m = |E|$. Vertices $u, v, w \in V$ are called *3-consecutive* if uv and vw are edges in E. A *3c-coloring* of G is a function $f : V \to \{1, 2, \ldots, k\}$ such that for any 3 consecutive vertices u, v and w, either $f(u) = f(v)$ or $f(w) = f(v)$ (or both). Equivalently, if $N(v) = \{u | uv \in E\}$ denotes the *open neighborhood* of a vertex v, then a 3c-coloring of G is an assignment of the colors $1, 2, \ldots, k$ to the vertices of G such that for each vertex v, at most one vertex in $N(v)$ is assigned a color different than the color assigned to v. Notice that the trivial function $f : V \to \{1\}$ is always a 3c-coloring. Thus, we are interested in the *3c-achromatic number* $\psi_{3c}(G)$, which equals the maximum number of colors that can be used in a 3c-coloring of G. This coloring number, called the *3-consecutive coloring number*, was first defined by E. Sampathkumar in (cf. [9]). This coloring number $\psi_{3c}(G)$ can also be defined as follows:

$\psi_{3c}(G)$ is the maximum order of a partition $\mathcal{P}$ of the vertex set V of G such that if u, v and w are 3-consecutive vertices, then for some set C_i in the partition $\mathcal{P}$ both $u, v \in C_i$ or both $v, w \in C_i$.

In this paper we initiate the study of this coloring number. We obtain exact values of $\psi_{3c}(G)$ for several classes of graphs, we give several upper and lower bounds for $\psi_{3c}(G)$, and we relate $\psi_{3c}(G)$ to the vertex covering number, $\alpha_0(G)$, the matching number $\beta_1(G)$ and the weakly connected domination number $\gamma_w(G)$.

For any definitions on graphs we refer the book [7].

Research supported by the Department of Science and Technology, Government of India, New Delhi under the grant SR/S4/MS:275/05.

1.1 Motivation

Consider a computer network $G = (V, E)$ in which processors are represented by vertices V and direct communication links between pairs of processors are represented by edges E. Each processor communicates with its direct neighbors using a common information protocol. But each processor also has a protocol translator that can be used to convert the protocol it uses to that of at most one neighbor that uses a different information protocol (for much more information on such communication technologies, the reader is referred to the ATIS website at: http://www.atis.org/about.shtml).

We seek to determine the maximum number of different protocols that can be used by the processors in a network, subject to the restrictions that (i) each processor can communicate with all but one of its neighbors using one common protocol, but (ii) can communicate with at most one of its neighbors using a different protocol through its protocol translator.

Translated into graph theory terminology, we assign colors (protocols) to the vertices of a graph subject to the restrictions that all but at most one neighbor of any vertex are assigned the same color, and at most one neighbor of any vertex v can be assigned a color different than that assigned to v. The maximum number of different colors that can be assigned to the vertices of a graph G subject to these restrictions is called the 3-consecutive achromatic number of G.

2. Exact values of $\psi_{3c}(G)$

Theorem 1. *For a connected graph $G = (V, E)$ of order $n \geq 3$*

(1) $\psi_{3c}(G) = n$ *if, and only if, each connected component of G is either K_1 or K_2.*
(2) *For a path P_n of order n,* $\psi_{3c}(P_n) = \lfloor n/2 \rfloor + 1$.
(3) *For a cycle C_n of order n,* $\psi_{3c}(C_n) = \lfloor n/2 \rfloor$.
(4) *For a complete bipartite graph $K_{m,n}$,*

$$\psi_{3c}(K_{m,n}) = \begin{cases} 2 & \text{when } m = 1 \text{ and } n \geq 1 \text{ or when } m = n = 2 \\ 1 & \text{when } m \geq 2 \text{ and } n \geq 3 \end{cases}$$

(5) *For the Petersen graph $\mathcal{P}$,* $\psi_{3c}(\mathcal{P}) = 2$.

It is of interest to note that the smallest bipartite graph of order at least 3 having $\psi_{3c}(G) = 1$ is the graph $K_{2,3}$.

3. Graphs for which $\psi_{3c}(G) = 1$

Surprisingly, it remains an open problem to characterize the graphs for which $\psi_{3c}(G) = 1$. For brevity let us call these $3\mathcal{C}$-1 graphs. In this section we provide several subclasses of this class of graphs.

Proposition 2. *Let G and H be two connected graphs of order at least 3 and let $G \odot H$ be the graph obtained by identifying any vertex in G with any vertex in H, thereby creating a cutvertex. Then if G and H are $3\mathcal{C}$-1 graphs, so is $G \odot H$.*

Proposition 3. *If G is a $3\mathcal{C}$-1 graph, then so is each block of G, and conversely, if every block of a graph G is a $3\mathcal{C}$-1 graph, then so is G.*

Proposition 4. *Let G be a connected graph of order $n \geq 3$ and let H be a graph obtained from G by either adding an edge or adding a new vertex and joining it to at least two vertices of G. Then if G is a $3\mathcal{C}$-1 graph, so is H.*

It is well known that all graphs in the class of k-trees, for $k \geq 2$, are obtained by starting with a copy of the complete graph K_k and repeatedly adding a new vertex and joining it to k vertices in a complete subgraph of order k. Therefore, we get the following corollary.

Corollary 5. *Every k-tree, for $k \geq 2$, is a $3\mathcal{C}$-1 graph.*

Proposition 6. *Let G be a connected graph of order $n \geq 3$. If G contains an edge that is not on a cycle, then $\psi_{3c}(G) \geq 2$.*

Proof. Suppose G contains an edge not on a cycle in G, then $G - x$ has two components say, G_1 and G_2. We can give color 1 to all vertices in G_1, and color 2 to all vertices in G_2. This coloring will give a 3-consecutive coloring of G, and hence $\psi_{3c}(G) \geq 2$. □

Corollary 7. *If G is a $3\mathcal{C}$-1 graph, then every edge of G lies on a cycle.*

Notice that the converse to Corollary 7 is not true; every edge of C_n lies on the cycle but C_n is not a $3\mathcal{C}$-1 graph.

Proposition 8. *Let G be a connected graph of order $n \geq 3$. If every edge of G lies on a triangle, then G is a $3\mathcal{C}$-1 graph.*

Proof. If T_1 is a triangle in G, then in any 3-consecutive coloring of G every vertex of T_1 should be colored by 1. Let T_2 be any other triangle in G having a common vertex or common edge with T_1. Clearly, in any 3-consecutive coloring of G all vertices of T_2 should also be colored by the color 1. Continuing this process we find that all vertices in G should be colored by the same color 1 in any 3-consecutive coloring of G, and hence $\psi_{3c}(G) = 1$. □

A wheel W_n is a graph obtained from a cycle C_n by adding a new vertex and joining it to every vertex on the cycle. Thus, every vertex of a wheel lies on a triangle.

Corollary 9. *Every wheel $W_n = K_1 + C_n$, for $n \geq 3$ is a $3\mathcal{C}$-1 graph.*

More generally, the *join* $G + H$ of two graphs is the graph obtained from graphs G and H by adding all possible edges between a vertex in G and a vertex in H. Thus, provided that either G or H contains no isolated vertices, every vertex of $G + H$ lies on a triangle.

Corollary 10. *The join $G + H$ of two graphs is a $3\mathcal{C}$-1 graph if either G or H contains no isolated vertex.*

Corollary 11. *The square G^2 of any connected graph G of order $n \geq 3$ is a $3\mathcal{C}$-1 graph.*

Corollary 12. *Every bridgeless chordal graph is a $3\mathcal{C}$-1 graph.*

Corollary 13. *Every maximal planar graph of order $n \geq 3$ is a $3\mathcal{C}$-1 graph.*

Corollary 14. *If every block of a connected graph G of order $n \geq 3$ is a complete subgraph of order at least 3, then G is a $3\mathcal{C}$-1 graph.*

One final class of $3\mathcal{C}$-1 graphs can be constructed as follows. Let G' and G'' be two copies of a connected graph G. To this union of two graphs add edges from every vertex $u' \in V'$ to each vertex in the open neighborhood $N(u'')$ of the corresponding vertex $u'' \in V''$. Let $G \bowtie G$ be this graph.

Proposition 15. *For any connected graph G of order $n \geq 3$, $G \bowtie G$ is a $3\mathcal{C}$-1 graph.*

Proof. We proceed by induction on the order n of G. It is easy to see that the result is true for both connected graphs of order $n = 3$. Assume that the result holds for all connected graphs G of order $n \leq k$, and let G be a connected graph of order $n = k+1$. Let G' and G'' be the two copies of G in $G \bowtie G$. Let u' be any non-cut vertex in G'; since G' is connected, any leaf of any spanning tree of G' will suffice as the vertex u'. Let u'' be the corresponding non-cut vertex in G''. The removal of vertices u' and u'' from $G \bowtie G$ results in the graph $G' - u' \bowtie G'' - u''$. By the induction hypothesis, $G' - u' \bowtie G'' - u''$ is a $3\mathcal{C}$-1 graph. But the graph $G \bowtie G$ can be obtained by two applications of adding a vertex to $G' - u' \bowtie G'' - u''$ and joining it to at least two vertices. Thus, by Proposition 4, $G' \bowtie G''$ is a $3\mathcal{C}$-1 graph. □

As we said earlier, it is an open problem to characterize $3\mathcal{C}$-1 graphs. Any attempt to characterize these graphs will have to determine what are the *minimal* $3\mathcal{C}$-1 graphs, that is, $3\mathcal{C}$-1 graphs G having the property that no proper subgraph G' of G is also a $3\mathcal{C}$-1 graph. These minimal $3\mathcal{C}$-1 graphs will include graphs that can be called 1-*minimal* $3\mathcal{C}$-1 graphs. These graphs have the property that if you remove any one vertex, the resulting graph is no longer a $3\mathcal{C}$-1 graph. Notice that the graph of order 5 consisting of two triangles joined at one vertex is a 1-minimal $3\mathcal{C}$-1 graph, but is not a minimal $3\mathcal{C}$-1 graph. Notice also that the complete bipartite graph $K_{2,3}$ is a 1-minimal $3\mathcal{C}$-1 graph.

4. Bounds for $\psi_{3c}(G)$

In this section we establish results for graphs that are not $3\mathcal{C}$-1 graphs, and we establish several upper and lower bounds for $\psi_{3c}(G)$.

The *Cartesian product* $G \square H$ of two graphs G and H is the graph with vertex set equal to the Cartesian product $V(G) \times V(H)$, where two vertices (u, v) and (w, x) are adjacent in $G \square H$ if, and only if, either u is adjacent to w in G or v is adjacent to x in H.

Theorem 16. *Let G be a connected graph of order $n \geq 1$. Then $\psi_{3c}(G \square K_2) \geq 2$, and if $\psi_{3c}(G) \geq 2$, then $\psi_{3c}(G \square K_2) = \psi_{3c}(G)$.*

Proof. Let G' and G'' be the two copies of G in $G \square K_2$. Give color 1 to all vertices in G' and color 2 to all vertices in G''. This gives a 3-consecutive coloring of G_2, and hence $\psi_{3c}(G_2) \geq 2$. Suppose now, $\psi_{3c}(G) \geq 2$. Consider a 3-consecutive coloring of G_2. We have two cases.

Case 1. Two corresponding vertices v and v' in G' and G'' respectively receive different colors, say 1 and 2. Noting that in any 3-consecutive coloring at most one neighbor of v may have a different color other than v, we see that, all neighbors of v except v' receive the color 1. Similarly, all neighbors of v' except v, receive the color 2. Let u be a neighbor of v in G', and u' be the corresponding neighbor of v' in G''. Since u has color 1 and u' has color 2, by the above argument we see that all neighbors of u except u' has color 1, all neighbors of u' except u has color 2. Continuing this way, we find that all vertices in G' have color 1 and all vertices in G'' have color 2 in this case.

Case 2. Corresponding vertices in G' and G'' have the same color. Let $\psi_{3c}(G) = k$ and consider a $\psi_{3c}(G)$ coloring of G' with colors $c_1, c_2, \ldots, c_k$. Color the vertices in G'' such that each vertex in G'' receives the color of the corresponding vertex in G'. One can easily verify that the coloring of $G \square K_2$ thus obtained is a 3-consecutive coloring.

Since k is the maximum number of colors that can be used in a 3-consecutive coloring of G', it follows that k is also maximum number of colors in a 3-consecutive coloring of $G \square K_2$. Thus $\psi_{3c}(G \square K_2) = \psi_{3c}(G)$. □

Corollary 17. *For the n-cube Q_n, $\psi_{3c}(Q_n) = 2$.*

In a graph $G = (V, E)$ a set $S \subseteq V$ is called a *neighbourhood set* or an *n-set* if $\cup G[N[v]] = G$, where $G[N[v]]$ denotes the subgraph of G induced by the vertices in $N[v]$. The *neighbourhood number* of a graph G, denoted $n_0(G)$, equals the minimum cardinality of an n-set in G (cf. [10]).

Theorem 18. *Let G be a connected graph. Then $\psi_{3c}(G) \leq n_0(G) + 1$, and this bound is sharp. Furthermore, for any tree T, $\psi_{3c}(T) = n_0(T) + 1$.*

Proof. Let S be an n-set of a graph G, with $|S| = n_0(G) = k$, and let $v_1 \in S$. Color the vertices in $N[v_1]$ using at most 2 colors. Next, choose a vertex $v_2 \in S$ such that $N[v_1] \cap N[v_2] \neq \emptyset$. Since at least one vertex in $N[v_2]$ has been colored, we can color the uncolored vertices in $N[v_2]$ using at most one new color. Next choose a vertex $v_3 \in S$ such that either $N[v_1] \cap N[v_3] \neq \emptyset$ or $N[v_2] \cap N[v_3] \neq \emptyset$. Since at least one vertex in $N[v_3]$ has already been colored, we can color the uncolored vertices in $N[v_3]$ using at most one new color. It follows that we can continue in this way through all of the vertices $v_i \in S$, coloring all of the vertices in $N[v_i]$ with at most one new color. When we are finished, we will have been able to produce a 3c-coloring of G with at most $k + 1 = n_0(G) + 1$ colors.

In the case of a tree T, one can show that at every step of the coloring process, one new color can always be used, and hence, a 3c-coloring can be produced that uses exactly $n_0(T) + 1$ colors. □

The *vertex covering number* $\alpha_0(G)$ equals the minimum cardinality of a set $S \subseteq V$ having the property that for every edge $uv \in E$, either $u \in S$ or $v \in S$ (or both). The *matching number* $\beta_1(G)$ equals the maximum cardinality of a set $F \subseteq E$ of edges having the property that no two edges in F have a vertex in common.

It is well known that: (i) for any graph G, $n_0(G) \leq \alpha_0(G)$, (ii) for any triangle-free graph G, $n_0(G) = \alpha_0(G)$ (cf. [10]), and (iii) for any bipartite graph G, $\alpha_0(G) = \beta_1(G)$. Thus, we have the following result.

Corollary 19. *For any connected graph G and any tree T,*

(i) $\psi_{3c}(G) \leq \alpha_0(G) + 1$.
(ii) $\psi_{3c}(T) = \alpha_0(T) + 1 = \beta_1(T) + 1$.

Proposition 20. *For any unicyclic graph G of order $n \geq 3$,*

$$\alpha_0(G) - 1 \leq \psi_{3c}(G) \leq \alpha_0(G) + 1,$$

and these bounds are sharp.

Proof. Inview of the Theorem 18 we need to establish only the lower bound. Let C be the cycle in G and x be an edge of C. Then $G - x$ is a tree and by Theorem 18,

$$\chi_{3c}(G - x) = \alpha_0(G - x) + 1 \qquad \text{(i)}$$

Since G is unicyclic, we observe that either $\alpha_0(G - x) = \alpha_0(G) - 1$, or $\alpha_0(G - x) = \alpha_0(G)$. In otherwords,

$$\alpha_0(G - x) \geq \alpha_0(G) - 1 \qquad \text{(ii)}$$

Also, $\chi_{3c}(G) = \chi_{3c}(G - x)$ or $\chi_{3c}(G) = \chi_{3c}(G - x) - 1$. In otherwords,

$$\chi_{3c}(G) \geq \chi_{3c}(G - x) - 1 \qquad \text{(iii)}$$

By (i), (ii) and (iii) we have $\chi_{3c}(G) \geq \alpha_0(G) - 1$.

The upper bound is attained if G is a cycle with exactly one pendant edge at each of its vertices. The lower bound is attained if G is an odd cycle. □

A set $S \subset V$ of vertices in a connected graph $G = (V, E)$ is called a *connected dominating set* if (i) every vertex $v \in V - S$ is adjacent to at least one vertex in S, and (ii) the subgraph $G[S]$ induced by S is connected. The minimum cardinality of a connected dominating set S is called the *connected domination number*, and is denoted $\gamma_c(G)$; such a set S is called a *γ_c-set* (cf. [11,8]).

By changing the first sentence in the proof of Proposition 18 to read: Let S be a γ_c-set of a graph G, with $|S| = \gamma_c(G) = k$, and let $v_1 \in S$, one can prove the following result.

Corollary 21. *For any connected graph G, $\psi_{3c}(G) \leq \gamma_c(G) + 1$.*

But an even stronger result is possible.

We say that a set of vertices $S \subset V$ in a graph G is called *weakly connected* if the subgraph $G_w[S] = (N[S], (S \times S \cap E) \cup (S \times V - S \cap E))$ is a connected subgraph. The edges in the subgraph $G_w[S]$ include all edges in E having at least one vertex in S.

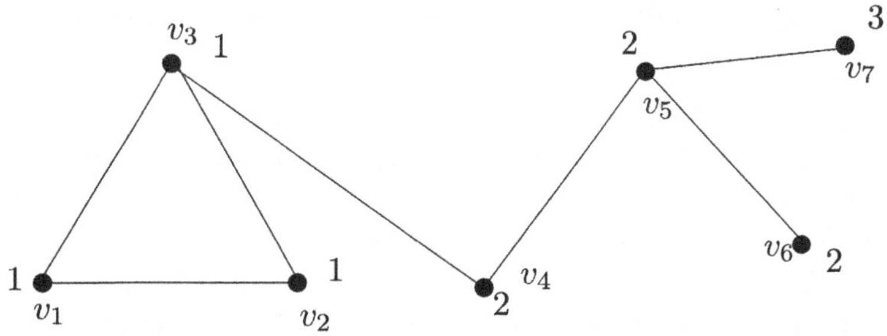

Figure 1.

A set $S \subset V$ of vertices in a connected graph $G = (V, E)$ is called a *weakly connected dominating set* if (i) every vertex $v \in V - S$ is adjacent to at least one vertex in S, and (ii) the subgraph $G_w[S]$ weakly induced by S is connected. The minimum cardinality of a weakly connected dominating set S is called the *weakly connected domination number*, and is denoted $\gamma_w(G)$; such a set S is called a *γ_w-set* (cf. [6,4,12,5]).

By changing the first sentence in the proof of Proposition 18 to read: Let S be a γ_w-set of a graph G, with $|S| = \gamma_w(G) = k$, and let $v_1 \in S$, one can prove the following.

Corollary 22. *For any connected graph G, $\psi_{3c}(G) \leq \gamma_w(G) + 1 \leq \gamma_c(G) + 1$.*

This bound is sharp. For example the bound is attained for the graph in Figure 1. For the graph G, $\gamma_w(G) = 2$, since the set $\{v_3, v_5\}$ is a weakly connected dominating set in G and is clearly minimum weakly connected dominating set and $\psi_{3c}(G) = 3 = \gamma_w(G) + 1$.

It is known that for a connected graph G of order n, $\gamma_c(G) \leq n - \Delta(G)$, where $\Delta(G)$ is the maximum degree of a vertex in G (cf. [8]). If $\chi(G)$ denotes the *chromatic number* of G, then it is clear that $\chi(G) \leq \Delta(G) + 1$. Therefore we have the following result.

Corollary 23. *For any connected graph G of order $n \geq 3$, $\psi_{3c}(G) \leq n - \Delta(G) + 1 \leq n - \chi(G) + 2$.*

Theorem 24. *For any connected graph G of order $n \geq 3$ and any spanning tree T of G, $\psi_{3c}(G) \leq \psi_{3c}(T)$.*

Proof. The proof is a consequence of the following two observations:

(i) Any 3c-coloring of G is also a 3c-coloring of T, but not conversely.
(ii) A maximum 3c-coloring of G need not be a maximum 3c-coloring of T. □

Consider, for example, the cycle C_5; $\psi_{3c}(C_5) = 2$, but for any edge x in C_5, $\psi_{3c}(C_5 - x) = 3$.

Corollary 25. *For any connected graph G of order $n \geq 3$, $\psi_{3c}(G) \leq \lfloor n/2 \rfloor + 1$.*

This bound is sharp. For example the bound is attained for paths.

Corollary 26. *If both* G *and its complement* $\bar{G}$ *are connected graphs of order* $n \geq 3$, *then* $\psi_{3c}(G) + \psi_{3c}(\bar{G}) \leq n + 2$.

This bound is sharp. For example the bound is attained for P_4, the path on 4 vertices, since $\bar{P}_4 = P_4$ and $\psi_{3c}(P_4) = 3$ and so $\psi_{3c}(P_4) + \psi_{3c}(\bar{P}_4) = 6 = 4 + 2$.

We now obtain a bound which appears to be better than the one given in Corollary 25. Let $G = (V, E)$ be a graph and let $v \in V$. Let C_i, $1 \leq i \leq r$ be the connected components in the subgraph $G[V - N[v]]$ obtained by removing the vertices in $N[v]$. Let c_i equal the number of vertices in the component C_i, for $1 \leq i \leq r$.

Proposition 27. *For any connected graph* G,

$$\psi_{3c}(G) \leq 2 + r + \sum_{i=1}^{r} \lfloor c_i/2 \rfloor.$$

Proof. In any 3c-coloring of G we can use at most two colors to color the vertices in $N[v]$. From Corollary 25 can use at most $\lfloor c_i/2 \rfloor + 1$ colors to color the vertices in each component C_i. □

Another bound involving the minimum degree $\delta(G)$ is the following.

Proposition 28. *For any connected graph* G *of order* $n \geq 3$, $\psi_{3c}(G) \leq \lfloor n/\delta \rfloor$.

Proof. Since every color class in a 3c-coloring has at least $\delta(G)$ vertices, the result follows. □

Corollary 29. *For any regular graph* G *of degree* r *and order* n, $\psi_{3c}(G) \leq \lfloor n/r \rfloor$.

The *corona* $G \circ K_1$ of a graph G is the graph obtained from G by adding a leaf adjacent to each vertex of G.

Proposition 30. *For any connected graph* G *of order* $n \geq 3$, $\psi_{3c}(G \circ K_1) = n + 1$.

Proof. It is easy to see that the order of the corona $G \circ K_1$ is $2n$ and that $\alpha_0(G \circ K_1) = n$. From Theorem 18 we know that $\psi_{3c}(G \circ K_1) \leq \alpha_0(G \circ K_1) + 1$. But this bound can be achieved by coloring each leaf of $G \circ K_1$ with a different color, and then coloring every vertex in G with one additional color. □

5. Neighbourhood-restricted colorings of graphs

The subject of 3c-colorings can be placed in a more general setting, in which one places a restriction on the number of different colors that can appear in any closed neighbourhood in a graph G. The focus of the *chromatic number* is to color the vertices of a graph with a minimum number of colors, but with no restrictions on the number of different colors that can be used to color the vertices in any closed neighbourhood $N[v]$. Here, by duality, we propose to color the vertices of a graph with a maximum number of colors, but subject to a restriction r on the number of different colors that can be used to color the vertices in any closed neighbourhood.

Definition 31. *An $[r]$-coloring of a graph G, of order k, is a function $f : V \to \{1, 2, \ldots, k\}$ having the property that for every vertex $v \in V$, $|f(N[v])| \leq r$, that is, at most r distinct colors appear in any closed neighbourhood $N[v]$ in G.*

Definition 32. *The $[r]$-achromatic number $\psi_{[r]}(G)$ equals the maximum integer k such that G has an $[r]$-coloring of order k.*

Notice that the simple function $f : V \to \{1\}$ is trivially an $[r]$-coloring for all values $r \geq 1$. Notice also that any function $f : V \to \{1, 2\}$ is automatically a $[2]$-coloring, and therefore, for any connected graph G of order $n \geq 2$, $\psi_{[2]}(G) \geq 2$.

The following observations follow immediately from the definitions.

Proposition 33. *Let $\pi = \{V_1, V_2, \ldots, V_k\}$ be an $[r]$-coloring of order k of a graph G. Then any partition π' obtained from π by forming the union $V_i \cup V_j$ of any two color classes is an $[r]$-coloring of G of order $k - 1$.*

Theorem 34. *$[r]$-coloring Interpolation Theorem. If a graph G has an $[r]$-coloring of order k, then for any integer j, $r \leq j \leq k$, G has an $[r]$-coloring of order j.*

Proposition 35. *For any connected graph G of order $n \geq 3$ and maximum degree Δ,*

$$1 = \psi_{[1]}(G) \leq \psi_{[2]}(G) \leq \psi_{[3]}(G) \leq \cdots \leq \psi_{[\Delta+1]}(G) = n.$$

It also follows that any 3c-coloring is automatically a [2]-coloring, and therefore we have the following.

Proposition 36. *For any connected graph G of order $n \geq 3$,*

$$\psi_{3c}(G) \leq \psi_{[2]}(G).$$

We have, however, not yet been able to construct a graph G for which the inequality above is strict. It is interesting, nevertheless, to observe that there can be a signficant difference between $\psi_{[2]}(G)$ and $\psi_{[3]}(G)$.

Proposition 37. *For the path P_n and cycle C_n of order $n \geq 3$,*

(i) $\psi_{3c}(P_n) = \psi_{[2]}(P_n) = \lfloor n/2 \rfloor + 1 < \psi_{[3]}(P_n) = n$.
(ii) $\psi_{3c}(C_n) = \psi_{[2]}(C_n) = \lfloor n/2 \rfloor < \psi_{[3]}(C_n) = n$.

A related coloring of a graph is as follows.

A *3-consecutive C-coloring* of graph $G = (V, E)$ is a mapping $\varphi : V \to \mathbb{N}$ such that every path on three vertices has at most two colors. The *3-consecutive c-coloring number number* $\bar{\chi}_{3CC}(G)$ is the maximum number of colors permitted in such a coloring of G. This coloring number is studied in (cf. [3]).

We discuss 3-*consecutive edge coloring* in (cf. [1] and [2]).

6. Open problems

A number of problems arising out of this study have not been resolved. Included among these are the following:

1. Characterize the class of $3\mathcal{C}$-1 graphs.
2. For any connected graph G of order $n \geq 3$, when is $\psi_{3c}(G) = \psi_{[2]}(G)$?
3. Is $\psi_{[2]}(G) \leq n_0(G) + 1$? Is $\psi_{[2]}(G) \leq \alpha_0(G) + 1$?
4. Is $\psi_{[2]}(G) \leq \lfloor n/2 \rfloor + 1$?

Acknowledgement

The authors express their deep sence of gratitude and appreciation to the referee for going through the manuscript very carefully and suggesting many changes. He has given a motivation to study the 3-consecutive coloring and suggested a new result (Corollary 22). Infact, he has retyped the entire manuscript in LaTex effecting the changes which is a very rare extraordinary help to get from a referee.

References

[1] Csilla Bujtás, E. Sampathkumar, Zsolt Tuza, Charles Dominic and L. Pushpalatha, 3-consecutive edge coloring of a graph (2010) Submitted.

[2] Csilla Bujtás, E. Sampathkumar, Zsolt Tuza, Charles Dominic and L. Pushpalatha, When the vertex coloring of a graph is an edge coloring of its line graph-A rare coincidence (2010) Submitted.

[3] Csilla Bujtás, E. Sampathkumar, Zsolt Tuza, M. S. Subramanya and Charles Dominic, 3-consecutive C-colorings of graphs, Discussiones Mathematicae Graph Theory (2010) to appear.

[4] G. S. Domke, J. H. Hattingh and L. R. Markus, On weakly connected domination in graphs II, *Discrete Math.*, **305** (2005) 112–122.

[5] W. Duckworth and B. Mans, Randomised algorithms for finding small weakly-connected dominating sets of regular graphs, *Algorithms and Complexity, LNCS*, **2653** 83–95, Springer, Berlin (2003).

[6] J. E. Dunbar, J. W. Grossman, J. H. Hattingh, S. T. Hedetniemi and A. A. McRae, On weakly-connected domination in graphs, *Discrete Math.*, **167/168** (1997) 261–269.

[7] F. Harary, Graph Theory, Addison-Wesley, Reading, MA (1969).

[8] S. Hedetniemi and R. Laskar, Connected domination in graphs, In Graph Theory and Combinatorics, B. Bollobás, Ed., Academic Press, London (1984) 209–218.

[9] E. Sampathkumar, New vertex and edge coloring numbers of a graph, DST Project Rept. No. SR/S4/MS.275/05 (2006).

[10] E. Sampathkumar and P. S. Neeralagi, The neighbourhood number of a graph, *Indian J. Pure Appl. Math.*, **16(2)** (1985) 126–132.

[11] E. Sampathkumar and H. B. Walikar, The connected domination number of a graph, *J. Math. Phys. Sci.*, **13** (1979) 607–613.

[12] L. A. Sanchis, On the number of edges in graphs with a given weakly connected domination number, *Discrete Math.*, **257** (2002) 111–124.

Proc. ICDM 2008, RMS-Lecture Notes Series No. 13, 2010, pp. 171–177.

Recent Developments on the Structure of Cartesian Products of Graphs

Sandi Klavžar

Faculty of Mathematics and Physics, University of Ljubljana, Jadranska 19, 1000 Ljubljana, Slovenia
and
Faculty of Natural Sciences and Mathematics, University of Maribor, Koroška 160, 2000 Maribor, Slovenia
e-mail: sandi.klavzar@fmf.uni-lj.si

Abstract. The Cartesian product of graphs was introduced more than 50 years ago and many fundamental results were obtained since then. Nevertheless, in the last years several basic problems on the Cartesian product were solved and interesting theorems proved on topics of contemporary interest in graph theory. Here we survey recent developments on the structure of the Cartesian product with emphasis on the connectivity, edge-connectivity and cancellation properties. Recognition algorithms, subgraph structure and the distinguishing number of Cartesian products are also mentioned.

Keywords. Cartesian product of graphs, connectivity, edge-connectivity, cancellation, recognition algorithm, distinguishing number.

1. Introduction

The Cartesian product of graphs is a straightforward operation, yet it is of utmost importance in graph theory and is definitely the most prominent graph product. Many important classes of graphs are Cartesian products (say hypercubes, Hamming graphs, grids, prisms), but it is even more important that Cartesian products are hosts for other graphs, say via isometric (that is, distance preserving) embeddings of graphs into products. For instance, graphs that allow such embeddings into hypercubes are known as partial cubes, one of the central classes of graphs in metric graph theory. Moreover, the Cartesian product is closely related to the most important other products of graphs and helps to understand them better. In particular, all known recognition algorithms for the strong and the direct product depend on that for the Cartesian one.

This graph product has been studied continuously since the 1950's and many structural theorems were proved along the way. From the 1960's we point out the unique prime factorization theorem [25,28]; in the (early) 1970's the structure of the automorphism group of the Cartesian product was clarified [10,23]; from the 1980's the canonical metric representation of a connected graph presents one of the highlights [9]; and from the 1990's we emphasize fixed box theorems [5,27]. We refer to the book [12] for more in depth information on the Cartesian product of graphs as well as to the forthcoming book [15] for a treatment of several recent results on Cartesian product graphs.

Despite all the research, many fundamental questions about Cartesian product of graphs remain unanswered or partially answered. For instance, how is the connectivity and the edge-connectivity of a Cartesian product graph dependent on invariants of its factors? Only in the last couple of years these two questions got their final answers. They are presented in Section 2..

From the algebraic point of view one of the natural questions about a graph product is whether the product obeys the cancellation law and if the rth root is well-defined. The answers for connected Cartesian products follow directly from the unique prime factorization theorem, but what about disconnected graphs? This was answered in the last year and is explained in Section 3..

Many additional new, deep, and fundamental results on the structure of the Cartesian products were obtained recently. We briefly mention them in the last section. The first such result is a linear recognition algorithm for Cartesian products. Then subgraphs of Cartesian products are characterized. Finally, good understanding of the structure of the automorphism group of Cartesian products enabled to determine the distinguishing number for all powers of graphs with respect to the Cartesian product.

In the rest of this section we formally introduce the Cartesian product of graphs.

For given graphs G and H, the *Cartesian product* $G \,\square\, H$ has $V(G) \times V(H)$ for the vertex set; the edge set consists of pairs $(g,h)(g',h')$ where either $g = g'$ and $hh' \in E(H)$, or $gg' \in E(G)$ and $h = h'$. The graphs G and H are called the *factors* of $G \,\square\, H$. The subgraph of $G \,\square\, H$ induced by $\{g\} \times V(H)$, where g is a vertex of G, is isomorphic to H and is called an *H-fiber*. Similarly one defines the *G-fibers*.

Since the Cartesian product is an associative operation, the product of several factors is well-defined. If the factors are the same we speak of powers with respect to the Cartesian product. More precisely, for a positive integer n, the nth *Cartesian power* of a graph G is $G^n = \square_{i=1}^{n} G$.

2. Connectivity

Clearly, $G \,\square\, H$ is connected if and only if the factors G and H are connected. But for a given (product) graph we also wish to know its vertex and edge connectivity. Hence it is natural to wonder if these connectivities of $G \,\square\, H$ can be expressed in terms of the corresponding invariants of the factors. Although this is a natural and basic question, it was answered only recently. The reason is that despite the simple definition of the Cartesian product the connectivity problem is far away from being trivial.

We start with the vertex connectivity $\kappa(G \,\square\, H)$. The first result on it goes back to 1957 when Sabidussi [24] noticed that for arbitrary graphs G and H, $\kappa(G \,\square\, H) \geq \kappa(G) + \kappa(H)$. Then, in 1978, Liouville [22] announced that for any graphs G and H on at least two vertices,

$$\kappa(G \,\square\, H) = \min\{\kappa(G)|V(H)|, \kappa(H)|V(G)|, \delta(G) + \delta(H)\}, \tag{1}$$

where δ denotes the minimum degree of a given graph.

Note that in (1) the "$\leq$" easily follows from the fiber structure of the Cartesian product and the fact that the connectivity cannot be larger that the minimum degree.

(Of course, $\delta(G \,\Box\, H) = \delta(G) + \delta(H)$.) But years have passed by and the announced proof of (1) never appeared. Finally, Špacapan gave a proof of it, so now the assertion (1) can be formulated as a theorem.

Theorem 2.1 ([26]). *For any graphs G and H on at least two vertices,*

$$\kappa(G \,\Box\, H) = \min\{\kappa(G)|V(H)|, \kappa(H)|V(G)|, \delta(G) + \delta(H)\}\,.$$

It is natural to expect that in "most" cases the minimum in Theorem 2.1 will be attained by the minimum degree. In fact, this holds for all powers of an arbitrary nontrivial graph.

Theorem 2.2 ([19]). *Let G be a connected graph on at least two vertices. Then for any $n \geq 2$, $\kappa(G^n) = n\,\delta(G)$.*

The edge-connectivity of the Cartesian product was determined in 2006 by Xu and Yang:

Theorem 2.3 ([30]). *For any graphs G and H on at least two vertices,*

$$\kappa'(G \,\Box\, H) = \min\{\kappa'(G)|V(H)|, \kappa'(H)|V(G)|, \delta(G) + \delta(H)\}\,.$$

Just as with Theorem 2.1, the "$\leq$" part of the proof is straightforward. To prove the other inequality Xu and Yang applied the edge version of Menger's theorem and by a careful analysis found enough edge-disjoint paths for any pair of vertices. Later, in [19] a short proof (independent of Menger's theorem) of Theorem 2.3 was given. Moreover, the structure of minimum edge cuts was clarified.

Theorem 2.4 ([19]). *Let S be a minimum edge cut of $G \,\Box\, H$. Then either S is induced by a minimum edge cut of a factor, or S is the set of edges incident to a vertex of $G \,\Box\, H$.*

Just as for the vertex connectivity, for the edge connectivity of powers of graphs the minimum degree always does the job.

Theorem 2.5 ([19]). *Let G be a connected graph on at least two vertices. Then for any $n \geq 2$, $\kappa'(G^n) = n\,\delta(G)$.*

3. Prime factorization and cancellation

We have already mentioned the following fundamental result on the Cartesian product.

Theorem 3.1 ([25,28]). *Every connected graph has a unique prime factor decomposition with respect to the Cartesian product.*

Theorem 3.1 cannot be extended to disconnected graphs which is demonstrated with the next example

$$(K_1 + K_2 + K_2^2) \,\Box\, (K_1 + K_2^3) = (K_1 + K_2^2 + K_2^4) \,\Box\, (K_1 + K_2)\,,$$

where the operation $+$ stands for the disjoint union of graphs.

Theorem 3.1 easily implies that for any connected graphs G and H and any positive integer r,

$$G^r = H^r \Rightarrow G = H\,, \tag{2}$$

and that for any connected graphs G, H and K,

$$G \,\square\, K = H \,\square\, K \Rightarrow G = H\,. \tag{3}$$

It is now natural to ask what can be said about the roof property (2) and the cancellation (3) property for arbitrary graphs. Contrary to the non-unique factorization of disconnected graphs, Fernández, Leighton and López-Presa in a recent paper [7] proved:

Theorem 3.2 ([7]). *Let G and H be arbitrary graphs and $r \geq 1$. Then $G^r = H^r$ implies $G = H$.*

The key idea of the proof of Theorem 3.2 is that finite graphs form a commutative semiring with respect to Cartesian multiplication and disjoint union. In this way finite graphs can be embedded into a polynomial ring with integer coefficients that is compatible with Cartesian multiplication and disjoint union. Then a given graph is uniquely embedded into the polynomial ring which in turn possesses the unique factorization property that is then lifted back to graphs.

Using the same algebraic approach one can also prove:

Theorem 3.3 ([14]). *Let G, H, K be arbitrary graphs such that $G \,\square\, K = H \,\square\, K$. Then $G = H$.*

4. Other recent results

4.1 *Recognizing Cartesian product graphs*

Another fundamental question on the Cartesian product is the following: given a connected graph G, is it a Cartesian product graph, and if so, find its unique prime factorization.

This question has a long history, the original question was whether the problem is polynomial or not. The problem was solved independently by Feigenbaum, Hershberger and Schäffer [6] and Winkler [29] by providing recognition algorithms of complexity $(n^{4.5})$ and (n^4), respectively. Here n stands for the number of vertices of a given graph, while m will denote the number of edges.

In 1992 Feder [4] followed with a new idea (to consider certain relations Θ and τ) and reduced the recognition complexity to $O(mn)$. At about the same time, Aurenhammer, Hagauer and Imrich [3] carefully considered the structure of Cartesian product graphs and further reduced the complexity to $O(m \log n)$. Finally, the story was rounded in 2007 when Imrich and Peterin [16] succeeded to recognize Cartesian products in linear time:

Theorem 4.1 ([16]). *Let G be a connected graph with m edges. Then its unique prime factor decomposition can be determined in $O(m)$ time.*

4.2 *Subgraphs of Cartesian products*

Given a graph G, we can ask the following question: Is G a subgraph of some Cartesian product graph? Of course, G is a subgraph of $G \square H$, so the question only makes sense if we require that G is a nontrivial subgraph. More precisely, we say that a subgraph of $G \square H$ is *nontrivial* if it intersects at least two G-fibers and at least two H-fibers.

A plotting of a graph is its drawing in the plane such that the endvertices of any edge have the same abscissa or the same ordinate. Now, calling a plotting to be trivial if all the vertices either have the same abscissa or the same ordinate, it is straightforward to observe that a graph is a nontrivial subgraph of a Cartesian product if and only if it has a nontrivial plotting [21].

In fact, a graph is a nontrivial subgraph of the Cartesian product of graphs if and only if it is a nontrivial subgraph of the Cartesian product of two complete graphs. But more can be said. Calling a labeled path, in which any two consecutive vertices receive different labels, a *properly colored path*, the following theorem was proved.

Theorem 4.2 ([17,18]). *Let X be a connected graph on at least three vertices. Then the following statements are equivalent.*

(i) *X is a nontrivial Cartesian subgraph.*
(ii) *$E(X)$ can be labeled with two labels such that on any induced cycle C of X that possesses both labels, the labels change at least three times while traversing the edges of C.*
(iii) *$V(X)$ can be labeled with k labels, $2 \leq k \leq |X| - 1$, such that the endvertices of any properly colored path receive different labels.*

We add that in [18] induced and isometric subgraphs of products of complete graphs are also characterized.

4.3 *Distinguishing Cartesian products*

The distinguishing number $D(G)$ of a graph G is a relatively recently introduced, yet already well-studied, graph invariant [2]. It is defined as the least integer d such that G has a labeling with d labels that is preserved only by a trivial automorphism.

Albertson [1] proved that for a connected prime graph G, $D(G^r) = 2$ for all $r \geq 4$, and, if $|V(G)| \geq 5$, then $D(G^r) = 2$ for all $r \geq 3$. In [20] the result was extended by showing that eventually $D(G^r) = 2$ for any connected graph $G \neq K_2$ and any $r \geq 3$. The story was completed by the following result.

Theorem 4.3 ([13]). *$D(G^k) = 2$ for $k \geq 2$ for all nontrivial, connected graphs $G \neq K_2, K_3$. Furthermore, $D(K_n^k) = 2$ for $n = 2, 3$, if $n + k \geq 6$ and $D(K_2^2) = D(K_2^3) = D(K_3^2) = 3$.*

The distinguishing number of the Cartesian product of two complete graphs was also determined independently by two groups of authors.

Theorem 4.4 ([8,11]). *Let k, n, d be integers so that $d \geq 2$ and $(d-1)^k < n \leq d^k$. Then*

$$D(K_k \,\Box\, K_n) = \begin{cases} d, & \text{if } n \leq d^k - \lceil \log_d k \rceil - 1; \\ d+1, & \text{if } n \geq d^k - \lceil \log_d k \rceil + 1. \end{cases}$$

If $n = d^k - \lceil \log_d k \rceil$ then $D(K_k \,\Box\, K_n)$ is either d or $d+1$ and can be computed recursively in $O(\log^*(n))$ time.

Acknowledgement

I wish to thank Wilfried Imrich for useful suggestions.

References

[1] M. O. Albertson, Distinguishing Cartesian powers of graphs, *Electron. J. Combin.*, **12** (2005) #N17.

[2] M. O. Albertson and K. L. Collins, Symmetry breaking in graphs, *Electron. J. Combin.*, **3** (1996) #R18.

[3] F. Aurenhammer, J. Hagauer and W. Imrich, Cartesian graph factorization at logarithmic cost per edge, *Comput. Complexity*, **2** (1992) 331–349.

[4] T. Feder, Product graph representations, *J. Graph Theory*, **16** (1992) 467–488.

[5] T. Feder, Stable Networks and Product Graphs, *Mem. Amer. Math. Soc.*, **116** (1995) no. 555.

[6] J. Feigenbaum, J. Hershberger and A. A. Schäffer, A polynomial time algorithm for finding the prime factors of Cartesian-product graphs, *Discrete Appl. Math.*, **12** (1985) 123–138.

[7] A. Fernández, T. Leighton and J. L. López-Presa, Containment propeties of product and power graphs, *Discrete Appl. Math.*, **155** (2007) 300–311.

[8] M. J. Fisher and G. Isaak, Distinguishing colorings of Cartesian products of complete graphs, *Discrete Math.*, **308** (2008) 2240–2246.

[9] R. L. Graham and P. M. Winkler, On isometric embeddings of graphs, *Trans. Amer. Math. Soc.*, **288** (1985) 527–536.

[10] W. Imrich, Automorphismen und das kartesische Produkt von Graphen, *Österreich. Akad. Wiss. Math.-Natur. Kl. S.-B. II*, **177** (1969) 203–214.

[11] W. Imrich, J. Jerebic and S. Klavžar, The distinguishing number of Cartesian products of complete graphs, *European J. Combin.*, **29** (2008) 922–929.

[12] W. Imrich and S. Klavžar, Product Graphs: Structure and Recognition, Wiley-Interscience, New York (2000).

[13] W. Imrich and S. Klavžar, Distinguishing Cartesian powers of graphs, *J. Graph Theory*, **53** (2006) 250–260.

[14] W. Imrich, S. Klavžar and D. F. Rall, Cancellation properties of products of graphs, *Discrete Appl. Math.*, **155** (2007) 2362–2364.

[15] W. Imrich, S. Klavžar and D. F. Rall, Topics in Graph Theory: Graphs and Their Cartesian Product, A K Peters, Ltd., Wellesley, MA (2008).

[16] W. Imrich and I. Peterin, Recognizing Cartesian products in linear time, *Discrete Math.*, **307** (2007) 472–483.

[17] S. Klavžar, A. Lipovec and M. Petkovšek, On subgraphs of Cartesian product graphs, *Discrete Math.*, **244** (2002) 223–230.

[18] S. Klavžar and I. Peterin, Characterizing subgraphs of Hamming graphs, *J. Graph Theory*, **49** (2005) 302–312.

[19] S. Klavžar and S. Špacapan, On the edge-connectivity of Cartesian product graphs, *Asian-European J. Math.*, **1** (2008) 93–98.

[20] S. Klavžar and X. Zhu, Cartesian powers of graphs can be distinguished by two labels, *European J. Combin.*, **28** (2007) 303–310.

[21] R. H. Lamprey and B. H. Barnes, A new concept of primeness in graphs, *Networks*, **11** (1981) 279–284.

[22] B. Liouville, Sur la connectivité des produits de graphes, *C. R. Acad. Sci. Paris Sér. A-B*, **286** (1978) A363–A365.

[23] D. J. Miller, The automorphism group of a product of graphs, *Proc. Amer. Math. Soc.*, **25** (1970) 24–28.

[24] G. Sabidussi, Graphs with given group and given graph-theoretical properties, *Canad. J. Math.*, **9** (1957) 515–525.

[25] G. Sabidussi, Graph multiplication, *Math. Z.*, **72** (1960) 446–457.

[26] S. Špacapan, Connectivity of Cartesian products of graphs, *Appl. Math. Lett.*, **21** (2008) 682–685.

[27] C. Tardif, A fixed box theorem for the Cartesian product of graphs and metric spaces, *Discrete Math.*, **171** (1997) 237–248.

[28] V. G. Vizing, Cartesian product of graphs, Vychisl. Sistemy (in Russian) **9** (1963) 30–43.

[29] P. M. Winkler, Factoring a graph in polynomial time, *European J. Combin.*, **8** (1987) 209–212.

[30] J.-M. Xu and C. Yang, Connectivity of Cartesian product graphs, *Discrete Math.*, **306** (2006) 159–165.

Proc. ICDM 2008, RMS-Lecture Notes Series No. 13, 2010, pp. 179–190.

Not so Easy Problems for Tree Decomposable Graphs

Stefan Szeider

Institute of Information Systems, Vienna University of Technology, A-1040 Vienna, Austria
e-mail: stefan@szeider.net

Abstract. We consider combinatorial problems for graphs that (a) can be solved in polynomial time for graphs of bounded treewidth and (b) where the order of the polynomial time bound is expected to depend on the treewidth of the considered graph. First we review some recent results for problems regarding list and equitable colorings, general factors, and generalized satisfiability. Second we establish a new hardness result for the problem of minimizing the maximum weighted outdegree for orientations of edge-weighted graphs of bounded treewidth.

Keywords. Treewidth, W[1]-hardness, graph coloring, general factors, generalized satisfiability, minimum maximum outdegree orientations.

1. Introduction

Treewidth is a graph invariant that indicates, in a certain sense, the global connectivity of a graph. Graphs of treewidth at most k are also known as partial k-trees and width-k tree decomposable graphs. Treewidth plays a central role in Robertson and Seymour's Graph Minors Project and has important algorithmic applications. Many hard graph problems are easy for graphs of small treewidth; for example, 3-COLORABILITY and HAMILTONICITY can be solved in linear time for graphs of treewidth bounded by a constant k (albeit with a running time containing a constant factor that is exponential in k). In fact, all problems that can be expressed in the formalism of Monadic Second Order Logic (that includes the two mentioned problems, but also linear optimization problems like DOMINATING SET) can be solved in linear time for graphs of bounded treewidth [8,2]. However, there are problems that are NP-hard for graphs of a certain fixed treewidth bound; for example BANDWIDTH is NP-hard for graphs of treewidth 1 [13] and $L(2, 1)$-COLORING is NP-hard for graphs of treewidth 2 [11] (to name an old and a new result).

In this paper we focus on problems that are, in a certain sense, neither very easy nor very hard for graphs of bounded treewidth and thus lie between the two extremes. More specifically, we focus on problems that can be solved in polynomial time for graphs of bounded treewidth, but where the order of the polynomial that bounds the running time necessarily depends on the treewidth bound. The theoretical framework of *parameterized complexity* provides the concepts and methods for providing evidence that a certain problem is of this type. The key method is to show that the problem at hand is W[1]-*hard under fpt-reductions* where W[1] is a complexity class that is considered as the parameterized analog of NP. As NP-hardness provides strong evidence that there is no polynomial-time algorithm for a problem, W[1]-hardness provides strong evidence that a problem cannot be solved in polynomial time for instances of bounded treewidth such that the order of the polynomial is independent of the treewidth bound.

We provide definitions and background information on treewidth and parameterized complexity in Section 2. In Section 3 we review some recent W[1]-hardness result for problems on graphs of bounded treewidth, including problems regarding list and equitable colorings, general factors, and generalized satisfiability. In Section 4 we establish a new W[1]-hardness result for the MINIMUM MAXIMUM OUTDEGREE problem for edge-weighted graphs.

2. Preliminaries

2.1 *Graphs and tree decompositions*

All considered graphs are finite, simple and undirected, unless stated otherwise. We denote the vertex set and the edge set of a graph G by $V(G)$ and $E(G)$, respectively, and an edge between vertices u and v by uv (or equivalently vu). Furthermore, we denote the subgraph of a graph G induced by a set $X \subseteq V(G)$ by $G[X]$; that is, $V(G[X]) = X$ and $E(G[X]) = \{ uv \in E(G) : u, v \in X \}$. We also write $G - X = G[V(G) \backslash X]$.

A *tree decomposition* of a graph G is a pair (T, χ) where T is a tree and χ is a mapping that assigns to each vertex $t \in V(T)$ a set $\chi(t) \subseteq V(G)$ such that the following conditions hold:

1. $V(G) = \bigcup_{t \in V(T)} \chi(t)$ and $E(G) \subseteq \bigcup_{t \in V(T)} \{ uv : u, v \in \chi(t) \}$.
2. The sets $\chi(t_1) \backslash \chi(t)$ and $\chi(t_2) \backslash \chi(t) = \emptyset$ are disjoint for any three vertices $t, t_1, t_2 \in V(T)$ such that t lies on a path from t_1 to t_2 in T.

The *width* of (T, χ) is $\max_{t \in V(T)} |\chi(t)| - 1$. The *treewidth* $\mathrm{tw}(G)$ of G is the smallest integer k such that G has a tree decomposition of width k. For more information on treewidth we refer to other sources [5,16].

We shall frequently use the following observation.

Observation 1. *Let G be a graph and $X \subseteq V(G)$. Then $tw(G) \leq tw(G - X) + |X|$.*

Proof. If (T, χ) is a tree decomposition of $G - X$, then (T, χ'), with $\chi'(t) = \chi(t) \cup X$ for $t \in V(T)$, is a tree decomposition of G. □

It is NP-hard to determine the treewidth of a graph [1]. However, for fixed $k \geq 1$, one can decide in linear time whether the treewidth of a graph is at most k, and if so, compute a tree decomposition of width k (Bodlaender's Theorem [4]).

2.2 *Parameterized complexity*

Let us first review some basic concepts of Parameterized Complexity; for more information we refer to the books of Downey and Fellows [9], Flum and Grohe [12], and

Niedermeier [22]. An instance of a parameterized problem is a pair (x, k), where x is the *main part* and k (usually a non-negative integer) is the *parameter*. A parameterized problem is *fixed-parameter tractable* if it can be solved in time $O(f(k)|x|^c)$ where f is a computable function and c is a constant independent of k. FPT denotes the class of all fixed-parameter tractable decision problems. A parameterized problem P *fpt-reduces* to a parameterized problem Q if we can transform an instance (x, k) of P into an instance $(x', g(k))$ of Q in time $O(f(k)|x|^c)$ (f, g are arbitrary computable functions, c is a constant) such that (x, k) is a yes-instance of P if and only if $(x', g(k))$ is a yes-instance of Q. This definition ensures that if there exists an fpt-reduction from P to Q and Q is fixed-parameter tractable, then so is P. A parameterized complexity class $\mathcal{C}$ is the class of parameterized decision problems fpt-reducible to a certain parameterized decision problem $Q_\mathcal{C}$. A parameterized problem P is $\mathcal{C}$-hard if $Q_\mathcal{C}$ (and so each problem in $\mathcal{C}$) can be fpt-reduced to P. A $\mathcal{C}$-hard problem that belongs to $\mathcal{C}$ is $\mathcal{C}$-complete. Of particular interest is the class W[1] that is considered as the parameterized analog to NP. It is believed that FPT $\neq$ W[1], and there is strong theoretical evidence that supports this belief; for example, FPT $=$ W[1] implies that the Exponential Time Hypothesis fails (cf. [12]). There are parameterized problems that are believed to be "harder" than problems in W[1]; indeed, there is an infinite hierarchy of parameterized complexity classes FPT $=$ W[0] $\subseteq$ W[1] $\subseteq$ W[2] $\subseteq$ W[3] $\subseteq \cdots$ where all inclusions are believed to be strict.

The following problem is well known to be W[1]-complete [9].

CLIQUE
Instance: A graph G and a non-negative integer k.
Parameter: The integer k.
Question: Does G contain a clique on k vertices?

As observed by Pietrzak [23] the problem remains W[1]-complete if the input graph is k-partite, which gives the following problem.

PARTITIONED CLIQUE
Instance: A k-partite graph G with partition $V_1, \ldots, V_k$ such that $|V_1| = \cdots = |V_k|$.
Parameter: The integer k.
Question: Does G contain a clique on k vertices?

PARTITIONED CLIQUE (also called MULTICOLORED CLIQUE) is particularly useful for reductions in the context of bounded treewidth. Several W[1]-hardness results that we consider in the sequel are obtained by fpt-reductions from this problem.

3. Some known W[1]-hardness results

3.1 *Coloring problems*

List coloring is an extensively studied variant of graph coloring [15,30,31].

LIST COLORING
Instance: A graph G and for each vertex $v \in V(G)$ a list $l(v)$ of allowed colors for v.
Question: Is there a proper coloring for G where each vertex is colored with a color from its list?

Theorem 1 ([10]). LIST COLORING *is* W[1]*-hard when parameterized by the treewidth of the instance graph.*

We sketch the proof as it is very simple and provides a good example for reductions from PARTITIONED CLIQUE.

Consider a k-partite graph G with partition $V_1, \ldots, V_k$. We construct a graph H as follows. Let $b : V(G) \to \{1, \ldots, |V(G)|\}$ be an arbitrary but fixed bijection. First we take new vertices $v_1, \ldots, v_k$, and set $l(v_i) = \{b(v) : v \in V_i\}$ $(1 \le i \le k)$. Second, for all $1 \le i < j \le k$ and each pair of *non*adjacent vertices $u \in V_i$, $v \in V_j$ we add a vertex v_{uv} and make it adjacent with v_i and v_j; we put $l(v_{uv}) = \{b(u), b(v)\}$. It is easy to verify that H has a proper list coloring if and only if G has a clique on k vertices. Note that $H - \{v_1, \ldots, v_k\}$ is edge-less and so of treewidth 1. Thus $\text{tw}(G) \le k + 1$ follows by Observation 1. So there is indeed an fpt-reduction from PARTITIONED CLIQUE to LIST COLORING parameterized by the treewidth of the instance graph.

Let us briefly mention a fixed-parameter tractability result that contrasts Theorem 1. A graph G is called *r-list-colorable* or *r-choosable* if for every list assignment l such that $|l(v)| \ge r$ for each vertex $v \in V(G)$, there exists a proper coloring for G where each vertex is colored with a color from its list. The *list-chromatic number* or *choice number* of G is the smallest integer r such that G is *r-list-colorable*. Now, as shown by Fellows *et al.* [10], determining the list chromatic number of a given graph is fixed-parameter tractable when parameterized by the treewidth of the graph.

Consider the following problem.

PRECOLORING EXTENSION
Instance: A graph G and a proper coloring c' of some induced subgraph G' of G using colors from $\{1, \ldots, r\}$.
Question: Is it possible to extend c' to a proper coloring c of G using only colors from $\{1, \ldots, r\}$?

One can fpt-reduce LIST COLORING to PRECOLORING EXTENSION by encoding the lists by means of precolored vertices of degree one, without increasing the treewidth.

Corollary 1 ([10]). PRECOLORING EXTENSION *is* W[1]*-hard when parameterized by the treewidth of the instance graph.*

The next problem was introduced by Meyer [21] motivated by a garbage truck scheduling problem; for history and recent results see [6,18].

EQUITABLE COLORING
Instance: A graph G and a positive integer r.
Question: Is there a proper coloring of G using colors from $\{1, \ldots, r\}$ such that the sizes of any two color classes differ at most by one?

Theorem 2 ([10]). EQUITABLE COLORING *is* W[1]*-hard when parameterized by the treewidth of the instance graph. The problem remains* W[1]*-hard when we parameterize simultaneously by the treewidth and the number r of colors.*

Theorem 2 can be shown by a reduction from PARTITIONED CLIQUE; the reduction is significantly more complicated than the one we sketched above.

For graphs of treewidth bounded by some arbitrary but fixed integer k, one can solve EQUITABLE COLORING in polynomial time, even when the number r of colors is not constant and given as part of the input (albeit the order of the polynomial depends on k). This was recently shown by Bodlaender and Fomin [6] using a combinatorial result of Kostochka, Nakprasit, and Pemmaraju [17].

3.2 *General factors*

Lovász [19,20] introduced the following problem.

> GENERAL FACTOR
> *Instance*: A graph G and for each vertex v of G a set $K(v) \subseteq \{0, \ldots, d(v)\}$; we call $K(v)$ the *cardinality set* of v.
> *Question*: Is there a subset $F \subseteq E(G)$ such that for each vertex $v \in V(G)$ the number of edges in F incident with v is an element of $K(v)$?

This problem clearly generalizes the polynomial-time solvable r-FACTOR problem where all cardinality sets are equal to $\{r\}$. However, GENERAL FACTOR is easily seen to be NP-hard, already if cardinality sets are restricted to $\{0, 3\}$ and $\{1\}$ (say, by reduction from 3-DIMENSIONAL MATCHING). Cornuéjols [7] gives a full classification of the complexity of GENERAL FACTOR when cardinality sets are restricted to some fixed class of sets (a dichotomy of NP-hard and polynomial-time solvable cases).

Theorem 3 ([26]). GENERAL FACTOR *is* W[1]*-hard when parameterized by the treewidth of the instance graph. The problem remains* W[1]*-hard when the given graph is bipartite and all cardinality sets for vertices of one side of the bipartition are equal to* $\{1\}$.

The proof of this result is, once again, obtained by an fpt-reduction from PARTITIONED CLIQUE.

GENERAL FACTOR can be solved in polynomial time for graphs of bounded treewidth where the order of the polynomial depends on the treewidth bound [29]. In fact, the main result of [29] is a meta-theorem that provides polynomial-time algorithms for a wide range of problems on graphs of bounded treewidth. Each of the covered problems asks for a given graph G with cardinality sets $K(v) \subseteq \{0, \ldots, |V(G)| + |E(G)| - 1\}$ whether there exists a set $X \subseteq V(G) \cup E(G)$ such that

1. for each vertex $v \in V(G)$, the number of vertices in X adjacent to v plus the number of edges in X incident with v belongs to $K(v)$,
2. X satisfies a fixed property $P(X)$ expressible in a certain formalism called "Monadic Second Order Logic."

For example $P(X)$ could state that X is a set of vertices that forms a color class for a proper 3-coloring of G. For GENERAL FACTOR the property $P(X)$ just states that X is a set of edges.

3.3 *Generalized satisfiability*

A *Boolean constraint* is a pair $C = ((x_1, \ldots, x_r), R)$ where $x_1, \ldots, x_r$ are distinct variables and $R \subseteq \{0, 1\}^r$ is a Boolean relation of arity $r > 0$. We write $\text{var}(C) = \{x_1, \ldots, x_r\}$ and say that C *is over* a set X of variables if $\text{var}(C) \subseteq X$. A mapping $\tau : X \to \{0, 1\}$ *satisfies* a Boolean constraint $C = ((x_1, \ldots, x_r), R)$ if C is over X and $(\tau(x_1), \ldots, \tau(x_r)) \in R$.

GENERALIZED SATISFIABILITY
Instance: A finite set X of variables and finite set S of Boolean constraints over X.
Question: Is there a mapping $\tau : X \to \{0, 1\}$ that satisfies all constraints in S?

Clearly GENERALIZED SATISFIABILITY is NP-complete, as, for example, it contains 3-SAT as the special case where all constraints use the same relation $R = \{0, 1\}^3 \setminus \{(0, 0, 0)\}$. Schaefer [27] classifies the complexity of GENERALIZED SATISFIABILITY problems for instances that use relations from a fixed class (a dichotomy of NP-hard and polynomial-time solvable cases).

By associating certain graphs to sets of Boolean constraints one can apply the treewidth parameter to the GENERALIZED SATISFIABILITY problem.

Consider an instance (X, S) of GENERALIZED SATISFIABILITY. The *primal graph* has vertex set X, two variables are adjacent if they occur together in a constraint. Symmetrically, the *dual graph* has as vertex set S, two constraints are adjacent if they share a variable. Finally, the *incidence graph* is the bipartite graph with vertex set $X \cup S$; a constraint and a variable are adjacent if the variable occurs in the constraint.

It is easy to see that GENERALIZED SATISFIABILITY is fixed-parameter tractable if parameterized by the treewidth of primal graphs [14]. However, regarding the treewidth of dual and incidence graphs we have the following negative results.

Theorem 4 ([24]). GENERALIZED SATISFIABILITY *is* W[1]*-hard when parameterized by the treewidth of the dual graph or by the treewidth of the incidence graph of the instance.*

We sketch the proof which uses an fpt-reduction from CLIQUE. Let G be a graph with $V(G) = \{v_1, \ldots, v_n\}$. We construct an instance (X, S) of GENERALIZED SATISFIABILITY as follows. First we construct a relation $R \subseteq \{0, 1\}^{2n}$ that encodes the edges of G using Boolean values 0 and 1. For each edge $v_p v_q$ of G, $1 \leq p < q \leq n$, we add to R the $2n$-tuple

$$(t_{p,1}, \ldots, t_{p,n}, t_{q,1}, \ldots, t_{q,n})$$

where $t_{p,i} = 1$ if and only if $p = i$, and $t_{q,i} = 1$ if and only if $q = i$, $1 \leq i \leq n$. We let S be the set of Boolean constraints

$$C_{i,j} = ((x_{i,1}, \ldots, x_{i,n}, x_{j,1}, \ldots, x_{j,n}), R)$$

and X the set of variables $x_{i,j}$, for $1 \leq i < j \leq k$. It is easy to verify that G contains a clique on k vertices if and only if S is satisfiable. Since there are exactly $\binom{k}{2}$ constraints in S, the treewidth of the dual graph is at most $\binom{n}{k} - 1$, thus bounded in terms of k. Using Observation 1 it is easy to see that the treewidth of the incidence graph is at most $\binom{k}{2}$. Hence Theorem 4 follows.

BOOLEAN SATISFIABILITY is defined similarly except that instead of Boolean constraints one considers *clauses* (disjunctions of variables or negated variables). Primal, dual, and incidence graphs are defined for sets of clauses in the obvious way. Interestingly, BOOLEAN SATISFIABILITY and GENERALIZED SATISFIABILITY are of different parameterized complexity: BOOLEAN SATISFIABILITY is fixed-parameter tractable when parameterized by the treewidth of any of the three associated graphs [25,28].

4. A new hardness result for the minimum maximum outdegree problem

A (positive integral) *edge weighting* of a graph G is a mapping w that assigns to each edge of G a positive integer. An *orientation* of G is a mapping $\Lambda : E(G) \to V(G) \times V(G)$ with $\Lambda(uv) \in \{(u, v), (v, u)\}$. The *weighted outdegree* of a vertex $v \in V(G)$ with respect to an edge weighting w and an orientation Λ is defined as

$$d^+_{G,w,\Lambda}(v) = \sum_{vu \in E(G) \text{ with } \Lambda(vu)=(v,u)} w(vu).$$

Asahiro, Miyano, and Ono [3] consider the following problem and discuss applications and related problems.

> MINIMUM MAXIMUM OUTDEGREE
> *Instance*: A graph G, an edge weighting w of G given in unary, and a positive integer r.
> *Question*: Is there an orientation Λ of G such that $d^+_{G,w,\Lambda}(v) \leq r$ for each $v \in V(G)$?

We assume that the edge weighting w is given in unary since otherwise the problem is already NP-complete for graphs of treewidth 2, as a simple reduction from PARTITION shows [3]. If all edge weights are identical, then MINIMUM MAXIMUM OUTDEGREE can be solved in polynomial time using network flows [3]. Furthermore, the problem can be solved for graphs of treewidth k in time bounded by a polynomial whose order depends on k [29]. The next theorem shows that this dependence is necessary, unless FPT = W[1].

Theorem 5. MINIMUM MAXIMUM OUTDEGREE *is* W[1]*-hard when parameterized by the treewidth of the instance graph.*

Proof. We use the following intermediate problem:

> CHOSEN MAXIMUM OUTDEGREE
> *Instance*: A graph G, an edge weighting w of G given in unary, and for each vertex $v \in V(G)$ a non-negative integer $\rho(v)$.
> *Question*: Is there an orientation Λ of G such that $d^{+}_{G,w,\Lambda}(v) \leq \rho(v)$ for each $v \in V(G)$? We call such an orientation *ρ-admissible.*

Claim 1. CHOSEN MAXIMUM OUTDEGREE fpt-reduces to MINIMUM MAXIMUM OUTDEGREE (both problems are parameterized by the treewidth of the instance graph).

To prove this claim, let G, w, ρ be an instance of CHOSEN MAXIMUM OUTDEGREE. We construct an edge weighted graph H from G as follows. Let $r = \max_{v \in V(G)} \rho(v)$. For each vertex $v \in V(G)$ with $\rho(v) < r$ we add to G two new vertices x_v, y_v and the edges vx_v, vy_v, and $x_v y_v$ with edge weights $r - \rho(v)$, $r - \rho(v)$, and r, respectively. It is easy to verify that H has an orientation with maximum weighted outdegree at most r if and only if G has a ρ-admissible orientation. Thus Claim 1 follows.

Next we give an fpt-reduction from PARTITIONED CLIQUE to CHOSEN MAXIMUM OUTDEGREE; the theorem will then follow by Claim 1.

Consider a k-partite graph G with partition $V_1, \dots, V_k$ with $|V_1| = \dots = |V_k| = n$. We write $V_i = \{v_i^1, \dots, v_i^n\}$ for $1 \leq i \leq k$. For $1 \leq i < i' \leq k$ let $E_{i,i'} = \{ (q, q') : 1 \leq q \leq n, 1 \leq q' \leq n, v_i^q v_{i'}^{q'} \in E(G) \}$. We are going to construct a graph H with edge weighting w and vertex weighting ρ.

The vertex set of H is obtained as follows:

1. For $1 \leq i \leq k$ and $1 \leq j \leq n$, we add to $V(H)$ three vertices u_i^j, x_i^j, and y_i^j.
2. For $1 \leq i \leq k$ we add to $V(H)$ a vertex a_i.
3. For $1 \leq i < i' \leq k$ we add to $V(H)$ vertices $b_{i,i'}$, $c_{i,i'}$, and $d_{i,i'}$.
4. For $1 \leq i < i' \leq k$ and each $(q, q') \in E_{i,i'}$ we add to $V(H)$ a vertex $e_{i,i'}^{q,q'}$.

The edge set of H is obtained as follows.

1. For $1 \leq i \leq k$ and $1 \leq j \leq n$ we add the edges $a_i u_i^j$, $u_i^j x_i^j$, and $u_i^j y_i^j$.
2. For $1 \leq i < i' \leq k$ and $(q, q') \in E_{i,i'}$ we add the edges $e_{i,i'}^{q,q'} d_{i,i'}$, $e_{i,i'}^{q,q'} b_{i,i'}$ and $e_{i,i'}^{q,q'} c_{i,i'}$.
3. For $1 \leq i < i' \leq k$, $1 \leq j \leq n$, and $1 \leq j' \leq n$ we add the edges $x_i^j b_{i,i'}$, $y_i^j c_{i,i'}$, and $x_{i'}^{j'} b_{i,i'}$, $y_{i'}^{j'} c_{i,i'}$.

We shall refer to the edges added in the last step as *special edges*.

Claim 2. The treewidth of H is at most $2\binom{k}{2} + 1$.

Indeed, the set $BC = \{b_{i,i'}, c_{i,i'} : 1 \leq i < i' \leq k\}$ is of cardinality $2\binom{k}{2}$ and $H - BC$ is a disjoint union of trees. Hence Claim 2 follows from Observation 1.

Let $N = n + 1$. We define the weights of special edges:

$$w(x_i^j b_{i,i'}) = w(y_i^j c_{i,i'}) = N^3 + j(i < i',\ 1 \leq j \leq n)$$

$$w(x_{i'}^j b_{i,i'}) = w(y_{i'}^j c_{i,i'}) = N^3 + jN(i < i',\ 1 \leq j \leq n)$$

Let $M(v)$ denote the sum of the weights of all special edges incident with vertex v. We set $M = k(N^3 + N^2)$ to ensure that we have $M(x_i^j) < M(y_i^j) < M$ for all $1 \leq i \leq k$ and $1 \leq j \leq n$.

We define further edge and vertex weights as follows.

For $1 \leq i \leq k$ we set $\rho(a_i) = 1$ and $w(a_i u_i^j) = 1$, for $1 \leq j \leq n$.

For $1 \leq i \leq k$ and $1 \leq i \leq n$ we set $w(u_i^j x_i^j) = \rho(x_i^j) = M$ and $w(u_i^j y_i^j) = \rho(u_i^j) = \rho(y_i^j) = M + 1$.

For $1 \leq i < i \leq k$ we set $\rho(d_{i,i'}) = |E_{i,i'}| - 1$, and for $(q, q') \in E_{i,i'}$ we define:

$$w(d_{i,i'} e_{i,i'}^{q,q'}) = 1$$

$$w(e_{i,i'}^{q,q'} b_{i,i'}) = w(x_i^q b_{i,i'}) + w(x_{i'}^{q'} b_{i,i'})$$

$$w(e_{i,i'}^{q,q'} c_{i,i'}) = w(y_i^q c_{i,i'}) + w(y_{i'}^{q'} c_{i,i'})$$

$$\rho(e_{i,i'}^{q,q'}) = w(e_{i,i'}^{q,q'} c_{i,i'}) \quad (> w(e_{i,i'}^{q,q'} b_{i,i'}))$$

For $1 \leq i < i \leq k$ we define:

$$\rho(b_{i,i'}) = \sum_{j=1}^{n} w(x_i^j b_{i,i'}) + \sum_{j=1}^{n} w(x_{i'}^j b_{i,i'})$$

$$\rho(c_{i,i'}) = \sum_{j=1}^{n} w(y_i^j c_{i,i'}) + \sum_{j=1}^{n} w(y_{i'}^j c_{i,i'})$$

Claim 3. If H has a ρ-admissible orientation then G has a clique on k vertices.

To prove this claim, let Γ be an admissible orientation. Let $A = \{\Lambda(e) : e \in E(G)\}$. We shall use terminology for directed graphs. For example, if $(x, y) \in A$ then we say that xy is an "outgoing edge" of x and an "incoming edge" of y.

Let $1 \leq i \leq k$. Observe that a_i has at most one outgoing edge. If it has no outgoing edges, then we can revert an arbitrarily chosen one maintaining a ρ-admissible orientation. Hence, without loss of generality, we may assume that a_i has exactly one outgoing edge, say $(a_i, u_i^{p(i)}) \in A$ for some $p(i) \in \{1, \ldots, n\}$. Consequently, for all $j \in \{1, \ldots, n\} \backslash \{p(i)\}$ we have $(y_i^j, u_i^j) \in A$, and in turn $(c_{i',i}, y_i^j) \in A$ for all $1 \leq i' \leq k$.

Let $1 \leq i < i' \leq k$. For similar reasons as in the previous paragraph we may assume, without loss of generality, that $d_{i,i'}$ has exactly one incoming edge, say $(e_{i,i'}^{q(i,i'),q'(i,i')} d_{i,i'}) \in A$ for $(q(i,i'),q'(i,i')) \in E_{i,i'}$. It follows that $(c_{i,i'}, e_{i,i'}^{q(i,i'),q'(i,i')}) \in A$. We have already concluded that $(c_{i,i'}, y_i^j) \in A$ for all $j \in \{1,\ldots,n\}\backslash\{p(i)\}$ and $(c_{i,i'}, y_{i'}^j) \in A$ for all $j \in \{1,\ldots,n\}\backslash\{p(i')\}$. Thus the number of outgoing edges from $c_{i,i'}$ is at least $2(n-1)+1 = 2n-1$. Observe that each edge incident with $c_{i,i'}$ has weight greater than N^3, the weight of $c_{i,i'}e_{i,i'}^{q(i,i'),q'(i,i')}$ is even greater than $2N^3$. Since $\lfloor \rho(c_{i,i'})/N^3 \rfloor = 2n$, we conclude that $c_{i,i'}$ has no further outgoing edges than the $2n-1$ edges identified so far. In particular $(e_{i,i'}^{q,q'}, c_{i,i'}) \in A$ for all $(q,q') \in E_{i,i'}\backslash\{(q(i,i'),q'(i,i'))\}$ and $(y_i^{p(i)} c_{i,i'}), (y_{i'}^{p(i')} c_{i,i'}) \in A$. The latter implies $(u_i^{p(i)}, y_i^{p(i)}) \in A$ and consequently $(x_i^{p(i)}, u_i^{p(i)}) \in A$ and $(b_{i,i'}, x_i^{p(i)}) \in A$; similarly $(b_{i,i'}, x_{i'}^{p'(i)}) \in A$. We concluded above that for $(q,q') \in E_{i,i'}\backslash\{(q(i,i'),q'(i,i'))\}$ we have $(e_{i,i'}^{q,q'}, c_{i,i'}) \in A$, thus $(b_{i,i}, e_{i,i'}^{q,q'}) \in A$. Hence the weighted outdegree of $b_{i,i'}$ is high enough to conclude, similarly as above for $c_{i,i'}$, that all edges incident with $b_{i,i'}$ that we have not yet identified as outgoing are incoming edges.

In view of $\rho(b_{i,i'})$ and $\rho(c_{i,i'})$ and the weights of the respective outgoing edges we conclude

$$\begin{aligned} w(b_{i,i'}x_i^{q(i,i')}) + w(b_{i,i'}x_{i'}^{q'(i,i')}) &= w(b_{i,i'}e_{i,i'}^{q(i,i'),q'(i,i')}) \\ &\geq w(b_{i,i'}x_i^{p(i)}) + w(b_{i,i'}x_{i'}^{p(i')}) \end{aligned}$$

and

$$\begin{aligned} w(c_{i,i'}y_i^{q(i,i')}) + w(c_{i,i'}y_{i'}^{q'(i,i')}) &= w(c_{i,i'}e_{i,i'}^{q(i,i'),q'(i,i')}) \\ &\leq w(c_{i,i'}y_i^{p(i)}) + w(c_{i,i'}y_{i'}^{p(i')}). \end{aligned}$$

The first inequality gives $q(i,i')+q'(i,i')N \geq p(i)+Np(i')$ and so $q'(i,i') \geq p(i')$; the second inequality gives $q'(i,i') \leq p(i')$; their combination gives $q'(i,i') = p(i')$. Using this identity to simplify the two inequalities we can finally obtain $q(i,i') = p(i)$. We conclude that $v_i^{p(i)}$ and $v_{i'}^{p(i')}$ are adjacent in G for all $1 \leq i < i' \leq k$. Consequently the vertices $v_1^{p(1)},\ldots,v_k^{p(k)}$ induce a clique in G, and Claim 3 follows.

Claim 4. If G has a clique on k vertices then H has a ρ-admissible orientation.

This is the easy direction. Assume there exists a clique on k vertices in G. We can write the vertices of the clique as $v_1^{p(1)},\ldots,v_k^{p(k)}$ where $p(i) \in \{1,\ldots,n\}$. Clearly $(p(i),p(i')) \in E_{i,i'}$ holds for $1 \leq i < i' \leq k$. We define a ρ-admissible orientation Λ es follows. Again we write $A = \{\,\Lambda(e) : e \in E(G)\,\}$.

For $1 \leq i \leq k$ we make $a_i u_i^{p(i)}$ the only outgoing edge of a_i; accordingly for $j = p(i)$ we set $(u_i^j, y_i^j), (x_i^j, u_i^j) \in A$, and $(y_i^j, c), (c, x_i^j) \in A$ for all $c = c_{i',i}$ $(1 \leq i' < i)$ and $c = c_{i,i'}$ $(i < i' \leq k)$; for $j \neq p(i)$ we take the inverse orientation of the mentioned edges.

For $1 \leq i < i' \leq k$ we we make $d_{i,i'}e_{i,i'}^{p(i),p(i')}$ the only incoming edge of $d_{i,i'}$; for $(p, p') \in E_{i,i'}$ we set $(e_{i,i'}^{p,p'}, b_{i,i'}) \in A$ exactly when $j = p(i)$, and we set $(e_{i,i'}^{p,p'}, c_{i,i'}) \in A$ exactly when $j \neq p(i)$.

This completes the definition of Λ. It is easy to verify that Γ is indeed ρ-admissible, hence Claim 4 follows.

It is evident that H can be computed in polynomial time from G. By Claim 2, the treewidth of H is a function of k, thus with Claims 3 and 4 we have established an fpt-reduction from PARTITIONED CLIQUE to CHOSEN MAXIMUM OUTDEGREE. In view of Claim 1 and the W[1]-hardness of PARTITIONED CLIQUE, Theorem 5 follows.

□

Acknowledgement

Research partially supported by the European Research Council, grant reference 239962.

References

[1] S. Arnborg, D. G. Corneil and A. Proskurowski, Complexity of finding embeddings in a k-tree, *SIAM J. Algebraic Discrete Methods 8*, **2** (1987) 277–284.
[2] S. Arnborg, J. Lagergren and D. Seese, Easy problems for tree-decomposable graphs, *J. Algorithms 12*, **2** (1991) 308–340.
[3] Y. Asahiro, E. Miyano and H. Ono, Graph classes and the complexity of the graph orientation minimizing the maximum weighted outdegree. In Proceedings of CATS 2008, Computing: The Australasian Theory Symposium, University of Wollongong, New South Wales, Australia, January 22–25, 2008, part of the Australasian Computer Society Week (ACSW 2008) (2008), J. Harland and P. Manyem, Eds., Vol. 77 of Conferences in Research and Practice in Information Technology, Australian Computer Society, pp. 97–106.
[4] H. L. Bodlaender, A linear-time algorithm for finding tree-decompositions of small treewidth, *SIAM J. Comput. 25*, **6** (1996) 1305–1317.
[5] H. L. Bodlaender, A partial k-arboretum of graphs with bounded treewidth, *Theoret. Comput. Sci. 209*, **1–2** (1998) 1–45.
[6] H. L. Bodlaender and F. V. Fomin, Equitable colorings of bounded treewidth graphs, *Theoret. Comput. Sci. 349*, **1** (2005) 22–30.
[7] G. Cornuéjols, General factors of graphs, *J. Combin. Theory Ser. B 45*, **2** (1988) 185–198.
[8] B. Courcelle, Graph rewriting: an algebraic and logic approach. In Handbook of Theoretical Computer Science, Vol. B. Elsevier Science Publishers, North-Holland, Amsterdam (1990) pp. 193–242.
[9] R. G. Downey and M. R. Fellows Parameterized Complexity, Monographs in Computer Science, Springer-Verlag (1999).
[10] M. R. Fellows, F. V. Fomin, D. Lokshtanov, F. A. Rosamond, S. Saurabh, S. Szeider and C. Thomassen, On the complexity of some colorful problems parameterized by treewidth. In Proceedings of COCOA 2007, Combinatorial Optimization and Applications, First International Conference, Xi'an, China, August 14–16, 2007 (2007), Vol. 4616 of Lecture Notes in Computer Science, Springer-Verlag, pp. 366–377.
[11] J. Fiala, P. A. Golovach and J. Kratochvíl, Distance constrained labelings of graphs of bounded treewidth. In Automata, Languages and Programming, Vol. 3580 of Lecture Notes in Computer Science. Springer-Verlag (2005) pp. 360–372.
[12] J. Flum and M. Grohe, Parameterized Complexity Theory, Vol. XIV of Texts in Theoretical Computer Science. An EATCS Series. Springer-Verlag (2006).

[13] M. R. Garey, R. L. Graham, D. S. Johnson and D. E. Knuth, Complexity results for bandwidth minimization, *SIAM J. Appl. Math. 34*, **3** (1978) 477–495.

[14] G. Gottlob, F. Scarcello and M. Sideri, Fixed-parameter complexity in AI and nonmonotonic reasoning, *Artificial Intelligence 138*, **1–2** (2002) 55–86.

[15] T. R. Jensen and B. Toft, Graph Coloring Problems. Wiley-Interscience Series in Discrete Mathematics and Optimization. John Wiley & Sons, New York (1995).

[16] T. Kloks, Treewidth: Computations and Approximations. Springer-Verlag (1994).

[17] A. V. Kostochka, K. Nakprasit and S. V. Pemmaraju, On equitable coloring of d-degenerate graphs, *SIAM J. Discrete Math. 19*, **1** (2005) 83–95.

[18] K.-W. Lih, The equitable coloring of graphs. In Handbook of Combinatorial Optimization, Vol. 3. Kluwer Academic Publishers, Dortrecht, Boston, MA (1998) pp. 543–566.

[19] L. Lovász, The factorization of graphs. In Combinatorial Structures and their Applications (Proc. Calgary Internat. Conf., Calgary, Alta., 1969). Gordon and Breach, New York (1970) pp. 243–246.

[20] L. Lovász, The factorization of graphs. II, *Acta Math. Acad. Sci. Hungar. 23*, (1972) 223–246.

[21] W. Meyer, Equitable coloring, *Amer. Math. Monthly 80*, (1973) 920–922.

[22] R. Niedermeier, Invitation to Fixed-Parameter Algorithms. Oxford Lecture Series in Mathematics and its Applications. Oxford University Press (2006).

[23] K. Pietrzak, On the parameterized complexity of the fixed alphabet shortest common supersequence and longest common subsequence problems, *J. of Computer and System Sciences 67*, **4** (2003) 757–771.

[24] M. Samer and S. Szeider, Constraint satisfaction with bounded treewidth revisited, *J. of Computer and System Sciences*, **76(2)** (2010) 103–114.

[25] M. Samer and S. Szeider, Algorithms for propositional model counting. In Proceedings of LPAR 2007, 14th International Conference on Logic for Programming, Artificial Intelligence and Reasoning, October 15–19, 2007 Yerevan, Armenia (2007), Vol. 4790 of Lecture Notes in Computer Science, Springer-Verlag, pp. 484–498. Full version to appear in *J. of Discrete Algorithms*, **8(1)** (2010) 50–64.

[26] M. Samer and S. Szeider, Tractable cases of the extended global cardinality constraint. In Proceedings of CATS 2008, Computing: The Australasian Theory Symposium, University of Wollongong, New South Wales, Australia, January 22–25, 2008 (2008), J. Harland and P. Manyem, Eds., Vol. 77 of Conferences in Research and Practice in Information Technology, Australian Computer Society, pp. 67–74. Full and extended version to appear in Constraints: an international journal.

[27] T. J. Schaefer, The complexity of satisfiability problems. In Conference Record of the Tenth Annual ACM Symposium on Theory of Computing (San Diego, Calif., 1978). ACM (1978) pp. 216–226.

[28] S. Szeider, On fixed-parameter tractable parameterizations of SAT. In Theory and Applications of Satisfiability, 6th International Conference, SAT 2003, Selected and Revised Papers (2004), E. Giunchiglia and A. Tacchella, Eds., Vol. 2919 of Lecture Notes in Computer Science, Springer-Verlag pp. 188–202.

[29] S. Szeider, Monadic second order logic on graphs with local cardinality constraints. In Mathematical Foundations of Computer Science 2008, 33rd International Symposium, MFCS 2008, Torun, Poland, August 25–29, 2008, Proceedings (2008), Vol. 5162 of Lecture Notes in Computer Science, Springer-Verlag, pp. 601–612. Full and extended version to appear in the ACM Transactions on Computational Logic.

[30] Z. Tuza, Graph colorings with local constraints—a survey, *Discuss. Math. Graph Theory 17*, **2** (1997) 161–228.

[31] D. R. Woodall, List colourings of graphs. In Surveys in Combinatorics, 2001 (Sussex), Vol. 288 of London Math. Soc. Lecture Note Ser., Cambridge University Press, Cambridge (2001) pp. 269–301.

Proc. ICDM 2008, RMS-Lecture Notes Series No. 13, 2010, pp. 191–205.

List Hereditary Colorings of Graphs (Extended Abstract)

C. R. Subramanian

The Institute of Mathematical Sciences, Taramani, Chennai 600 113, India
e-mail: crs@imsc.res.in

Abstract. We study the list variant of a generalized notion of proper vertex coloring of graphs which is defined as follows. For an (induced) hereditary property of labeled undirected graphs $\mathcal{P}$, the $\mathcal{P}$-chromatic number $\chi_{\mathcal{P}}(G)$ of a graph $G = (V, E)$ is the minimum k such that there exists a labeled partition $(V_1, \ldots, V_k)$ of V into k (possibly empty) subsets such that for each i, the induced subgraph $G[V_i]$ is a member of $\mathcal{P}$. We present a survey of some known and some new upper bounds on $\chi_{\mathcal{P}}(G)$ in terms of other invariants like maximum degree and degeneracy of G and also on a parameter which depends only on $\mathcal{P}$.

In the list variant, each vertex $u \in V$ is assigned a list L_u of colors. For $k \geq 1$, G is $(k, \mathcal{P})$-choosable if *for every* list assignment $\{L_u : u \in V\}$ of lists of colors such that $|L_u| \geq k$ for all $u \in V$: there exists a coloring $\{f(u)\}_{u \in V}$ such that (i) $f(u) \in L_u$ for each $u \in V$ and (ii) for each color $x \in \cup_{u \in V} L_u$, the subgraph induced by V_x, the set of vertices u such that $f(u) = x$, is a member of $\mathcal{P}$. The $\mathcal{P}$-choice number of G, denoted by $ch_{\mathcal{P}}(G)$, is the minimum value of k such that G is $(k, \mathcal{P})$-choosable.

It is shown that for any graph G, $ch_{\mathcal{P}}(G) \leq c\chi_{\mathcal{P}}(\ln(n/\chi_{\mathcal{P}})) + 1)$ where c is an absolute positive constant and $\chi_{\mathcal{P}} = \chi_{\mathcal{P}}(G)$ is the $\mathcal{P}$-chromatic number of G. Here, $n = |V(G)|$. We also show that the bounds are constructive leading to efficient algorithms for list $\mathcal{P}$-coloring provided we are also given an exact or an approximate $(\chi_{\mathcal{P}}, \mathcal{P})$-coloring of G.

Keywords. Graph colorings, list colorings, hereditary properties, choice number.

1. Introduction

We consider only finite, undirected and simple graphs. Throughout the paper, we use $G = (V, E)$ to denote a graph on $n = |V|$ vertices. Let $\mathcal{P}$ be a hereditary property of labeled undirected graphs. That is, if $G \in \mathcal{P}$ and H is an induced subgraph of G, then $H \in \mathcal{P}$. Also, if G and H are isomorphic, then $G \in \mathcal{P}$ if and only if $H \in \mathcal{P}$. Such properties are also known as induced hereditary properties in the literature. But, for the sake of brevity, we refer to them simply as hereditary properties. For an introduction to the hereditary properties and related notions, see the comprehensive survey by Borowiecki, Broere, Frick, Mihok and Semanisin [10] which also mentions some weaker versions of the bounds presented in this abstract.

1.1 *$\mathcal{P}$-chromatic numbers*

A $(k, \mathcal{P})$-coloring of $G = (V, E)$ is a labeled partition $(V_1, \ldots, V_k)$ of V into k (possibly empty) subsets such that for each i, the induced subgraph $G[V_i]$ is a member of $\mathcal{P}$. The minimum value of k such that G admits a $(k, \mathcal{P})$-coloring is known as the $\mathcal{P}$-chromatic number of G which we denote by $\chi_{\mathcal{P}}(G)$. When $\mathcal{P}$ is the set of all labeled graphs having no edges, $(k, \mathcal{P})$-coloring specializes to the standard k-coloring and $\chi_{\mathcal{P}}(G)$ is the same as the standard chromatic number $\chi(G)$. Hereditary colorings are also referred to as Generalized colorings in the literature. In Section 2 of this paper, we state some easy-to-verify observations (see Theorem 2.1) about hereditary colorings which we have not seen mentioned in the literature.

Various special instances of hereditary colorings have been previously studied (see the work of Harary [23]) and estimates or bounds on the corresponding chromatic numbers have been obtained for various special classes of graphs. For example, Sampathkumar and Venkatachalam [35] consider the case when $\mathcal{P}$ is the set of all c-colorable graphs for some fixed $c \geq 0$. Albertson, Jamison, Hedetniemi and Locke [2] consider the case when $\mathcal{P}$ is the collection of all graphs each of which is a vertex disjoint union of complete graphs, with the corresponding $\chi_{\mathcal{P}}(G)$ known as the subchromatic number of G. Gimbel and Straight [20] and Gimbel, Kratsch and Stewart [21] consider the case when $\mathcal{P}$ is the set of all graphs each of which is either an independent set or a complete graph, with the corresponding $\chi_{\mathcal{P}}(G)$ known as the cochromatic number of G. When $\mathcal{P}$ is the set of all forests, the chromatic number $\chi_{\mathcal{P}}(G)$ is known as the vertex arboricity of G and has been studied in the literature, see [11,1,22,12]. When $\mathcal{P}$ is the set of all perfect graphs, $\mathcal{P}$-colorings are referred to as perfect colorings (see [13,8]). When $\mathcal{P}$ is the set of all P_4-free graphs, $\mathcal{P}$-colorings are referred to as P_4-free colorings (see [25]).

General non-trivial upper bounds which are applicable to any hereditary $\mathcal{P}$ have been few. To the best of our knowledge, the only such upper bound is due to Cockayne [14] who obtains an upper bound on $\chi_{\mathcal{P}}$ in terms of a parameter which can thought of as a $\mathcal{P}$-dependant generalization of the well-known degeneracy of a graph. In fact, we can generalize the standard definition of degrees to their $\mathcal{P}$ analogues, namely, the $\mathcal{P}$-degrees. Further details on this generalization will appear in the full version [39] of this extended abstract. However, there are properties $\mathcal{P}$ for which it is likely to be impossible to efficiently compute this parameter for an arbitrary G.

In [10], a general upper bound on $\chi_{\mathcal{P}}$ which is expressed as a function of $\Delta(G)$ (maximum degree) and another parameter which depends only on $\mathcal{P}$, is mentioned. In Section 3 (see Theorem 3.2), we improve this bound by replacing $\Delta(G)$ with $d(G)$, the degeneracy of G (defined as the maximum value of the minimum degree of any induced subgraph). Hence, for any fixed $\mathcal{P}$, it depends only on $d(G)$ and hence the bound is very efficiently computable for any G since the other parameter needs to be determined theoretically only once. It is also shown that this bound is tight (that is, there are infinitely many graphs achieving the bound) for several types of properties. We also show that a $(k, \mathcal{P})$-coloring (k stands for the bound) is computable using only $O(d|V|)$ number of $\mathcal{P}$-membership tests. Hence, for any $\mathcal{P}$ admitting an efficient membership testing algorithm, we have a polynomial time algorithm for computing such a $(k, \mathcal{P})$-coloring. Examples of properties admitting polynomial time membership tests are forests, partial k-trees (for fixed k), planar graphs, etc.

1.2 $\mathcal{P}$-choice numbers

We also consider the list versions of hereditary colorings in this abstract. In the list variant, each vertex $u \in V$ is assigned a list L_u of colors. Given a list assignment $\mathcal{L} = \{L_u : u \in V\}$, we say that G is $(\mathcal{L}, \mathcal{P})$-choosable if there exists a coloring $\{f(u)\}_{u \in V}$ such that (i) $f(u) \in L_u$ for each $u \in V$ and (ii) for each color $x \in \cup_{u \in V} L_u$, the subgraph induced by V_x, the set of vertices u such that $f(u) = x$, is a member of $\mathcal{P}$. For $k \geq 1$, G is $(k, \mathcal{P})$-choosable if *for every* list assignment $\{L_u : u \in V\}$ of lists of colors such that $|L_u| \geq k$ for all $u \in V$, G is $(\mathcal{L}, \mathcal{P})$-choosable. The $\mathcal{P}$-choice number of G, denoted by $ch_{\mathcal{P}}(G)$, is the minimum value of k such that G is $(k, \mathcal{P})$-choosable. When $\mathcal{P}$ is the set of independent sets, the $(k, \mathcal{P})$-choosability specializes to the k-choosability notion and the $\mathcal{P}$-choice number $ch_{\mathcal{P}}(G)$ specializes to the standard choice number $ch(G)$.

In Section 4, we present (see Theorem 4.1) and prove some useful observations about the numbers $ch_{\mathcal{P}}(G)$. These can be thought of as list analogues of similar observations about $\chi_{\mathcal{P}}(G)$. While special types of hereditary colorings have been studied and reported in the literature under various names, we are not aware of any systematic research work on the list versions of hereditary colorings reported in the literature. The work reported in this abstract seems to be one of the first attempts at studying these notions.

Note that $ch_{\mathcal{P}}(G) \geq \chi_{\mathcal{P}}(G)$ for any graph G since $(k, \mathcal{P})$-choosability specializes to $(k, \mathcal{P})$-colorability when $L_u = L_v$ for each pair u, v of vertices. Also, the difference $ch_{\mathcal{P}}(G) - \chi_{\mathcal{P}}(G)$ is not bounded and can be arbitrarily large as shown in [9].

The notion of k-choosability was introduced by Vizing in [42] and also independently by Erdős, Rubin and Taylor in [17]. The notion of choice number has received a lot of attention recently and interesting results have been obtained relating $ch(G)$ with various other graph invariants, see [5,4,3,32,33,24,34,7,27,37]. For some recent work on the admissibility of a proper coloring when the lists are randomly and independently chosen for each vertex, see [28], [29] and [38].

It is also well-known that $ch(G)$ cannot be bounded from above by any function $f(\chi(G))$ since $ch(K_{n,n}) = \Theta(\log n)$ whereas $\chi(K_{n,n}) = 2$. However, by extending the work of Alon in [3], the present author obtained the following upper bound in [37]:

Theorem 1.1 ([37]). *There exists a constant $c > 0$ such that for every graph G,*

$$ch(G) \leq c\chi\left(\ln\left(\frac{n}{\chi}\right) + 1\right)$$

where $\chi = \chi(G)$ denotes the chromatic number of G.

This bound is tight up to a constant factor for every n, χ and is achieved by any complete χ-partite graph on n vertices where the size of each part is either $\lfloor n/\chi \rfloor$ or $\lceil n/\chi \rceil$ as explained in [37]. In other words, $f(n, \chi) = \Theta\big(\chi\big(\ln\big(\frac{n}{\chi}\big) + 1\big)\big)$ where $f(n, \chi) = \max\{ch(G) : \chi(G) = \chi,\ |V(G)| = n\}$.

This leads to the natural problem of determining a tight upper bound on $ch_{\mathcal{P}}(G)$. In Section 4, we show that the degeneracy based upper bound on $\chi_{\mathcal{P}}(G)$ (obtained in Section 3) is also a constructive upper bound for $ch_{\mathcal{P}}(G)$. But this bound can be far away from $ch_{\mathcal{P}}(G)$ and this is illustrated later. It leads to the natural question of whether a relationship (analogous to the one in Theorem 1.1) exists between $ch_{\mathcal{P}}(G)$ and $\chi_{\mathcal{P}}(G)$ where $\mathcal{P}$ can be any arbitrary hereditary property. Surprisingly, there is indeed such a relationship and the proof arguments of [37] can be adapted to work for any hereditary $\mathcal{P}$. This leads us to the following result.

Theorem 1.2. *There exists a constant $c > 0$ such that for every hereditary property $\mathcal{P}$ and for every graph G with $n = |V(G)|$,*

$$ch_{\mathcal{P}}(G) \leq c\chi_{\mathcal{P}}\left(\ln\left(\frac{n}{\chi_{\mathcal{P}}}\right)+1\right)$$

where $\chi_{\mathcal{P}} = \chi_{\mathcal{P}}(G)$ denotes the $\mathcal{P}$-chromatic number of G. Also, given lists $\mathcal{L} = \{L_u\}_u$ of this size and also an optimal $\mathcal{P}$-coloring of G, one can efficiently find a $\mathcal{P}$-coloring permitted by $\mathcal{L}$ in polynomial time.

In Section 5, we present a proof of the above theorem. A careful examination of the proof of Theorem 1.2 reveals that this proof is in fact applicable to a more general scenario, namely, the *choice numbers of k-uniform hypergraphs*. Further details on this generalization will appear in the full version [39] of this paper. Finally, in Section 6, we conclude with some remarks.

2. Hereditary colorings

Note that by allowing $\mathcal{P}$ to be different hereditary properties, we specialize to other chromatic parameters studied in the literature. For example, when $\mathcal{P}$ is the set of labeled complete graphs, $\chi_{\mathcal{P}}(G)$ is the same as the clique covering number of G. Also, this is the same as $\chi(\bar{G})$ where $\bar{G}$ is the complement of the graph G.

When $\mathcal{P}$ is the set of all labeled forests, $\chi_{\mathcal{P}}(G)$ is the same as the vertex arboricity of G. When $\mathcal{P}$ is the set of all labeled linear forests (a linear forest is a vertex disjoint collection of paths on one or more vertices), $\chi_{\mathcal{P}}(G)$ is the same as the vertex linear arboricity of G. When $\mathcal{P}$ is the set of all star forests (a star forest is a vertex disjoint collection of stars), $\chi_{\mathcal{P}}(G)$ is the same as the vertex star arboricity of G.

The following observations are easy to establish for the hereditary properties.

Theorem 2.1. *Let $\mathcal{P}$ be any hereditary property. Then, for any two graphs G and H, we have*

(0) *If $G \in \mathcal{P}$, then $\chi_{\mathcal{P}}(G) = 1$.*
(1) *G always has a $\mathcal{P}$-coloring using at most $|V(G)|$ colors, namely the trivial coloring which colors each vertex in $V(G)$ with a distinct color. That is, $\chi_{\mathcal{P}}(G) \leq |V(G)|$.*
(2) *$\chi_{\mathcal{P}}(G \backslash S) \leq \chi_{\mathcal{P}}(G)$ for any $S \subseteq V(G)$.*
(3) *If H is obtained from G by adding a new vertex (with some incident edges) to G, then $\chi_{\mathcal{P}}(H) \leq \chi_{\mathcal{P}}(G) + 1$.*

(4) *Suppose $\mathcal{P}$ is such that: For any two $G, H \in \mathcal{P}$ with $V(G) \cap V(H) = \emptyset$, the graph K obtained by taking the disjoint union of G and H with no edges (in K) between G and H, is also in $\mathcal{P}$. Then, for any graph G having k connected components $G_1, \ldots, G_k$, we have $\chi_{\mathcal{P}}(G) = \max_i \chi_{\mathcal{P}}(G_i)$.*

(5) *Suppose $\mathcal{P}$ is such that: For any two $G, H \in \mathcal{P}$ with $V(G) \cap V(H) = \emptyset$, the graph K obtained by taking the disjoint union of G and H with all edges (in K) joining $V(G)$ and $V(H)$, is also in $\mathcal{P}$. Then, for any graph G such that $\bar{G}$ has k connected components $\overline{G_1}, \ldots, \overline{G_k}$, we have $\chi_{\mathcal{P}}(G) = \max_i \chi_{\mathcal{P}}(G_i)$.*

(6) *Suppose $\mathcal{P}_1$ and $\mathcal{P}_2$ are two hereditary properties such that $\mathcal{P}_1 \subseteq \mathcal{P}_2$. That is, any G satisfies $\mathcal{P}_2$ whenever it satisfies $\mathcal{P}_1$. Then, for any graph G, $\chi_{\mathcal{P}_2}(G) \leq \chi_{\mathcal{P}_1}(G)$.*

(7) *Suppose $\mathcal{P}$ is such that: (i) For each $n \geq 1$, there is a $G \in \mathcal{P}$ with $n = |V(G)|$ and (ii) $G - e \in \mathcal{P}$ whenever $G \in \mathcal{P}$ and $e \in E(G)$. Then, $\chi_{\mathcal{P}}(G) \leq \chi(G)$ where $\chi(G)$ denotes the standard chromatic number of G.*

(8) *Suppose $\mathcal{P}$ is such that: (i) For each $n \geq 1$, there is a $G \in \mathcal{P}$ with $n = |V(G)|$ and (ii) $G + e \in \mathcal{P}$ whenever $G \in \mathcal{P}$ and $e \notin E(G)$. Then, $\chi_{\mathcal{P}}(G) \leq \chi(\bar{G})$ where $\chi(\bar{G})$ denotes the standard chromatic number of $\bar{G}$.*

3. An upper bound on $\chi_{\mathcal{P}}(G)$

It is well-known [10,15] that hereditary properties are characterized by forbidden induced subgraphs. More formally, given a collection of graphs M which is closed under isomorphism, let us denote by $Free(M)$ the class of graphs each containing no induced subgraphs isomorphic to graphs in M. In [15], the following theorem is proved.

Theorem 3.1 ([15]). *An isomorphism closed class $\mathcal{P}$ of graphs is hereditary if and only if $\mathcal{P} = Free(M)$ for a set M. Moreover, the minimal set M with this property is unique. Also, M is precisely the set of all minimal ($H \notin \mathcal{P}$ but $H - v \in \mathcal{P}$ for every $v \in V(H)$) graphs which are not in $\mathcal{P}$.*

Thus, every hereditary property $\mathcal{P}$ is characterized by a unique class of minimal forbidden subgraphs M, that is, $\mathcal{P} = Free(M)$. For example, independent sets are characterized by the absence of a K_2, cliques are characterized by the absence of an induced $\overline{K_2}$, forests are characterized by the absence of induced cycles, bipartite graphs are characterized by the absence of induced odd cycles, chordal graphs are characterized by the absence of induced cycles of length 4 or more, perfect graphs are characterized by the absence of induced odd cycles and induced complements of odd cycles of length 5 or more, and so on.

A graph $G = (V, E)$ is k-degenerate ($k \geq 0$) if there exists a linear ordering $(v_1, \ldots, v_n)$ of $V(G)$ such that for each $i \leq n$, v_i has at most k neighbors in $\{v_{i+1}, \ldots, v_n\}$. We call such a linear ordering a k-degenerate order of G. Such graphs have been studied in [30]. The degeneracy of a graph, denoted by $d(G)$, is the minimum $k \geq 0$ for which G is k-degenerate. It is well-known [40] that $d(G)$ is the maximum value of $\delta(H)$ where the maximum is taken over all induced subgraphs H of G and $\delta(H)$ is the minimum degree of any vertex in H. It is also well-known that a linear ordering which certifies that G is $d(G)$-degenerate can be found in $O(|V| + |E|)$ time for any graph G.

The degeneracy is an important invariant of graphs and it is easy to verify that for any graph G, we always have $\delta(G) \le d(G) \le \Delta(G)$ where $\delta(G)$ and $\Delta(G)$ denote respectively the minimum and maximum degree of any vertex in G. For regular graphs (where every vertex has the same common degree), we have $\delta(G) = d(G) = \Delta(G)$.

A simple greedy coloring using a $d(G)$-degenerate ordering of G shows that any graph can be colored properly using at most $d(G) + 1$ colors. This leads to the following well-known observation.

Fact 3.1. [40] For any graph G, $\chi(G) \le d(G) + 1$.

However, it is also well-known that $d(G) + 1$ could be significantly larger than $\chi(G)$. For example, $\chi(K_{n,n}) = 2$ but $d(K_{n,n}) = n$. Similarly, $d(G)$ could be significantly smaller than $\Delta(G)$. For example, for any tree T, $d(T) = 1$ but $\Delta(T)$ could be as large as $n-1$. Also, planar graphs are 5-degenerate but can have Δ as large as $n-1$. For such graphs, $d(G)+1$ is a better bound on $\chi(G)$ compared to $\Delta(G)+1$. In addition, this bound is computable in linear time. Polynomial time computable bounds are important since computing $\chi(G)$ exactly is a notoriously hard computational problem [19].

Surprisingly, a similar bound based on degeneracy can be obtained for $\chi_{\mathcal{P}}(G)$ of any graph G provided every minimal forbidden graph for $\mathcal{P}$ has no isolated vertex. We present it below. In [9], a weaker bound (using $\Delta(G)$ instead of $d(G)$) was obtained. The improvement is due to using a $d(G)$-degenerate ordering of $V(G)$ instead of an arbitrary order of V. In the final version of the present abstract, we present bounds which are better than the following one.

For a class M of graphs, let $\delta(M)$ be defined by $\delta(M) = \min\{\delta(H) : H \in M\}$.

Theorem 3.2. *Let $\mathcal{P} = Free(M)$ be any hereditary class of graphs where M is the set of minimal forbidden graphs of $\mathcal{P}$. Suppose $\delta(M) > 0$. Then, for any graph G with degeneracy $d = d(G)$, we have*

$$\chi_{\mathcal{P}}(G) \le \left\lfloor \frac{d}{\delta(M)} \right\rfloor + 1.$$

Moreover, a $(k, \mathcal{P})$-coloring of G can be computed in polynomial time if there is a polynomial time algorithm for testing membership in $\mathcal{P}$. Here k refers to the upper bound specified above.

Proof. Let $(v_1, \ldots, v_n)$ be a d-degenerate ordering of G. For each i, let V_i denote the set $\{v_i, \ldots, v_n\}$ and G_i denote the subgraph of G induced by V_i. We color vertices inductively in the reverse order $(v_n, \ldots, v_1)$. Since $\mathcal{P}$ is hereditary, every trivial graph on one vertex is always a member of $\mathcal{P}$. Hence, we color v_n with color 1.

Suppose we have obtained a $(k, \mathcal{P})$-coloring $(C_1, \ldots, C_k)$ of G_{i+1} where k is defined before. Let $\{u_1, \ldots, u_l\}$, $l \le d$, be the neighbors of v_i in V_{i+1}. We claim that there are at most $k-1$ colors j such that $G[C_j \cup \{v_i\}]$ is not in $\mathcal{P}$. Suppose the claim is true. Then, there exists some color $j \in \{1, \ldots, k\}$ such that $G[C_j \cup \{v_i\}]$ is in $\mathcal{P}$.

Thus, we extend a $(k, \mathcal{P})$-coloring of G_{i+1} to a $(k, \mathcal{P})$-coloring of G_i. Iterating this, we obtain a $(k, \mathcal{P})$-coloring of G, thereby proving the upper bound specified above.

To complete the arguments, we only need to prove the claim mentioned before. Suppose, for each of the k color classes C_j, the subgraph $G[C_j \cup \{v_i\}]$ is not in $\mathcal{P}$. This means that there are k minimal forbidden subgraphs $H_1, \ldots, H_k$ such that

(i) $v_i \in V(H_j)$ for each j,
(ii) H_j is an induced subgraph of $G[C_j \cup \{v_i\}]$ for each j,
(iii) for each j, $H_j \notin \mathcal{P}$ but $H_j - u \in \mathcal{P}$ for each $u \in V(H_j)$.

Since $\mathcal{P} = Free(M)$, each of the subgraphs H_j should be isomorphic to some member of M. Hence, from the definition of $\delta(M)$, it follows that $deg_{H_j}(v_i) \geq \delta(H_j) \geq \delta(M)$ for each j. Since the subgraphs H_j are vertex disjoint except for v_i and each H_j is an induced subgraph of G_{i+1}, we have that

$$k\delta(M) \leq \sum_j deg_{H_j}(v_i) \leq d$$

and hence

$$k = \left\lfloor \frac{d}{\delta(M)} \right\rfloor + 1 \leq d/\delta(M)$$

which is impossible. This contradiction establishes the claim.

A $d(G)$-degenerate ordering of $V(G)$ can be computed in linear time. For each v_i, coloring of v_i only involves k membership tests in $\mathcal{P}$ resulting in a total of nk $\mathcal{P}$-membership tests. Hence, if there is a polynomial time algorithm for $\mathcal{P}$-membership, then we will have a polynomial time algorithm for computing a $(k, \mathcal{P})$-coloring of G. ■

We obtain the following consequences immediately.

Corollary 3.1. *Let G be any graph of degeneracy d. Then,*

(1) $\chi_{\mathcal{P}}(G) \leq \lfloor d/2 \rfloor + 1$ *if $\mathcal{P}$ is the set of all forests.*
(2) $\chi_{\mathcal{P}}(G) \leq \lfloor d/2 \rfloor + 1$ *if $\mathcal{P}$ is the set of all partial 2-trees.*
(3) $\chi_{\mathcal{P}}(G) \leq \lfloor d/(k+1) \rfloor + 1$ *if $\mathcal{P}$ is the set of all k-degenerate graphs. Also, for each $d, k \geq 0$, there exists an infinite family $\{G_l\}_{l \geq 1}$ of d-degenerate graphs achieving this upper bound.*
(4) $\chi_{\mathcal{P}}(G) \leq \lfloor d/(k-1) \rfloor + 1$ *if $\mathcal{P}$ is the set of all K_k-free graphs for some $k \geq 2$. Also, for each $d \geq 0, k \geq 2$, there exists an infinite family of d-degenerate graphs achieving this upper bound.*
(5) $\chi_{\mathcal{P}}(G) \leq \lfloor d/2 \rfloor + 1$ *if $\mathcal{P}$ is the set of all planar graphs.*
(6) $\chi_{\mathcal{P}}(G) \leq \lfloor d/2 \rfloor + 1$ *if $\mathcal{P}$ is the set of all perfect graphs.*

Proof.

1. When $\mathcal{P}$ is the set of all forests, $\mathcal{P} = Free(M)$ where M is the collection of all cycles and hence $\delta(M) = 2$.

2. When $\mathcal{P}$ is the set of all partial 2-trees, $\mathcal{P} = Free(M)$ where M is the collection of all minimal graphs which are not partial 2-trees. Any such minimal forbidden subgraph H should have $\delta(H) \geq 2$ obviously. Also, each subdivision of K_4 is a minimal forbidden subgraph. Hence $\delta(M) = 2$.
3. When $\mathcal{P}$ is the set of all k-degenerate graphs, $\mathcal{P} = Free(M)$ where M is the collection of all minimal graphs of degeneracy at least $k+1$. Note that any such graph H should have $\delta(H) \geq k+1$. Also, K_{k+1} is a minimal forbidden graph. Hence $\delta(M) = k+1$.
   For each $d, k \geq 0$, the upper bound of (3) is tight for the complete graph K_{d+1}. This is because the degeneracy of K_{d+1} is d and in any $(l, \mathcal{P})$-coloring of K_{d+1}, each color class cannot have more than $k+1$ vertices and hence l is at least $(d+1)/(k+1) \geq \lfloor d/(k+1) \rfloor + 1$. In fact, there exists an infinite family $\mathcal{C} = \{G_l : l \geq 1\}$ of graphs such that for each $G_l \in \mathcal{C}$, the degeneracy of G_l is d and $\chi_{\mathcal{P}}(G_l)$ is exactly the upper bound mentioned in (3). The graphs G_l are obtained by taking l vertex disjoint copies of K_{d+1} and joining these l copies in a path like fashion by adding an edge between two consecutive copies. It can be verified that the degeneracy of each G_l is d.
4. In this case, $\mathcal{P} = Free(M)$ where $M = \{K_k\}$ and $\delta(M) = k-1$. For the tightness, note that degeneracy of K_{d+1} is d and $\chi_{\mathcal{P}}(G) \geq (d+1)/(k-1) \geq \lfloor d/(k-1) \rfloor + 1$. The infinite family achieving the tightness is the same as the family $\mathcal{C} = \{G_l\}_{l \geq 1}$ described before.
5. When $\mathcal{P}$ is the set of all planar graphs, $\mathcal{P} = Free(M)$ where M is the collection of all minimal non-planar graphs H. For any such H, $\delta(H) \geq 2$ since otherwise $H - u$ is non-planar for some $u \in V(H)$. Also, each non-trivial subdivision of K_5 and $K_{3,3}$ is a minimal non-planar graph with minimum degree 2. Hence $\delta(M) = 2$.
6. When $\mathcal{P}$ is the set of all perfect graphs, $\mathcal{P} = Free(M)$ where M is the collection of all minimal non-perfect graphs, namely, odd cycles C_k or complements of odd cycles C_k^c for some odd $k \geq 5$. Obviously, $\delta(H) \geq 2$ for any such $H \in M$. Also, each odd cycle is regular of degree 2. Hence $\delta(M) = 2$. ■

Since planar graphs are 5-degenerate, the upper bound of the following result from [22] now follows from the previous corollary.

Corollary 3.2 ([22]). *Let G be any planar graph. Then, $\chi_{\mathcal{P}}(G) \leq 3$ if $\mathcal{P}$ is the set of all forests. Also, $\chi_{\mathcal{P}}(G) = 3$ if and only if G has a cycle and its planar dual G^* does not contain a connected Eulerian spanning subgraph.*

However, if we define $\mathcal{P}$ to be the set of all graphs having no induced cycles of length 4 or more, then it is known [41] that $\chi_{\mathcal{P}}(G) \leq 2$ for each planar graph G.

Tightness. The tightness of the above-given upper bounds vary widely depending on restrictions on G and $\mathcal{P}$. For example, the bound given in (3) is exact if G is restricted to be a k-degenerate graph since for such graphs, we require only one color. On the other hand, the bound given in (4) is 3 if G is restricted to be planar but only one color is required for planar graphs. This contrasts with the tightness of the bound obtained in (3).

Remarks. Note that for each fixed hereditary property $\mathcal{P}$ for which the corresponding $\delta(M)$ is known, the bound of Theorem 3.2 is a linear $O(|V| + |E|)$ time computable bound. Also, such a $(k, \mathcal{P})$-coloring achieving the chromatic bound of Theorem 3.2 can also be constructed using only a $O(d(G)|V|)$ number of $\mathcal{P}$-membership tests.

4. List hereditary colorings

Each of the observations on $\chi_{\mathcal{P}}(G)$ mentioned in Theorem 2.1 has an analogue with respect to its list variant $ch_{\mathcal{P}}(G)$. The following theorem specifies all of them.

Theorem 4.1. *Let $\mathcal{P}$ be any hereditary property. Then, for any two graphs G and H, we have*

(0) *If $G \in \mathcal{P}$, then $ch_{\mathcal{P}}(G) = 1$.*
(1) *G always has a $\mathcal{P}$-coloring using at most $|V(G)|$ colors, namely the coloring which colors each vertex in $V(G)$ with a distinct color. That is, $ch_{\mathcal{P}}(G) \leq |V(G)|$.*
(2) *$ch_{\mathcal{P}}(G \backslash S) \leq ch_{\mathcal{P}}(G)$ for any $S \subseteq V(G)$.*
(3) *If H is obtained from G by adding a new vertex (with some incident edges) to G, then $ch_{\mathcal{P}}(H) \leq ch_{\mathcal{P}}(G) + 1$.*
(4) *Suppose $\mathcal{P}$ is such that: For all $G, H \in \mathcal{P}$ with $V(G) \cap V(H) = \emptyset$, the graph K obtained by taking the disjoint union of G and H with no edges (in K) between G and H, is also in $\mathcal{P}$. Then, for any graph G having k connected components $G_1, \ldots, G_k$, we have $ch_{\mathcal{P}}(G) = \max_i ch_{\mathcal{P}}(G_i)$.*
(5) *Suppose $\mathcal{P}$ is such that: For all $G, H \in \mathcal{P}$ with $V(G) \cap V(H) = \emptyset$, the graph K obtained by taking the disjoint union of G and H with all edges (in K) joining $V(G)$ and $V(H)$, is also in $\mathcal{P}$. Then, for any graph G such that $\bar{G}$ has k connected components $\overline{G_1}, \ldots, \overline{G_k}$, we have $ch_{\mathcal{P}}(G) = \max_i ch_{\mathcal{P}}(G_i)$.*
(6) *Suppose $\mathcal{P}_1$ and $\mathcal{P}_2$ are two hereditary properties such that $\mathcal{P}_1 \subseteq \mathcal{P}_2$. That is, any G satisfies $\mathcal{P}_2$ whenever it satisfies $\mathcal{P}_1$. Then, for any graph G, $ch_{\mathcal{P}_2}(G) \leq ch_{\mathcal{P}_1}(G)$.*
(7) *Suppose $\mathcal{P}$ is such that: (i) For each $n \geq 1$, there is a $G \in \mathcal{P}$ with $n = |V(G)|$ and (ii) $G - e \in \mathcal{P}$ whenever $G \in \mathcal{P}$ and $e \in E(G)$. Then, $ch_{\mathcal{P}}(G) \leq ch(G)$ where $ch(G)$ denotes the standard choice number of G.*
(8) *Suppose $\mathcal{P}$ is such that: (i) For each $n \geq 1$, there is a $G \in \mathcal{P}$ with $n = |V(G)|$ and (ii) $G + e \in \mathcal{P}$ whenever $G \in \mathcal{P}$ and $e \notin E(G)$. Then, $ch_{\mathcal{P}}(G) \leq ch(\bar{G})$ where $ch(\bar{G})$ denotes the standard choice number of $\bar{G}$.*

Proof. We provide below a proof for each of these eight observations. Suppose $\mathcal{L} = \{L_u : u \in V(G)\}$ be any arbitrary assignment of lists of colors to vertices. Note that we can assume, without loss of generality, that $|L_u| = |L_v|$ for each pair of vertices u and v. In a $\mathcal{P}$-coloring admitted by $\mathcal{L}$ (if there is any), for any color x, let V_x denote the set of vertices assigned the color x.

(0) Suppose $G \in \mathcal{P}$ and each L_u is of size exactly one. Each u is then forced to be colored with the unique color in its list. Clearly, for any color x, the subgraph

induced by all vertices colored x is a member of $\mathcal{P}$ and we get $\mathcal{P}$-coloring of G. This shows that $ch_{\mathcal{P}}(G) = 1$.

(1) Suppose each L_u is of size n where $n = |V(G)|$. Fix an ordering $\sigma = \langle u_1, \ldots, u_n \rangle$ of V. Color the vertices inductively in the order of σ such that each vertex is given a color which is distinct from the colors used so far. When we come to color u_{i+1}, at most i colors from $L_{u_{i+1}}$ have been used for the vertices in $\{u_1, \ldots, u_i\}$ and hence there is always a color in $L_{u_{i+1}}$ with which we can color u_{i+1}. Now, each non-empty color class is a singleton set and hence is a member of $\mathcal{P}$. This proves that $ch_{\mathcal{P}}(G) \leq |V(G)|$.

(2) Since each L_u is of size $ch_{\mathcal{P}}(G)$, we see that $\mathcal{L}$ admits a $\mathcal{P}$-coloring of G. For each color class V_x, the subgraph induced by $V_x \cap (V(G) \backslash S)$ is also a member of $\mathcal{P}$. Thus, $\mathcal{L}$ (restricted to $G \backslash S$) also admits a $\mathcal{P}$-coloring and hence $ch_{\mathcal{P}}(G \backslash S) \leq ch_{\mathcal{P}}(G)$.

(3) Suppose H is obtained by adding a vertex $v \notin V(G)$. Given a list assignment $\mathcal{L}' = \{L_u : u \in V(H)\}$ such that each L_u is of size $ch_{\mathcal{P}}(G) + 1$, we choose an arbitrary color $x \in L_v$ and assign it to v and remove this color from every other list L_u (if it is there). Each truncated list has size at least $ch_{\mathcal{P}}(G)$ and hence admits a $\mathcal{P}$-coloring of G. This coloring together with the singleton class $\{v\}$ yields a $\mathcal{P}$-coloring of H. This shows that $ch_{\mathcal{P}}(H) \leq ch_{\mathcal{P}}(G) + 1$.

(4) Let m denote the quantity $\max_i ch_{\mathcal{P}}(G_i)$. From statement (2), it follows that $ch_{\mathcal{P}}(G_i) \leq ch(G)$ for each i and hence $m \leq ch_{\mathcal{P}}(G)$. To prove the converse, suppose that each L_u is of size m. Then, for each $i \leq k$ and each $u \in V(G_i)$, we have $|L_u| = m \geq ch_{\mathcal{P}}(G_i)$ and hence $\mathcal{L}$ admits a $\mathcal{P}$-coloring for each G_i. Now, for each color x, if V_x^i denotes the set of vertices in G_i colored x, then the subgraph induced by $V_x = \cup_i V_x^i$ is the vertex disjoint union of the k induced subgraphs $G_i[V_x^i]$ each of which is a member of $\mathcal{P}$. Hence, by the nature of $\mathcal{P}$, $G[V_x]$ is also a member of $\mathcal{P}$ for each color x. This proves that $ch_{\mathcal{P}}(G) \leq m$ and hence $ch_{\mathcal{P}}(G) = \max_i ch_{\mathcal{P}}(G_i)$.

(5) This can be proved using arguments similar to those employed in the proof of (4) given above.

(6) If each L_u is of size $ch_{\mathcal{P}_1}(G)$, then $\mathcal{L}$ admits a $\mathcal{P}_1$-coloring of G in which, for every color x, the induced subgraph $G[V_x]$ is a member of $\mathcal{P}_1$ and hence is also a member of $\mathcal{P}_2$. This establishes the inequality.

(7) Let $\mathcal{Q}$ denote the collection of all empty graphs. It follows from (i) that $\mathcal{P}$ contains an empty graph (that is, an independent set) on n vertices for every $n \geq 1$. That is, $\mathcal{Q} \subseteq \mathcal{P}$. By applying (6) to $\mathcal{Q}$ and $\mathcal{P}$, it follows that $ch_{\mathcal{P}}(G) \leq ch(G)$ since $ch(G) = ch_{\mathcal{Q}}(G)$.

(8) Let $\mathcal{Q}$ denote the collection of all complete graphs. It follows from (i) that $\mathcal{P}$ contains a complete graph on n vertices for every $n \geq 1$. That is, $\mathcal{Q} \subseteq \mathcal{P}$. By applying (6) to $\mathcal{Q}$ and $\mathcal{P}$, it follows that $ch_{\mathcal{P}}(G) \leq ch(\bar{G})$ since $ch(\bar{G}) = ch_{\mathcal{Q}}(G)$.

5. Bounds for the $\mathcal{P}$-choice number

The argument used to prove Theorem 3.2 essentially verifies the following theorem as well.

Theorem 5.1. *Let $\mathcal{P} = Free(M)$ be any hereditary class of graphs where M is the set of minimal forbidden graphs of $\mathcal{P}$. Suppose $\delta(M) > 0$. Then, for any graph G with degeneracy $d = d(G)$, we have*

$$ch_{\mathcal{P}}(G) \leq \left\lfloor \frac{d}{\delta(M)} \right\rfloor + 1.$$

Moreover, given a list collection $\mathcal{L} = \{L_u : u \in V\}$ such that $|L_u| \geq k$ for all u, a $\mathcal{P}$-coloring of G admitted by $\mathcal{L}$ can be computed in polynomial time if there is a polynomial time algorithm for testing membership in $\mathcal{P}$. Here k refers to the upper bound specified above.

Thus, we have a constructive bound in terms of degeneracy of the given graph. But this bound could be arbitrarily large compared to the actual value of $ch_{\mathcal{P}}(G)$ in some cases. For example, when $\mathcal{P}$ is any fixed hereditary property that contains all independent sets, then by Theorem 4.1:(7), we have $\chi_{\mathcal{P}}(K_{n,n}) \leq 2$, $ch_{\mathcal{P}}(K_{n,n}) \leq ch(K_{n,n}) = \Theta(\ln n)$, but $d(K_{n,n}) = n$.

But, Theorem 1.2 presents an upper bound for $ch_{\mathcal{P}}(G)$ in terms of the corresponding number $\chi_{\mathcal{P}}(G)$. This shows that $ch_{\mathcal{P}}(G)$ can be larger than $\chi_{\mathcal{P}}(G)$ by at most a multiplicative factor of $(\ln n)$ even if $d(G)$ grows linearly with the size of the graph. In this section, we prove this theorem.

Proof (of Theorem 1.2). Actually, we prove the more general statement that $ch_{\mathcal{P}}(G) \leq ck(\ln(n/k) + 1)$ for each $k \geq \chi_{\mathcal{P}}(G)$. The validity of the case $k = \chi_{\mathcal{P}}$ then follows as a special case and leads to the smallest upper bound since $k(\ln(n/k) + 1)$ is an increasing function for $k < n$.

The proof of this theorem is based on probabilistic arguments. In fact, a careful study of the proof of Theorem 1.1 described in [37] shows that it can be adapted to work for the more general $ch_{\mathcal{P}}(G)$. Hence we proceed by only providing an outline of the major steps of that proof adapted to our new context of list $\mathcal{P}$-coloring, without (due to lack of space) elaborating on further details. Full details will appear in the journal version. This establishes the proof of Theorem 1.2.

Outline of the Proof. We outline the major steps in the proof.

- S1: It suffices to prove the theorem for all values of k which are powers of 2. Henceforth, k is a power of 2.
- S2: The proof is by induction on the value of k.
- S3: The basis of induction is any value of k such that either $k \geq n/e^2$ or $k \leq n^{0.9}$. For such cases, one can prove the bound directly using either deterministic arguments (for $k \geq n/e^2$) or probabilistic arguments (for $k \leq n^{0.9}$). Henceforth, we assume that $n^{0.9} \leq k \leq n/e^2$.
- S4: Assume that we are given a $(k, \mathcal{P})$-coloring $\langle V_1, \ldots, V_k \rangle$ of G. The problem is now to prove the existence of a $\mathcal{P}$-coloring admitted by $\mathcal{L}$. We reduce this problem to two subproblems of list $\mathcal{P}$-coloring two vertex disjoint induced subgraphs G_1 and G_2 such that (i) each subgraph is induced by the union of $k/2$ color classes and hence has a $(k/2, \mathcal{P})$-coloring, (ii) The list of colors for each vertex

in the two subgraphs has been truncated randomly to a smaller list. The random truncation is based on randomly bipartitioning the universe $\mathcal{U}$ of all colors in all lists into two subsets $(\mathcal{U}_1, \mathcal{U}_2)$. It is explained below.

S4a: For each $u \in V(G)$, the associated list L_u is truncated to $L'_u = L_u \cap \mathcal{U}_i$ where $i = 1$ if $u \in V(G_1)$ and $i = 2$ if $u \in V(G_2)$. We apply the inductive hypothesis (since they both admit a $(k/2, \mathcal{P})$-coloring each) to both G_1 and G_2 and deduce that both have a $\mathcal{P}$-coloring admitted by their respective L'_u lists. Note that $L'_u \cap L'_v = \emptyset$ for any $u \in V(G_1)$ and any $v \in V(G_2)$. Hence, any $\mathcal{P}$-coloring of G_1 admitted by its truncated lists can be combined with any $\mathcal{P}$-coloring of G_2 admitted by its truncated lists to get a $\mathcal{P}$-coloring of G admitted by $\mathcal{L}$.

S4b: To apply the inductive hypothesis to G_1 and G_2, we need to show that each randomly truncated list L'_u is of sufficient size (required by the hypothesis). Let n_1 (respectively n_2) denote the size $|V(G_1)|$ (resp., $|V(G_2)|$). More precisely, we need to show that the following holds with positive probability:

$$\forall\, j \in \{1, 2\},\ \forall\, u \in V(G_j),\quad |L'_u| \geq c\left(\frac{k}{2}\right)\left(\ln\left(\frac{2n_j}{k}\right) + 1\right). \qquad \text{(A)}$$

In fact, using known upper bounds on the tail probabilities of Bernoulli trials, it will be shown that (A) holds with probability more than 0.5. Thus, we complete the inductive proof.

S5: How do we determine the pair of $k/2$ color classes each, whose induced subgraphs are G_1 and G_2? These are determined as follows. We first look at the difference $D = n_2 - n_1$ where n_1 (resp., n_2) is the total size of the smallest (resp., the largest) $k/2$ color classes. If D is NOT "sufficiently small", say, $D \geq n^{0.8}$, then G_1 (resp., G_2) is defined to be the subgraph induced by the $k/2$ smallest (resp., largest) color classes. Otherwise (that is, $D < n^{0.8}$), it can be shown that there are $k/2$ color classes all of which are of the same size which is either $\lfloor n/k \rfloor$ or $\lceil n/k \rceil$. In this case, we define G_1 to be the subgraph induced by these $k/2$ equal sized classes and G_2 to be the subgraph induced by the remaining $k/2$ classes. In this case, it can be shown that $(n-k)/2 \leq |V(G_1)|, |V(G_2)| \leq (n+k)/2$. How do we randomly split the universe $\mathcal{U}$ of all colors into two disjoint parts?.

S6: When $D \geq n^{0.8}$, we put each color $x \in \mathcal{U}$ independently into $\mathcal{U}_1$ with probability p and into $\mathcal{U}_2$ with probability $1 - p$. p is defined by

$$p = 0.5\left(\frac{\ln\left(\frac{2n_1}{k}\right) + 1}{\ln\left(\frac{n}{k}\right) + 1}\right) + 3\sqrt{\frac{\ln k}{ck(\ln(n/k) + 1)}}$$

where n_1 (n_2) is the total size of all $k/2$ smallest (largest) color classes. For this choice of p, using known bounds on the tail probabilities of Bernoulli trials, it can be shown (see [37] for details) that Statement (A) holds with probability more than 0.5.

S7: When $D < n^{0.8}$, we put each color $x \in \mathcal{U}$ independently into $\mathcal{U}_1$ with probability $1/4$ and into $\mathcal{U}_2$ with probability $3/4$. Note that, in this case, G_1 is a special type of graph possessing a $(k/2, \mathcal{P})$-coloring in which all color classes are of the same size which itself is either $\lfloor n/k \rfloor$ or $\lceil n/k \rceil$. For such special type of graphs, one can show, by generalizing an upper bound of Alon [3] on the choice numbers of such graphs, that the following is true.

Theorem 5.2. *There exists an absolute positive constant c_2 such that: for every hereditary $\mathcal{P}$, every $m \geq 2$ and for every $r \geq 2$, for every graph G on mr vertices having a $(r, \mathcal{P})$-coloring in which each color class has exactly m vertices,*

$$ch_{\mathcal{P}}(G) \leq c_2 r \ln m.$$

Hence, by choosing c sufficiently larger than c_2, we can show using known bounds on Bernoulli trials that the following statement is true, with probability more than 0.5.

$$|L'_u| \geq c_2 \left(\frac{k}{2}\right)\left(\ln\left(\frac{n+k}{k}\right)+1\right) \qquad \text{for all } u \in V(G_1)$$

$$|L'_u| \geq c \left(\frac{k}{2}\right)\left(\ln\left(\frac{n+k}{k}\right)+1\right) \qquad \text{for all } u \in V(G_2)$$

S8: Thus, in both cases of $D \geq n^{0.8}$ and $D < n^{0.8}$, we arrive at two list $\mathcal{P}$-coloring problems on vertex disjoint induced subgraphs each possessing a $(k/2, \mathcal{P})$-coloring and each vertex in these graphs having a truncated list (of colors) of sufficiently large size which enable us to apply the inductive hypothesis and obtain a $\mathcal{P}$-coloring each admitted by the truncated lists in each of the two graphs. This shows that $\mathcal{L}$ admits a $\mathcal{P}$-coloring of G and proves the bound.

Given a $(k, \mathcal{P})$-coloring of G and also a list assignment $\mathcal{L}$, a $\mathcal{P}$-coloring admitted by $\mathcal{L}$ can be found using essentially the two algorithms described in [37]. In fact, an efficient $\mathcal{P}$-membership algorithm is not required for this purpose. ■

6. Conclusions

An immediate question that arises is that of determining whether the upper bound of Theorem 1.2 is tight. In other words, are there graphs for every $\chi_{\mathcal{P}} \leq n$ whose $ch_{\mathcal{P}}$ value equals or is within a constant multiplicative factor of the upper bound? As mentioned in Section 1, the upper bound (on choice numbers) of Theorem 1.1 is tight up to a constant multiplicative factor.

Some issues on the computational complexity of standard and list $\mathcal{P}$-coloring problems are also being looked into. One can also similarly obtain results about the edge coloring analogue of the $\mathcal{P}$-coloring of vertices (usual or list variant) and obtain bounds and algorithms for the corresponding chromatic parameter. These notions and related results will appear in the full version of this abstract.

A careful examination of the proof of Theorem 1.2 reveals that this proof is in fact applicable to a more general scenario, namely, the choice numbers of k-uniform hypergraphs. Further details on this generalization will appear in the full version [39] of this paper.

References

[1] Y. Aifeng and Y. Jinjiang, On the vertex arboricity of planar graphs of diameter two, *Discrete Mathematics*, **307** (2007) 2438–2447.

[2] M. O. Albertson, R. E. Jamison, S. T. Hedetniemi and S. C. Locke, The subchromatic number of a graph, *Discrete Mathematics*, **74** (1989) 33–49.

[3] N. Alon, Choice numbers of graphs; a probabilistic approach, *Combinatorics, Probability and Computing*, **1** (1992) 107–114.

[4] N. Alon, Restricted colorings of graphs in Surveys in Combinatorics 1993, London Math. Soc. Lecture Notes Series 187 (K. Walker, ed.), Cambridge University Press (1993) 1–33.

[5] N. Alon, Degrees and choice numbers, *Random Structures & Algorithms*, **16** (2000) 364–368.

[6] N. Alon and J. Spencer, The Probabilistic Method, John Wiley & Sons (1992).

[7] B. Bollobas and A. J. Harris, List-colourings of graphs, *Graphs and Combinatorics*, **1(2)** (1985) 115–127.

[8] J. I. Brown and D. G. Corneil, Perfect Colorings, *Ars Combinatoria*, **30** (1990) 141–159.

[9] M. Borowiecki, E. Drgas-Burchardt and P. Mihk, Generalized list colouring of graphs, *Discussiones Mathematicae Graph Theory*, **15** (1995) 185–193.

[10] M. Borowiecki, I. Broere, M. Frick, P. Mihok and G. Semanisin, A Survey of Heriditary Properties of Graphs, *Discussiones Mathematicae Graph Theory*, **17(1)** (1997) 5–51.

[11] P. A. Catlin and H-J. Lai, Vertex arboricity and maximum degree, *Discrete Mathematics*, **41** (1995) 37–46.

[12] G. Chartrand, H. V. Kronk and C. E. Wall, The point-arboricity of a graph, *Israel Journal of Mathematics*, **6** (1968) 169–175.

[13] S. A. Choudum, T. Karthick and M. A. Shalu, Perfect coloring and Linearly χ-Bound P_6-Free Graphs, *Journal of Graph Theory*, **54** (2007) 293–206.

[14] E. J. Cockayne, Colour classes for r-graphs, *Canadian Mathematical Bulletin*, **15** (1972) 349–354.

[15] L. S. Chandran, V. V. Lozin and C. R. Subramanian, Graphs of low chordality, *Discrete Mathematics and Theoretical Computer Science*, **7** (2005) 25–36.

[16] P. Erdős and L. Lovász, Problems and results on 3-chromatic hypergraphs and some related questions, in A. Hajnal *et al.* eds., Infinite and Finite Sets, North-Holland, Amsterdam, 609–628.

[17] P. Erdős, A. L. Rubin and H. Taylor, Choosability in graphs, *Proc. West Coast Conf. on Combinatorics, Graph Theory and Computing, Congressus Numerantium XXVI* (1979) 125–157.

[18] J. Folkman, Graphs with monochromatic complete subgraphs in every edge colouring, *SIAM Journal on Applied Mathematics*, **18(1)**, January (1970) 19–24.

[19] U. Feige and J. Kilian, Zero Knowledge and the Chromatic Number, *Proceedings of the 11th Annual IEEE Conference on Computational Complexity*, Philadelphia (1996) 278–287.

[20] J. Gimbel and H. Joseph Straight, Some topics in cochromatic theory, *Graphs and Combinatorics*, **3** (1987) 255–265.

[21] J. Gimbel, D. Kratsch and L. Stewart On cocolourings and cochromatic numbers, *Discrete Applied Mathematics*, **48** (1994) 111–127.

[22] S. L. Hakimi and E. F. Schmeichel, A note on the vertex arboricity of a graph, *SIAM Journal of Discrete Mathematics*, **2(1)** February (1989) 64–67.

[23] F. Harary, Conditional colorability in graphs, *Graphs and Applications* (F. Harary and J. Maybee, Editors), Wiley, New York (1985) 127–136.

[24] P. E. Haxell, A Note on Vertex List Colouring, *Combinatorics, Probability & Computing*, **10(4)** (2001) 345–347.

[25] C. T. Hoang and V. B. Le, P_4-free colorings and P_4-bipartite graphs, *Discrete Mathematics and Theoretical Computer Science*, **4** (2001) 109–122.

[26] J. Kahn, Asymptotically Good List-Colorings, *Journal of Combinatorial Theory, Series A*, **73** 1–59.

[27] J. H. Kim, On Brooks' Theorem for Sparse Graphs, *Combinatorics, Probability & Computing*, **4** (1995) 97–132.

[28] M. Krivelevich and A. Nachmias, Colouring powers of cycles from random lists, *European Journal of Combinatorics*, **25** (2004) 961–968.

[29] M. Krivelevich and A. Nachmias, Colouring complete bipartite graphs from random lists, *Random Structures and Algorithms*, **29** (2006) 436–449.

[30] D. R. Lick and A. T. White, k-degenerate graphs, *Canadian Journal of Mathematics*, **22** (1970) 1082–1096.

[31] M. Molloy and B. Reed, Near-optimal List Colourings, *Random Structures & Algorithms*, **17** 376–402.

[32] K. Ohba, On chromatic-choosable graphs, *Journal of Graph Theory*, **40(2)** 130–135.

[33] B. Reed and B. Sudakov, List coloring when the chromatic number is close to the order of the graph, To appear in Combinatorica.

[34] B. Reed and B. Sudakov, Asymptotically the List Colouring Constants are 1, *Journal of Combinatorial Theory, Series B*, **86** (2002) 27–37.

[35] E. Sampathkumar and C. V. Venkatachalam, Chromatic partitions of a graph, *Discrete Mathematics*, **74** (1989) 227–239.

[36] C. R. Subramanian, Coloring sparse random graphs in polynomial average time, *Proceedings of the 8th Annual European Symposium on Algorithms (ESA'00)*, (2000) 415–426.

[37] C. R. Subramanian, List set coloring: algorithms and bounds, *Combinatorics, Probability & Computing*, **16** (2007) 145–158.

[38] C. R. Subramanian, Coloring graphs using random lists of colors, manuscript (2007).

[39] C. R. Subramanian, List Hereditary Colorings of Graphs and Hypergraphs, manuscript in preparation (2009).

[40] G. Szekers and H. S. Wilf, An inequality for the chromatic number of a graph, *Jounal of Combinatorial Theory*, **4** (1968) 1–3.

[41] C. Thomassen, Every planar graph is 5-choosable, *Jounal of Combinatorial Theory, Ser. B*, **62** (1994) 180–181.

[42] V. G. Vizing, Coloring the vertices of a graph in prescribed colors, Diskret. Analiz. No. 29, Metody Diskret. Anal. v. Teroii Kodov i Shem, **101** (1976) 3–10.

Proc. ICDM 2008, RMS-Lecture Notes Series No. 13, 2010, pp. 207–229.

Matrices in the Theory of Signed Simple Graphs

Thomas Zaslavsky

Department of Mathematical Sciences, Binghamton University (SUNY), Binghamton, NY 13902-6000, U.S.A.
zaslav@math.binghamton.edu

Abstract. I discuss the work of many authors on various matrices used to study signed graphs, concentrating on adjacency and incidence matrices and the closely related topics of Kirchhoff ('Laplacian') matrices, line graphs, and very strong regularity.

Keywords. Signed graph, bidirected graph, line graph, incidence matrix, adjacency matrix, Seidel adjacency matrix, Abelson–Rosenberg adjacency matrix, Kirchhoff matrix, graph Laplacian, very strong regularity, weighting matrix, very strong regularity.

2010 Subject Classification: Primary 05C22, 05C50; Secondary 05C62, 05C76.

Introduction

This article is a survey of the uses of matrices in the theory of simple graphs with signed edges. A great many authors have contributed ideas and results to this field; but amongst them all I have felt to be exceptionally inspiring and important the relevant works of J. J. Seidel and G. R. Vijayakumar.

A *signed simple graph* is a graph, without loops or parallel edges, in which every edge has been declared positive or negative. Such a signed graph is illustrated in Figure 1(a). For many purposes the most significant thing about a signed graph is not the actual edge signs, but the sign of each circle (or 'cycle' or 'circuit'), which is the product of the signs of its edges. This fact is manifested in simple operations on the matrices I will present.

I treat three kinds of matrix of a signed graph, all of them direct generalisations of familiar matrices from ordinary, unsigned graph theory.

The first is the adjacency matrix. The adjacency matrix of an ordinary graph has 1 for adjacent vertices; that of a signed graph has $+1$ or -1, depending on the sign of the connecting edge. The adjacency matrix leads to questions about eigenvalues and strong regularity.

The second matrix is the vertex-edge incidence matrix. There are two kinds of incidence matrix of an unsigned graph. The unoriented incidence matrix has two 1's in each column, corresponding to the endpoints of the edge whose column it is. The oriented incidence matrix has a $+1$ and a -1 in each column. For a signed graph, there are both kinds of column, the former corresponding to a negative edge and the latter to a positive edge.

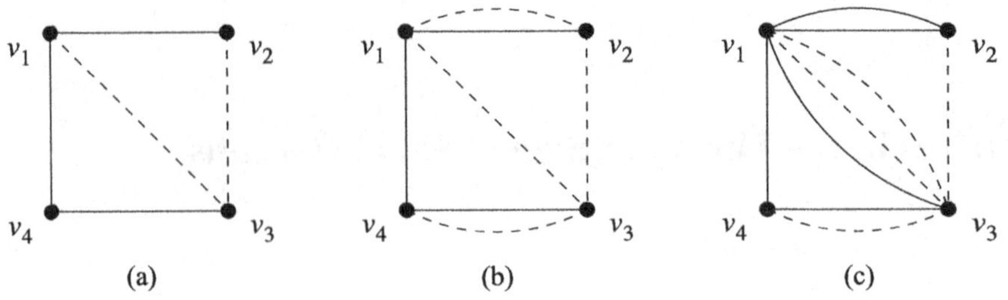

Figure 1. A signed simple graph Σ_4 (a), a simply signed multigraph (b), and a signed multigraph that is not simply signed (c).

Finally, there is the Kirchhoff matrix.[1] This is the adjacency matrix with signs reversed, and with the degrees of the vertices inserted in the diagonal. The Kirchhoff matrix equals the incidence matrix times its transpose. If we multiply in the other order, the transpose times the incidence matrix, we get the adjacency matrix of the line graph, but with 2's in the diagonal.

All this generalises ordinary graph theory. Indeed, much of graph theory generalises to signed graphs, while much—though certainly not all—signed graph theory consists of generalising facts about unsigned graphs.

As this article is expository I will give only elementary proofs, to illustrate the ideas.

1. Fundamentals of signed graphs

1.1 *Definitions about vectors and matrices*

A vector of all 1's is denoted by $\mathbf{j}$. The matrix J consists of all 1's.

1.2 *Definitions about graphs*

A graph is $\Gamma = (V, E)$ with vertex set V and edge set E. All graphs will be undirected, finite, and *simple*—without loops or multiple edges—except where explicitly stated otherwise (though most of what I say works well when loops and multiple edges are allowed). The *order* of the graph is $n := |V|$. $c(\Gamma)$ is the number of connected components of Γ. Γ^c is the complement of Γ.

An edge with endpoints v, w may be written vw or e_{vw}. An edge with endpoints v_i, v_j may also be written e_{ij}.

The *degree* $d_\Gamma(v)$ of a vertex v in a graph Γ is the number of edges incident with v. If every vertex has the same degree k, we say Γ is *regular* of degree k, or briefly, *k-regular*.

A *walk* in Γ is a sequence $W = e_{01}e_{12}\cdots e_{l-1,l}$ of edges, where the second endpoint v_i of $e_{i-1,i}$ is the first endpoint of $e_{i,i+1}$. Its *length* is l. Vertices and edges may be repeated. A *path* is a walk with no repeated vertices or edges. A *closed path* is a

[1] The Kirchhoff matrix is sometimes called the 'Laplacian', but other matrices are also called 'Laplacian'. I adopt the unambiguous name.

walk of positive length in which $v_0 = v_l$, but having no other repeated vertices or edges.

Important subgraphs

A subgraph of Γ is *spanning* if it contains all the vertices of Γ. A *circle* (or circuit, cycle, polygon) is the graph of a closed path; that is, it is a 2-regular connected subgraph. A *theta graph* consists of three internally disjoint paths joining two vertices. A *pseudoforest* is a graph in which every component is a tree or a 1-tree (a tree with one extra edge forming a circle). A *block* of Γ is a maximal 2-connected subgraph, or an isthmus or an isolated vertex. A *cutset* is the set of edges between a vertex subset and its complement, except that the empty edge set is not considered a cutset.

Adjacency matrix

The *adjacency matrix* of Γ is the $n \times n$ matrix $A(\Gamma)$ in which $a_{ij} = 1$ if $v_i v_j$ is an edge and 0 if not. The *Seidel adjacency matrix* of Γ is the $n \times n$ matrix $S(\Gamma)$ in which $s_{ij} = 0$ if $i = j$ and otherwise is -1 if $v_i v_j$ is an edge, $+1$ if it is not. Seidel introduced this matrix in [20]; his very successful use of it in many papers led to his name's becoming firmly attached. I will have much to say about the Seidel matrix in terms of signed graphs.

1.3 *Definitions about signed graphs*

A *signed graph* Σ is a pair $(|\Sigma|, \sigma)$ where $|\Sigma| = (V, E)$ is a graph, called the *underlying graph*, and $\sigma : E \to \{+1, -1\}$ is the *sign function* or *signature*. The sign group $\{+1, -1\}$ can also be written $\{+, -\}$; I shall treat the two notations as equivalent.[2] Often, we write $\Sigma = (\Gamma, \sigma)$ to mean that the underlying graph is Γ. E^+ and E^- are the sets of positive and negative edges. The *positive* and *negative subgraphs*,

$$\Sigma^+ = (V, E^+) \quad \text{and} \quad \Sigma^- = (V, E^-),$$

are unsigned graphs.

A signed graph is *simply signed* if it has no loops[3] and no parallel edges with the same sign; but it may have two edges, one positive and one negative, joining a pair of vertices. We shall be concerned mostly with signed simple graphs (those with no parallel edges) but occasionally simply signed graphs, and even loops, will play a role.

Signed graphs Σ and Σ' are *isomorphic* if there is a graph isomorphism $f : |\Sigma| \to |\Sigma'|$ that preserves edge signs.

Define Σ to be *regular* if both Σ^+ and Σ^- are regular graphs.

1.4 *Examples*

1. We say Σ is *homogeneous* if all edges have the same sign, and *heterogeneous* otherwise. (I got this handy terminology from M. Acharya and co-authors.) It is

---

[2] The important thing is that the signs form a multiplicative group of two elements that acts on numbers. For matrix theory the notation is simpler if they are themselves numbers.

[3] Usually, negative loops would be allowed; but that is not so suitable to the matrix theory.

all positive or *all negative* if all edges are positive or negative, respectively. $+\Gamma$ denotes Γ with all positive signs. $-\Gamma$ is Γ with all negative signs.

2. K_{Γ_0} denotes a complete graph K_n with vertex set $V = V(\Gamma_0)$, whose edges are negative if they belong to Γ_0 and positive otherwise. That is, $(K_{\Gamma_0})^- = \Gamma_0$ and $(K_{\Gamma_0})^+ = {\Gamma_0}^c$.

1.5 *Walks, circles, and their signs*

The sign of a walk $W = e_1 e_2 \cdots e_l$ is the product of its edge signs:

$$\sigma(W) := \sigma(e_1)\sigma(e_2)\cdots\sigma(e_l).$$

Thus, a walk is either positive or negative, depending on whether it has an even or odd number of negative edges, counted with their multiplicity in W if there are repeated edges.

A circle, being the graph of a closed path, has a definite sign, either positive or negative. The class of positive, or negative, circles in Σ is denoted by $\mathcal{C}^+(\Sigma)$, or $\mathcal{C}^-(\Sigma)$, respectively. One can characterise the class of negative circles by the following property. A *theta graph* is the union of three paths joining the same two vertices, but otherwise disjoint.

Proposition 1.1 (cf. [34]). *A class $\mathcal{B}$ of circles in a graph Γ is the class of negative circles of a signed graph (Γ, σ) if and only if every theta subgraph contains an even number of circles in $\mathcal{B}$.*

A negative circle of length 2—which consists of one positive edge and one negative edge and can only occur in a signed multigraph—is called a *negative digon*.

1.6 *Balance*

A signed graph Σ, or a subgraph or edge set, is called *balanced* if every circle in it is positive. $b(\Sigma)$ is the number of connected components of Σ that are balanced. For $S \subseteq E$, $b(S)$ is the number of balanced components of (V, S).

A circle is balanced iff it is positive. A walk is called balanced when its underlying graph is balanced; thus, a positive walk may be balanced or unbalanced, and the same holds for a negative walk.

It is easy to see that Σ is balanced if and only if every block is balanced.

Theorem 1.2 (Harary's Balance Theorem [10]). *A necessary and sufficient condition for Σ to be balanced is that there be a bipartition of V into X and Y such that an edge is negative precisely when it has one endpoint in X and one in Y. (X or Y may be empty.)*

In other words, Σ is balanced just when E^- is empty or a cutset. When Σ is balanced, a *Harary bipartition* is any bipartition $\{X, Y\}$ as in the theorem. It is unique if and only if Σ is connected. It was not so easy to prove this theorem at the time, but with switching it becomes simple; see later.

Corollary 1.3. $-\Gamma$ *is balanced iff* Γ *is bipartite.*

This means balance is a kind of generalisation of biparticity. That turns out to be a valid insight in a number of ways; e.g., in regard to the unoriented incidence matrix of a graph (Example 4.6.2).

We call Σ *antibalanced* if $-\Sigma$ is balanced; equivalently, if all even circles are positive and all odd circles are negative. The quick way to get an antibalanced signed graph is to give negative signs to all edges of a graph. Corresponding to Harary's Balance Theorem is the following result:

Corollary 1.4. *A necessary and sufficient condition for* Σ *to be antibalanced is that there be a bipartition of* V *into* X *and* Y *such that an edge is positive precisely when it has one endpoint in* X *and one in* Y. *(*X *or* Y *may be empty.)*

Thus, Σ is antibalanced iff E^+ is empty or a cutset. Correspondingly to Corollary 1.3, $+\Gamma$ is antibalanced iff Γ is bipartite.

1.7 *Switching*

Switching Σ means reversing the signs of all edges between a vertex set X and its complement. X may be empty. We say X is *switched* in Σ. The switched graph is written Σ^X. *Vertex switching* means switching a single vertex. Switching a set X has the same effect as switching all the vertices in X, one after another.

Another version of switching, which is equivalent to the preceding and is very useful, is in terms of a function $\theta : V \to \{+, -\}$, called a *switching function*. *Switching* Σ *by* θ means changing σ to σ^θ defined by

$$\sigma^\theta(vw) := \theta(v)\sigma(vw)\theta(w).$$

The switched graph is written $\Sigma^\theta := (|\Sigma|, \sigma^\theta)$.

If Σ can be switched to become Σ', we say Σ and Σ' are *switching equivalent*. Switching equivalence is an equivalence relation on signatures of a fixed graph. An equivalence class is called a *switching class*. $[\Sigma]$ denotes the switching class of Σ.

If Σ' is isomorphic to a switching of Σ, we say Σ and Σ' are *switching isomorphic*. Switching isomorphism is an equivalence relation on all signed graphs. Often in the literature switching isomorphism is not distinguished from switching equivalence, but I prefer to separate the two concepts.

Lemma 1.5 (Switching Lemma [35, Corollary 3.3]). Σ *is balanced if and only if it switches to an all-positive signature, and it is antibalanced if and only if it switches to an all-negative signature.*

Proof. To prove the first statement one can assume Σ is connected. Take a spanning tree, rooted at any vertex v, and switch Σ so the tree is all positive. The switching function for this is $\theta(w) := \sigma(T_{vw})$, where T_{vw} is the unique vw-path in T. Σ is balanced if and only if there are no remaining negative edges.

The second statement follows by a similar proof, or by negation from the first part. □

A useful way to think of the Switching Lemma without actually switching is that Σ is balanced iff there is a function $\mu : V \to \{+, -\}$ such that $\sigma(e_{vw}) = \mu(v)\mu(w)$. This amounts to saying that σ has a *potential function*, i.e., μ.

Properties preserved by switching are the signs of circles, and balance or imbalance of Σ and of any subgraph.

The proof technique of the Switching Lemma yields a valuable insight into equivalence of signed graphs.

Theorem 1.6 (Switching Equivalence [28], [35, Proposition 3.2]). *Two signed graphs with the same underlying graph are switching equivalent if and only if they have the same class of positive circles.*

Proof. As in the previous proof we may assume the graphs, Σ_1 and Σ_2, are connected. Switch both signed graphs so a fixed spanning tree T is all positive. Call the switched graphs Σ_1' and Σ_2'.

If Σ_1 and Σ_2 are switching equivalent, so are Σ_1' and Σ_2', but as they agree on a spanning tree, one can only be switched to the other by no switching at all or by switching every vertex. Thus, $\Sigma_1' = \Sigma_2'$, whence $\mathcal{C}^+(\Sigma_1) = \mathcal{C}^+(\Sigma_1') = \mathcal{C}^+(\Sigma_2') = \mathcal{C}^+(\Sigma_2)$.

If $\mathcal{C}^+(\Sigma_1) = \mathcal{C}^+(\Sigma_2)$, then after switching $\mathcal{C}^+(\Sigma_1') = \mathcal{C}^+(\Sigma_2')$. Now, with an all-positive spanning tree T, the only possible difference between Σ_1' and Σ_2' is in the signs of the non-tree edges. But the sign of an edge $e \notin T$ is the sign of the unique circle in $T \cup \{e\}$, which is the same in Σ_1' and Σ_2'. Therefore, $\Sigma_1' = \Sigma_2'$. As Σ_1 and Σ_2 switch to the same signed graph Σ_1', they are switching equivalent. □

Corollary 1.7 (Switching Isomorphism [28]). *Two signed graphs are switching isomorphic if and only if there is an isomorphism of underlying graphs that preserves the signs of circles.*

With switching I can give short proofs of such results as Harary's fundamental theorem.

Proof of Harary's Balance Theorem. If there is such a bipartition, every circle has an even number of negative edges, so Σ is balanced.

If Σ is balanced, switch it to be all positive. Letting X be the set of switched vertices, the bipartition is $\{X, V \backslash X\}$. □

1.8 *History*

Signed graphs and balance were invented by Harary by 1953 [10] to treat a question in social psychology [4]. They have since been reinvented over and over in many contexts in physics, geometry, economics, and more—thus showing that they are a natural concept—but it was Harary who first had the idea of labelling with the 2-element group by putting signs on the edges and multiplying them. Harary also introduced antibalance, in [11].

It is remarkable that, years before Harary, König [19, Section X.3] had the idea of a graph with a distinguished subset of edges, in a way we now recognise as equivalent

$$\begin{pmatrix} 0 & 1 & -1 & 1 \\ 1 & 0 & -1 & 0 \\ -1 & -1 & 0 & 1 \\ 1 & 0 & 1 & 0 \end{pmatrix} \qquad \begin{pmatrix} 0 & 0 & -1 & 1 \\ 0 & 0 & -1 & 0 \\ -1 & -1 & 0 & 0 \\ 1 & 0 & 0 & 0 \end{pmatrix} \qquad \begin{pmatrix} 0 & 2 & -1 & 1 \\ 2 & 0 & -1 & 0 \\ -1 & -1 & 0 & 0 \\ 1 & 0 & 0 & 0 \end{pmatrix}$$

(a) $A(\Sigma_4)$ (b) (c)

Figure 2. The adjacency matrices of the signed graphs in Figure 1. Note the cancellations due to simultaneous adjacencies with opposite signs in (b, c) and the 2 for double adjacency in (c).

to signed graphs, and even proved Harary's Balance Theorem and defined switching in the form of taking the set sum of a cutset with the set of negative edges. Despite all this, he failed to notice the value of edge labels that one can multiply, which I regard as the crucial step in the invention of signed graphs.

The first to think of switching as an operation on signed graphs were the social psychologists Abelson and Rosenberg [1]. However, their formalism for switching, in terms of a Hadamard product operation on their adjacency matrix R (see Section 2.6), was awkward and ungraphical. I arrived at switching of signed graphs as an adaptation and generalisation of graph switching, introduced by Seidel and applied with great effect in many papers (cf. [20,25]). Switching, like signed graphs themselves, has been reinvented several times, with varying names and notation.

The observation following the proof of the Switching Lemma was published, independently of other work on switching, by Sampathkumar [23].

2. Adjacency matrices

2.1 *Definition*

The adjacency matrix $A = A(\Sigma)$ is an $n \times n$ matrix in which $a_{ij} = \sigma(v_i v_j)$ (the sign of the edge $v_i v_j$) if v_i and v_j are adjacent, and 0 if they are not. Thus A is a symmetric matrix with entries $0, \pm 1$ and zero diagonal, and conversely, any such matrix is the adjacency matrix of a signed simple graph. The absolute value matrix, $|A(\Sigma)|$, equals $A(|\Sigma|)$, the adjacency matrix of the underlying unsigned graph.

When Σ has multiple edges, the (i, j) entry of A is the sum of the signs of all $v_i v_j$ edges. Positive and negative edges cancel each other. This will become important to us in the treatment of line graphs.

2.2 *Walks and neighbors*

Powers of A count walks in a signed way. Let $w_{ij}^{+}(l)$ be the number of positive walks of length l from v_i to v_j (that is, the sign product of the edges in W is positive), and let $w_{ij}^{-}(l)$ be the number of negative walks.

Theorem 2.1. *The (i, j) entry of A^l is $w_{ij}^{+}(l) - w_{ij}^{-}(l)$.*

I omit the proof, which is not difficult.

A noteworthy special case is the square of A. Let p_{ij}^{+} denote the number of common positive neighbors of distinct vertices v_i and v_j, p_{ij}^{-} the number of their common negative neighbors, and $p_{ij}^{\pm}$ the number of neighbors that are positive neighbors of one and negative neighbors of the other.

Corollary 2.2. *In* $A(\Sigma)^2$ *the* (i, i) *entry is* $d_{|\Sigma|}(v_i)$ *and the* (i, j) *entry, for* $i \neq j$, *is* $p_{ij}^{+} + p_{ij}^{-} - p_{ij}^{\pm}$.

Proof. The diagonal entry is the number of walks $v_i v_j v_i$, since every such walk has the positive sign $\sigma(v_i v_j)^2$. This number equals the number of neighbors of v_i. As for the off-diagonal entry, $p_{ij}^{+} + p_{ij}^{-} = w_{ij}^{+}(2)$ and $p_{ij}^{\pm} = w_{ij}^{-}(2)$. □

2.3 *Examples*

1. If Σ is complete, i.e., $|\Sigma| = K_n$, then $A(\Sigma)$ is the Seidel adjacency matrix $S(\Sigma^{-})$ of the negative subgraph Σ^{-}. That means the Seidel matrix of a graph is the adjacency matrix of a signed complete graph. This fact inspired my work on adjacency matrices of signed graphs.
2. If Σ is bipartite, so that $|\Sigma| \subseteq K_{r,s}$, then $A(\Sigma) = \begin{pmatrix} O & B \\ B^{\mathrm{T}} & O \end{pmatrix}$ where B is an $r \times s$ matrix of 0's, +1's, and −1's. If Σ is complete bipartite, B has no 0's.

2.4 *Switching*

Switching has a simple effect on $A(\Sigma)$. Given $X \subseteq V$, switching by X negates both the rows and columns of vertices in X. Given a function $\theta : V \to \{+1, -1\}$, switching by θ negates both the rows and columns of vertices with $\theta(v_i) = -1$. In matrix terms, let $\mathrm{Diag}(\theta)$ be the diagonal $(0, \pm 1)$-matrix with $\theta(v_i)$ in the ith diagonal position. Then switching by θ conjugates $A(\Sigma)$ by $\mathrm{Diag}(\theta)$; that is,

$$A(\Sigma^{\theta}) = \mathrm{Diag}(\theta)^{-1} A(\Sigma)\, \mathrm{Diag}(\theta).$$

(Note that $\mathrm{Diag}(\theta)^{-1} = \mathrm{Diag}(\theta)$; I inserted the inversion in order to show that this is truly conjugation.) We may conclude that:

Proposition 2.3. *Signed graphs* Σ_1 *and* Σ_2 *on the same vertex set are switching equivalent if and only if their adjacency matrices satisfy* $A(\Sigma_2) = D^{-1} A(\Sigma_1) D$ *for some diagonal* $(0, \pm 1)$*-matrix* D *whose diagonal has no zeros.*

We can strengthen this to a criterion for switching *isomorphism* (if not one that is computationally practical). By 'similarly rearranging' the rows and columns of a square matrix A, I mean that one applies the same permutation to the rows and to the columns.

Corollary 2.4. *Signed graphs* Σ_1 *and* Σ_2 *are switching isomorphic if and only if the rows and columns of* $A(\Sigma_1)$ *can be similarly rearranged so that* $A(\Sigma_2) = D^{-1} A(\Sigma_1) D$ *for some diagonal* $(0, \pm 1)$*-matrix* D *whose diagonal has no zeros.*

I omit the simple verifications.

Similarly rearranging the rows and columns is represented with matrices as conjugation by a permutation matrix P, i.e., A is changed to $P^{-1}AP$. Thus, in strictly matrix language, $A(\Sigma_2) = (PD)^{-1}A(\Sigma_1)(PD)$.

If Σ is bipartite, as in Example 2.3.2, with bipartition $V = V_1 \cup V_2$, we can examine the effect of switching X on B. Say the rows of B are indexed by V_1 and the columns by V_2. Switching negates each row of B corresponding to a vertex in $X \cap V_1$ and each column corresponding to a vertex in $X \cap V_2$.

2.5 *Eigenvalues and eigenvectors*

An *eigenvalue of* Σ is an eigenvalue of its adjacency matrix. The *spectrum* of Σ is the list of its eigenvalues with their multiplicities. Individual eigenvalues and the spectrum may give information about Σ. I know of nothing that has been done on this aside from special cases: signed graphs whose underlying graph is regular, which can be treated through the Kirchhoff matrix (see Section 4.5); those with no eigenvalues greater than 2 (they are, with a few exceptions, line graphs of signed graphs; see Section 5.5); and Acharya's matricial criterion for balance. There is one fundamental result that must be the starting point for all investigations:

Proposition 2.5. *Switching a signed graph does not change its spectrum.*

Proof. This is a corollary of Proposition 2.3, because conjugating a matrix does not change its spectrum. □

Proposition 2.6 (B.D. Acharya [2]). Σ *is balanced if and only if* $A(\Sigma)$ *has the same eigenvalues (with multiplicities) as does* $A(|\Sigma|)$.

I omit the proof. Acharya's criterion is the more interesting because it cannot be strengthened to an eigenvalue criterion for switching isomorphism. Although switching-isomorphic signed graphs obviously must have the same eigenvalues, the converse is false, as one can see from the facts that nonisomorphic unsigned graphs can have the same eigenvalues and that $A(+\Gamma) = A(\Gamma)$.

A regular signed graph has $d^{\pm}(\Sigma)$ as an eigenvalue, associated to the eigenvector $\mathbf{j}$.

Proposition 2.7. Σ *is regular if and only if* $\mathbf{j}$ *is an eigenvector of both* $A(|\Sigma|)$ *and* $A(\Sigma)$.

Proof. The term of v_i in $A(|\Sigma|)\mathbf{j}$ is $d_{|\Sigma|}(v_i) = d_{\Sigma^+}(v_i) + d_{\Sigma^-}(v_i)$, the sum of degrees in the positive and negative subgraphs. The term of v_i in $A(\Sigma)\mathbf{j}$ is $d_{\Sigma^+}(v_i) - d_{\Sigma^-}(v_i)$, the difference of degrees. A necessary and sufficient condition for $\mathbf{j}$ to be an eigenvector of both is that the sum and difference be independent of i; equivalently, that the positive and negative subgraphs be regular; thus, by definition, that Σ be regular. □

Although it does not directly provide information about the signed graph itself, from matrix theory we do know something about the relationship between eigenvalues of Σ and those of induced subgraphs.

Proposition 2.8. *If* Σ_1 *is an induced subgraph of* Σ_2, *then the largest eigenvalues satisfy* $\lambda^1(\Sigma_1) \leq \lambda^1(\Sigma_2)$.

Proof. This is an instance of the general interlacing theorem for eigenvalues of real, symmetric matrices [9, Theorem 9.1.1]. □

McKee and Smyth [21] start with an eigenvalue question and arrive at signed graphs. If a symmetric integer matrix A has all its eigenvalues (which are necessarily real) in the interval $[-2, 2]$, then it is the adjacency matrix of a signed simple graph except for having possibly nonzero elements on the diagonal. The signed graphs that can appear are subgraphs of members of three infinite families and a handful of sporadic examples. Their result is one of several reasons to think that ± 2 are especially significant bounds on eigenvalues. I refer the reader to their long paper for the details.

As I mentioned, the eigenvalue properties of signed graphs are an open field of inquiry.

2.6 *The Abelson–Rosenberg adjacency matrix*

An oddity from early in signed graph theory is the adjacency matrix formulated by the inventive social psychologists Abelson and Rosenberg [1]. Their matrix, R, has entries **o**, **p**, **n**, **a** (and these symbols have addition and multiplication operations), where **o** stands for nonadjacency, **p** and **n** are for positive and negative adjacency, and **a** is for vertices that are adjacent by both a positive and a negative edge; and it has **p** on the diagonal instead of **o**. Abelson and Rosenberg employed their symbols to stand for an 'unrelated', 'positive', 'negative', and 'ambivalent' relationship between the vertices, which represented persons in a social group. The symbol **a**, which can occur with a simply signed graph but not a signed simple graph, was necessary for the psychology.

Harary, Norman, and Cartwright [14, Theorem 13.8] used R to show the existence of walks of given length and sign between two specified vertices. Their theorem says there is a positive (resp., negative) walk of length l from v_i to v_j iff the (i, j) position in $(R - \mathbf{p}I)^l$ contains **p** or **a** (resp., **n** or **a**).

2.7 *Degrees*

The degree of a vertex has several generalisations to signed graphs. The *underlying degree* $d_{|\Sigma|}(v)$ is the degree in the underlying graph, that is, the total number of edges incident with v. The *positive* or *negative degree*, $d^+_\Sigma(v)$ or $d^-_\Sigma(v)$, is the degree in the positive subgraph Σ^+ or the negative subgraph Σ^-. The *net degree* is

$$d^\pm_\Sigma(v) := d^+_\Sigma(v) - d^-_\Sigma(v).$$

If Σ is regular, we can speak of its *degree* (of any kind): it is the degree of any vertex. For example, the net degree of Σ is $d^\pm(\Sigma) := d^\pm(v)$ for every vertex v. Σ is regular if and only if $|\Sigma|$ is regular and $A\mathbf{j} = r\mathbf{j}$ for some real number r, which necessarily equals $d^\pm(\Sigma)$.

2.8 *Very strong regularity*

Seidel discovered that a strongly regular graph Γ_0 has a nice definition in terms of its Seidel adjacency matrix $S = S(\Gamma_0)$, namely, that

$$S^2 - tS - kI = p(J - I) \quad \text{and} \quad S\mathbf{j} = \rho_0 \mathbf{j}$$

for some constants t, k, p, ρ_0 (thus in particular $\mathbf{j}$ is an eigenvector of S). (See [24,25].) The constants have combinatorial interpretations. For a start, $k = n - 1$, the degree of K_n. Given an edge vw, the number of vertices adjacent to exactly one of v and w equals $\frac{1}{2}(n-2-p+t)$. Given a nonadjacent pair v, w, that number is $\frac{1}{2}(n-2-p-t)$. If $p \neq 0$, the second defining equation $S\mathbf{j} = \rho_0\mathbf{j}$ is superfluous.

Contemplating the fact that $S(\Gamma_0)$ is the same matrix as $A(K_{\Gamma_0})$, I was somehow led to the following definition: A signed graph is *very strongly regular* [40] if its adjacency matrix satisfies

$$A^2 - tA - kI = p\bar{A} \quad \text{and} \quad A\mathbf{j} = \rho_0\mathbf{j} \tag{2.1}$$

for some constants t, k, p, ρ_0. Here $\bar{A}$ is the adjacency matrix of the complement of $|\Sigma|$. The combinatorial interpretation of these parameters is:

a. $|\Sigma|$ is k-regular.
b. $\rho_0 = d^{\pm}(\Sigma)$, the net degree of every vertex (hence it is an integer). Hence Σ is regular.
c. $t = t_{ij}^{+} - t_{ij}^{-}$ where t_{ij}^{+}, t_{ij}^{-} are the numbers of positive and negative triangles on an edge e_{ij}.
d. $p = p_{ij}^{+} - p_{ij}^{-}$ for any pair v_i, v_j of nonadjacent vertices, where p_{ij}^{+}, p_{ij}^{-} are the numbers of positive and negative length-2 paths joining the vertices.
e. t and p are independent of the choices of adjacent or nonadjacent vertices.

The big problem is to, in some sense, classify very strongly regular signed graphs. I am currently working on this. Various simplifications are helpful. For instance, $-\Sigma$, whose signature is $-\sigma$, behaves just like Σ except that t and ρ_0 are negated. Thus, one may assume that t is nonnegative, or that Σ is not all negative, when it is convenient to do so.

The most important factor in the classification is which of p and t are 0. As is usual in such problems, there are strong numerical restrictions on the values of t, k, p, ρ_0. It is interesting that some of the types include kinds of matrices that have already been studied for many years. I will run down the possibilities. I write $d^{\pm}$ for $d^{\pm}(\Sigma)$. A *weighing matrix* is a $(0, \pm 1)$-matrix W such that $W^{\mathrm{T}}W$ is a multiple of I [17].

(a) Homogeneous signed graphs can be assumed (by negation) to be all positive. $+\Gamma$ is very strongly regular if and only if Γ is a strongly regular unsigned graph. This demonstrates that we have a true generalisation of strong regularity.

For the rest of the cases I assume Σ is inhomogeneous.

(b) When $p = t = 0$, the defining equations are $A^2 = kI$ and $A\mathbf{j} = d^{\pm}\mathbf{j}$. The eigenvalues are $\pm\sqrt{k}$, one of which is $d^{\pm}$. Consequently, k is a square and A is a symmetric weighing matrix with zero diagonal and in each row $\binom{d^{\pm}+1}{2}$ entries equal to $+1$ and $\binom{d^{\pm}}{2}$ entries equal to -1.

(c) When $p = 0$ but $t \neq 0$, the defining equations are $A^2 - tA - kI = O$ and $A\mathbf{j} = d^{\pm}\mathbf{j}$. Solving for the eigenvalues shows there is an integer $s \equiv t \pmod 2$ such that

$$k = \frac{s-t}{2} \cdot \frac{s+t}{2} \quad \text{and} \quad d^{\pm} = \frac{s+t}{2}.$$

(d) When $p \neq 0$, from the eigenvector $\mathbf{j}$ we deduce that $p = [d^{\pm}(d^{\pm} + t) - k]/[n - 1 - k]$, which significantly constrains the numbers. This case is too complicated for further description here. For instance, the eigenvalues of Σ depend on those of $|\Sigma|^c$.

Much more can be said, but not here; see [40].

2.9 *History*

To my knowledge, the first adjacency matrix of a signed graph was that of Abelson and Rosenberg. The standard adjacency matrix, $A(\Sigma)$, appeared soon after, but I am not sure exactly when and where. Harary certainly used it early on.

Switching of the standard adjacency matrix is implicit in [20] and explicit in a form that is equivalent to switching signed complete graphs in [24] but was not made explicitly and generally signed-graphic until later (I am not sure just when). Abelson and Rosenberg, as I mentioned, had the concept of switching but they did not develop it far, and I think that would have been difficult given their mathematical formulation.

3. Orientation

3.1 *Bidirected graphs*

In a *bidirected graph*, every edge has an independent orientation at each end. (This concept is due to Jack Edmonds; cf. [7].) We think of these in two ways: as an arrow at each end, which may point towards or away from the endpoint, and as a sign $\eta(v, e)$ on the end of e that is incident with v, which is $+1$ if the arrow points to the endpoint, -1 if the arrow is directed away from the endpoint.

3.2 *Oriented signed graphs*

A bidirected graph is naturally signed by the formula

$$\sigma(e) = -\eta(v, e)\eta(w, e) \tag{3.1}$$

for an edge e_{vw}. An edge is negative if its arrows both point toward their corresponding endpoints (I call such edges *extraverted*), or both away from their endpoints (I call these *introverted* edges). An edge is positive if one arrow points at its endpoint while the other is directed away from its endpoint. Thus, a positive edge is just like an ordinary directed edge.

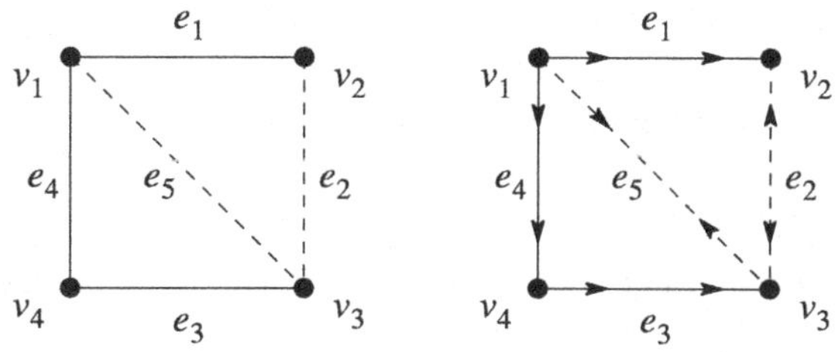

Figure 3. The signed graph Σ_4 and an orientation of it.

Conversely, an *orientation* of a signed graph Σ is a bidirection η of $|\Sigma|$ that satisfies the sign formula (3.1).

Reorienting an edge e_{vw} means replacing $\eta(v, e)$ and $\eta(w, e)$ by their negatives. In terms of arrows, it means reversing the arrows at both endpoints of the edge. This does not change the sign of the edge.

3.3 *Switching*

One switches a vertex set X in a bidirected graph by changing the signs of all the edge ends incident with a vertex in X. In terms of a switching function $\theta : V \to \{+, -\}$, one has $\eta^\theta(v, e) := \theta(v)\eta(v, e)$.

Recall that switching the entire vertex set V does not change a signed graph; but it does have an effect on an orientation: it reverses all the arrows.

4. Incidence matrices

4.1 *Definition*

An *incidence matrix* of Σ is a $V \times E$ matrix in which the column of edge e has two entries ± 1, one in the row of each endpoint of e, and 0's elsewhere. The two nonzero entries must have product equal to $-\sigma(e)$; that is, they are equal if e is negative, but if e is positive, one is $+1$ and the other is -1.

The notation I use for an incidence matrix of a signed graph is $\mathrm{H}(\Sigma)$ (read 'Eta') $= (\eta_{ve})_{v \in V, e \in E}$.

The incidence matrix is not unique. The choice of signs in each column reflects a choice of orientation η of Σ; in fact, the (v, e) entry in $\mathrm{H}(\Sigma)$ is equal to

$$\begin{cases} \eta(v, e) & \text{if } e \text{ is incident with } v, \\ 0 & \text{if } e \text{ is not incident with } v. \end{cases}$$

(Thus, we may define $\eta(v, e) := 0$ in the latter case. Then $\eta(v, e) = \eta_{ve}$; the only difference between the two is the point of view: the former is η regarded as a bidirection and the latter is a matrix entry.) Conversely, the entries of an incidence matrix of Σ determine an orientation η.

The incidence matrix can be treated as a matrix over any field $\mathbf{k}$, or indeed any ring, in particular over the ring of integers, or the 2-element field.

4.2 *Switching and reorientation*

The effect on $\mathrm{H}(\Sigma)$ of switching $X \subseteq V$ is to negate the rows corresponding to the vertices in X. In terms of a switching function θ and the matrix $\mathrm{Diag}(\theta)$, $\mathrm{H}(\Sigma)$ switches to $\mathrm{Diag}(\theta)\mathrm{H}(\Sigma)$.

The effect of reorienting edges is to negate the corresponding columns of $\mathrm{H}(\Sigma)$.

4.3 *Rank*

The rank of $\mathrm{H}(\Sigma)$ is a basic fact that generalises more widely known but limited results about incidence matrices.

Theorem 4.1 (Jeurissen [18], Zaslavsky [35, Section 8A]). *If* $\mathbf{k}$ *has characteristic* 2, *then* $\mathrm{H}(\Sigma)$ *has rank* $n - c(\Sigma)$. *Otherwise,* $\mathrm{H}(\Sigma)$ *has rank* $n - b(\Sigma)$.

Proof. As $\mathrm{H}(\Sigma)$ can be written in the block form

$$\begin{pmatrix} \mathrm{H}(\Sigma_1) & O & \cdots & O \\ O & \mathrm{H}(\Sigma_2) & \cdots & O \\ & & \cdots & \\ O & O & \cdots & \mathrm{H}(\Sigma_k) \end{pmatrix},$$

where $\Sigma_1, \Sigma_2, \ldots, \Sigma_k$ are the connected components of Σ, it suffices to prove the rank formula for a connected signed graph. For $S \subseteq E$, write $\mathrm{H}(S)$ for the submatrix of $\mathrm{H}(\Sigma)$ that consists of the columns corresponding to S.

Let T be a spanning tree of Σ and switch so T is all positive. If Σ is balanced, then after switching it is all positive. Then its incidence matrix is identical with that of $|\Sigma|$, which is well known to have rank $n - 1$ (not n, because the rows sum to 0; not less than $n - 1$, because $\mathrm{H}(T)$ after dropping one row has determinant ± 1).

If $2 = 0$ in $\mathbf{k}$, then the definition shows that $\mathrm{H}(\Sigma) = \mathrm{H}(|\Sigma|)$ over $\mathbf{k}$, and by the preceding argument the rank is $n - 1$.

If Σ is unbalanced, there is a negative edge e. Consider the submatrix $\mathrm{H}(T \cup \{e\})$ of $\mathrm{H}(\Sigma)$ that consists of the columns corresponding to T and e. It has the form $(\mathrm{H}(T)\ \mathrm{H}(e))$. The last column has two $+1$'s or two -1's, so its sum is ± 2. The only linear dependence among the rows of $\mathrm{H}(T)$ is that they sum to 0; but the sum of the rows of the matrix $(\mathrm{H}(T)\ \mathrm{H}(e))$ has ± 2 in its last column. Therefore, if $2 \neq 0$ in $\mathbf{k}$ the rows of $(\mathrm{H}(T)\ \mathrm{H}(e))$ are linearly independent; this submatrix has rank n. Thus, $\mathrm{H}(\Sigma)$ has rank n. □

Since the incidence matrix of $|\Sigma|$ has rank $n - c(\Sigma)$, and $b(\Sigma) = c(\Sigma)$ when, and only when, Σ is balanced, $\mathrm{H}(\Sigma)$ has the same rank as $\mathrm{H}(|\Sigma|)$ if and only if Σ is balanced.

4.4 *The Kirchhoff matrix and matrix-tree theorems*

The *Kirchhoff matrix* (also called the *Laplacian*) is

$$K(\Sigma) := \mathrm{H}(\Sigma)\mathrm{H}(\Sigma)^{\mathrm{T}} = D(|\Sigma|) - A(\Sigma), \tag{4.1}$$

$$\mathrm{H}(\Sigma_4, \eta) = \begin{pmatrix} -1 & 0 & 0 & -1 & -1 \\ +1 & +1 & 0 & 0 & 0 \\ 0 & +1 & +1 & 0 & -1 \\ 0 & 0 & -1 & +1 & 0 \end{pmatrix}$$

Figure 4. The incidence matrix of Σ_4 corresponding to the orientation η in Figure 3.

$$K(\Sigma_4) = \mathrm{H}(\Sigma_4, \eta)\mathrm{H}(\Sigma_4, \eta)^{\mathrm{T}} = \begin{pmatrix} 3 & -1 & 1 & -1 \\ -1 & 2 & 1 & 0 \\ 1 & 1 & 3 & -1 \\ -1 & 0 & -1 & 2 \end{pmatrix}$$

Figure 5. The Kirchoff matrix of Σ_4 from Figure 3.

where $D(|\Sigma|)$ is the degree matrix of $|\Sigma|$, that is, the diagonal matrix whose diagonal entries are the degrees of the vertices in the underlying graph.

Lemma 4.2. *The rank of K (over the real numbers) is $n - b(\Sigma)$. Its nullity is $b(\Sigma)$. Its eigenvalues are non-negative.*

Proof. A matrix of the form MM^{T} is positive semidefinite and has rank equal to the rank of M. By standard matrix theory, its eigenvalues are all non-negative. In our case, $M = \mathrm{H}(\Sigma)$ has rank $n - b(\Sigma)$ by Theorem 4.1. □

The classical Matrix-Tree Theorem expresses the number of spanning trees of a graph in terms of the Kirchhoff matrix: the number is $\det K_{ij}$ where K_{ij} is K with any one row and column deleted. I state two signed-graphic analogs. For proofs see the references; the first is not hard (it uses the Binet–Cauchy theorem in the standard way), but the second is rather complicated both to state fully and to prove.

Theorem 4.3 (Zaslavsky [35, Section 8A]). *The determinant* $\det K$ *is the sum, over all pseudoforests F with n edges and with no positive circles, of $4^{c(F)}$, where $c(F)$ is the number of components of F.*

The reason is that $\det \eta(F) = \pm 2^{c(F)}$. Note that the pseudoforests in the theorem, because they have as many edges as vertices, can have no tree components.

Theorem 4.4 (Chaiken's All-Minors Matrix-Tree Theorem [5]). *From K delete k rows, corresponding to vertices $r_1, \ldots, r_k$, and k columns, corresponding to $c_1, \ldots, c_k$, and then take the determinant. The resulting number is the sum of $\pm 4^q$, where q is the number of circles in F, over all $n - k$-edge spanning pseudoforests without positive circles such that each tree component of F contains exactly one r_i and one c_j. The sign of the term is given by a complicated rule.*

Chaiken's theorem is even more general: the edges can be weighted.

4.5 *Eigenvalues of the Kirchhoff matrix*

The eigenvalues of the Kirchhoff matrix are a natural topic of study. As a simple example, they constrain the eigenvalues of Σ when the underlying graph is regular, in just the same way as with ordinary graphs.

Theorem 4.5. *If the underlying graph of Σ is regular of degree k, then all eigenvalues of Σ are $\leq k$. The value k is an eigenvalue of multiplicity $b(\Sigma)$; thus, k is an eigenvalue if and only if Σ has a balanced component.*

Proof. If $|\Sigma|$ is regular of degree k, then $D(|\Sigma|) = kI$. Thus, $K = kI - A$. The Kirchhoff matrix is positive semidefinite and has nullity $b(\Sigma)$ by Lemma 4.2. Therefore, all eigenvalues of K are at least 0. The eigenvalues λ_i^A of A are determined by $\lambda_i^K = k - \lambda_i^A$ where $\lambda_1^K, \ldots, \lambda_n^K$ are the eigenvalues of K. It follows that all $\lambda_i^A \leq k$ and that the eigenvalue k of A has the same multiplicity as the corresponding eigenvalue 0 of A. The multiplicity of 0 as an eigenvalue of K is the nullity of K. That gives the multiplicity of k as an eigenvalue of A. □

Hou, Li, and Pan [16] studied the eigenvalues of K in general, without assuming regularity of the underlying graph. They have two kinds of results: upper and lower bounds on the largest eigenvalue of K (mostly in terms of the underlying graph), and an interlacing theorem for all eigenvalues. Write the eigenvalues in decreasing order as $\lambda^1 \geq \lambda^2 \geq \cdots \geq \lambda^n$. Here are some of their results:

Theorem 4.6 (Hou, Li, and Pan [16]). *Let Γ be a connected simple graph and let Σ be a signed simple graph.*

(1) $\lambda^1(\Gamma, \sigma) \leq \lambda^1(-\Gamma)$, *with equality iff σ is antibalanced (Lemma 3.1).*
(2) $\lambda^1(\Sigma) \leq 2(n-1)$, *with equality iff Σ switches to $-K_n$ (Theorem 3.4).*
(3) $\lambda^1(\Sigma^+) + \lambda^1(\Sigma^-) \geq \lambda^1(\Sigma) \geq \lambda^1(\Sigma^+), \lambda^1(\Sigma^-)$ *(Corollary 3.8).*
(4) $\lambda^1(\Sigma) \geq 1 + \max_{v \in V} d_{|\Sigma|}(v)$ *(Theorem 3.10).*

Theorem 4.7 (Hou, Li, and Pan [16, Lemma 3.7]). $\lambda^i(\Sigma) \geq \lambda^i(\Sigma \backslash e) \geq \lambda^{i+1}(\Sigma)$.

4.6 *Examples*

1. An *oriented incidence matrix* of the unsigned graph Γ is the same as an incidence matrix $\mathrm{H}(+\Gamma)$ of the all-positive signed graph $+\Gamma$. Since $+\Gamma$ is balanced, the rank given by our formula equals $n - c(\Gamma)$, as is well known.
   The Kirchhoff matrix of $+\Gamma$ is simply that of Γ, i.e., $D(\Gamma) - A(\Gamma)$. Its determinant is zero, which is consistent with Theorem 4.3 because there are no negative circles in $+\Gamma$, so the number of pseudoforests of the kind counted by Theorem 4.3 is zero.
2. The *unoriented incidence matrix of* Γ is a (0, 1)-matrix which has 1 in position (v, e) if v is an endpoint of e and has 0 otherwise. (This matrix is often, though I think unfortunately, called simply 'the incidence matrix' of Γ.) It is an incidence

matrix of the all-negative signed graph, $-\Gamma$. Since an all-negative graph is balanced if and only if it is bipartite (Corollary 1.3), the rank of the matrix (except in characteristic 2) equals $n - b$ where b is the number of bipartite components of Γ. This result was previously obtained by ad hoc methods (originally by van Nuffelen [22]), but it is really a special case of the general rank theorem for signed graphs.
The Kirchhoff matrix $K(-\Gamma)$ equals $D(\Gamma) + A(\Gamma)$ (now often called the 'signless Laplacian' of Γ). Its determinant equals the sum of $4^{c(F)}$ over all n-edge pseudoforests F of which no component is bipartite.

3. The signed graphs of Figure 1 are connected and unbalanced. Therefore, each of their incidence matrices has rank $|V|$, which is 4.

5. Line graphs

5.1 *Unsigned line graphs*

The line graph of an unsigned graph Γ is denoted by $L(\Gamma)$. Its vertex set is $E(\Gamma)$, and two edges are adjacent if they have a common endpoint in Γ. $L(\Gamma)$ has two kinds of distinguished circles: *vertex triangles* are formed by three edges incident with a common vertex, and *derived circles* are the line graphs of circles in Γ. Every circle in $L(\Gamma)$ is known to be a set sum (i.e., symmetric difference of edge sets) of vertex triangles and derived circles.

5.2 *Signed line graphs*

The line graph $\Lambda(\Sigma)$ of Σ is a switching class, not a single signed graph [39]. Its underlying graph is the line graph $L(|\Sigma|)$ of the underlying graph. To define $\Lambda(\Sigma)$ we may take the approach of edge orientation or a direct definition of the circle signs.

1. *Definition by Orientation [39].*
Choose an orientation η of Σ. We define a bidirection η' of $L(|\Sigma|)$ and therefore an edge signature, thus forming the line signed graph $\Lambda(\Sigma)$. Two Σ-edges e_{vw}, e_{vu} incident with a vertex v form an edge $e := e_{vw}e_{vu}$ in Λ, whose vertices are e_{vw} and e_{vu}. An end of e may therefore be written (e_{vw}, e), corresponding to the end (v, e_{vw}) in Σ. Define $\eta'(e_{vw}, e) = \eta(v, e_{vw})$.
In terms of arrows, bidirect each edge of Σ with two arrows as indicated by η, and let the arrow on (e_{vw}, e) point into the vertex e_{vw} iff the arrow on (v, e_{vw}) points into the vertex v in Σ.
Reorienting an edge in Σ corresponds to switching the corresponding vertex in Λ. Thus, $\Lambda(\Sigma)$ is well defined only up to switching. That is, it is a well defined switching class.
2. *Definition by Circle Signs [36,39].*
Make every vertex triangle negative and give to every derived circle the same sign as the circle in Σ it derives from. Other circles get signed by the following *sum*

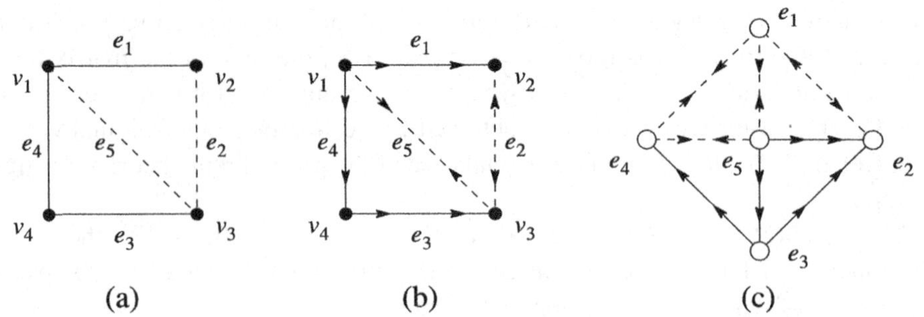

Figure 6. A signed simple graph (a), an orientation (b), and the oriented line graph (c).

rule: If C is the set sum of certain vertex triangles and derived circles, its sign is the product of the signs of those vertex triangles and derived circles. This rule is a consequence of Proposition 1.1. One has to prove that the sum rule gives the same sign no matter how it is applied, which is most easily achieved by showing that the definition by circle signs agrees with that by edge orientation, the latter being obviously well defined.

5.3 *Reduced line graphs*

If we allow Σ to have parallel edges—though only with opposite sign, so that Σ is simply signed but two vertices can be joined by both a positive and a negative edge—there is a definition of line graph similar to the preceding one. In the line graph there are negative digons derived from those of Σ. In the line-graph adjacency matrix $A(\Lambda)$, these digon edges cancel. To represent that phenomenon accurately in the line graph we should *reduce* the line graph by deleting pairs of parallel edges, $+ef$ and $-ef$, of opposite sign. The *reduced line graph*, $\bar{\Lambda}(\Sigma)$, is what results. It is a signed simple graph.

The importance of reduction is that, even though the underlying graph of Σ itself may not be simple, that of $\bar{\Lambda}(\Sigma)$ is. More precisely, if the underlying graph $|\Sigma|$ is simple, $|\Lambda(\Sigma)|$ is already simple and does not need to be reduced. If $|\Sigma|$ is not simple, then $|\Lambda(\Sigma)|$ is not simple but the underlying graph of the reduced line graph is simple. Thus, amongst the signed simple graphs, there are graphs that are reduced line graphs of simply signed graphs but not line graphs of simply signed graphs. One is shown in Figure 7; since $|\bar{\Lambda}(\Sigma_5)|$ is not a line graph, $\bar{\Lambda}(\Sigma_5)$ cannot be an unreduced line graph.

5.4 *Geometry*

This larger class, the reduced line graphs, has been studied by Vijayakumar and his coworkers, under the name of 'signed graphs represented by D_∞' [30,6]. From our line-graphic point of view, the name comes from the fact that

$$D_n := \{\pm \mathbf{b}_i \pm \mathbf{b}_j : 0 < i < j \le n\},$$

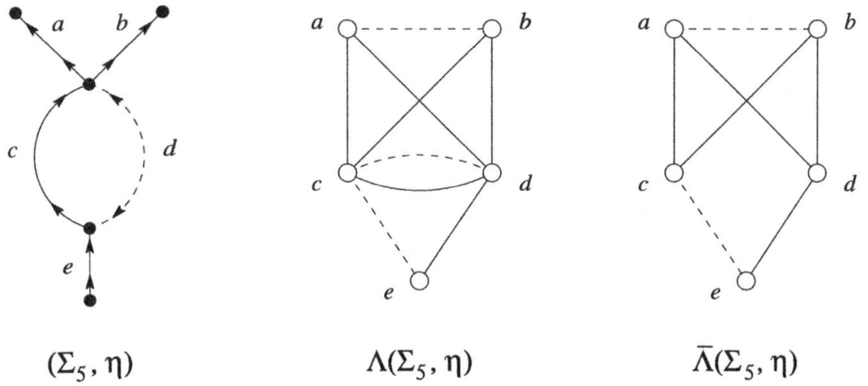

Figure 7. An oriented signed graph (Σ_5, η), its line graph $\Lambda(\Sigma_5)$ and its reduced line graph $\bar{\Lambda}(\Sigma_5)$, whose underlying graph $|\bar{\Lambda}(\Sigma_5)|$ is not a line graph.

where $\{\mathbf{b}_1, \ldots, \mathbf{b}_n\}$ is the standard orthonormal basis of $\mathbb{R}^n$, is the set of all those vectors that can be the column of an edge in the incidence matrix of a signed graph with n vertices; the line graph corresponds to taking dot products of these vectors. Allowing n to be arbitrarily large, one has D_∞.

D_n is best known as one of the classical root systems in Lie algebra. The other important root system for us is E_8, an exceptional root system in $\mathbb{R}^8$ whose exact definition is unnecessary here—but see [31,32].

We need to define representation by $W \subseteq \mathbb{R}^n$ since it depends on the adjacency matrix and is at the heart of the treatment of line graphs. A signed simple graph Σ is *represented by* W if there is an injection $f : V \to W$ such that the dot product $f(v) \cdot f(w) = \sigma(e_{vw})$ if v and w are adjacent and 0 if they are not, and all $\|f(v)\| = \sqrt{2}$. (That ought to have been $-\sigma(e_{vw})$ to be consistent with our definition of line graph but the negative sign was omitted by Vijayakumar, who defined this terminology.)

5.5 *Adjacency matrix and eigenvalues*

The adjacency matrix of $\Lambda(\Sigma)$ is the $E \times E$ matrix given by

$$A(\Lambda(\Sigma)) = 2I - \mathrm{H}(\Sigma)^{\mathrm{T}}\mathrm{H}(\Sigma). \tag{5.1}$$

Theorem 5.1. *The largest eigenvalue of $A(\Lambda(\Sigma))$ is at most 2. Moreover, 2 is an eigenvalue of $A(\Lambda(\Sigma))$ with multiplicity $|E| - n + b(\Sigma)$; in particular, it is an eigenvalue if and only if Σ has a connected component that is neither a tree nor an unbalanced 1-tree.*

Proof. A matrix product of the form $M^{\mathrm{T}}M$ is positive semidefinite, so all its eigenvalues are at least 0. An eigenvalue β of $M^{\mathrm{T}}M$ corresponds to the eigenvalue $\alpha - \beta$ of $\alpha I - M^{\mathrm{T}}M$. Letting $M = \mathrm{H}$ and $\alpha = 2$, we deduce that the eigenvalues of the right side of Equation (5.1) are not greater than 2.

By matrix theory, the rank of $\mathrm{H}^{\mathrm{T}}\mathrm{H}$ equals that of H, which is $n - b(\Sigma)$. It has 0 as an eigenvalue of multiplicity (order − rank). The order of $\mathrm{H}^{\mathrm{T}}\mathrm{H}$ is $|E|$. Thus, it has 0 as an eigenvalue of multiplicity $|E| - n + b(\sigma)$. The corresponding eigenvalue 2 of $A(\Lambda(\Sigma))$ has the same multiplicity. □

Theorem 5.2 ([3], [6, Theorem 1.1]). *If Σ is a signed simple graph whose eigenvalues are ≤ 2, then $-\Sigma$ is represented by D_n or E_8.*

That is, the connected signed simple graphs whose eigenvalues are at most 2 are the reduced line graphs of simply signed graphs (represented by D_n) and only a few additional sporadic examples (represented by E_8) which are neither line graphs nor reduced line graphs. This fact was explicitly recognised by Chawathe and Vijayakumar in [6, Theorem 1.1] (or see [33, the first Theorem 2.4, on p. 214]), although the proof is actually in the classic paper of Cameron, Goethals, Seidel, and Shult [3].

The essential observation behind Theorem 5.2 is that, if Σ has all eigenvalues ≤ 2, then $2I - A(\Sigma)$ is a positive semidefinite matrix and therefore is the matrix of inner products of a set W of vectors in $\mathbb{R}^m$ for some m. The proof involves classifying the possible sets W, which turn out to be the subsets of D_n and E_8.

Singhi and Vijayakumar showed that having an eigenvalue > 2 implies an induced subgraph with all eigenvalues ≤ 2; indeed, their result is stronger:

Theorem 5.3 (Singhi and Vijayakumar [27]). *If a signed simple graph Σ has an eigenvalue greater than or equal to 2, then it contains an induced subgraph whose largest eigenvalue is exactly 2.*

This result is the harder converse of the following corollary of Proposition 2.8.

Proposition 5.4. *If Σ has an induced subgraph whose largest eigenvalue is 2, then the largest eigenvalue of Σ is at least 2.*

5.6 *Examples*

1. The line graph of the all-negative signed graph $-\Gamma$ is $[-L(\Gamma)]$, the switching class of the ordinary line graph with all negative signs. From the standpoint of line graphs, therefore, ordinary graphs should be considered as all-negative signed graphs, instead of all-positive as in other parts of signed graph theory.
2. Hoffman's generalised line graph [15] is the reduced line graph of a signed graph; specifically, of $-\Gamma(m_1, \ldots, m_n)$, which consists of $-\Gamma$ together with m_i negative digons attached at the vertex v_i. See Figure 8.
3. The line graphs $\Lambda(\Sigma)$ that are antibalanced are those of the form $[-\Gamma']$ where Γ_0 is an ordinary line graph or a generalised line graph; that is:

   Theorem 5.5 (Cameron, Goethals, Seidel, and Shult [3]). *The all-negative signed graphs that are reduced line graphs of signed graphs are precisely the all-negative generalised line graphs $-\Gamma(m_1, \ldots, m_n)$.*

   (See the statement in the introduction to [30] [where 'The family of sigraphs represented by D_∞' should be 'The family of *graphs* ...'].) Again, we see that the usual line graphs are all-negative signed graphs. (This result was not originally stated explicitly in terms of signed graphs, but in an equivalent fashion in terms of adjacency matrices.)

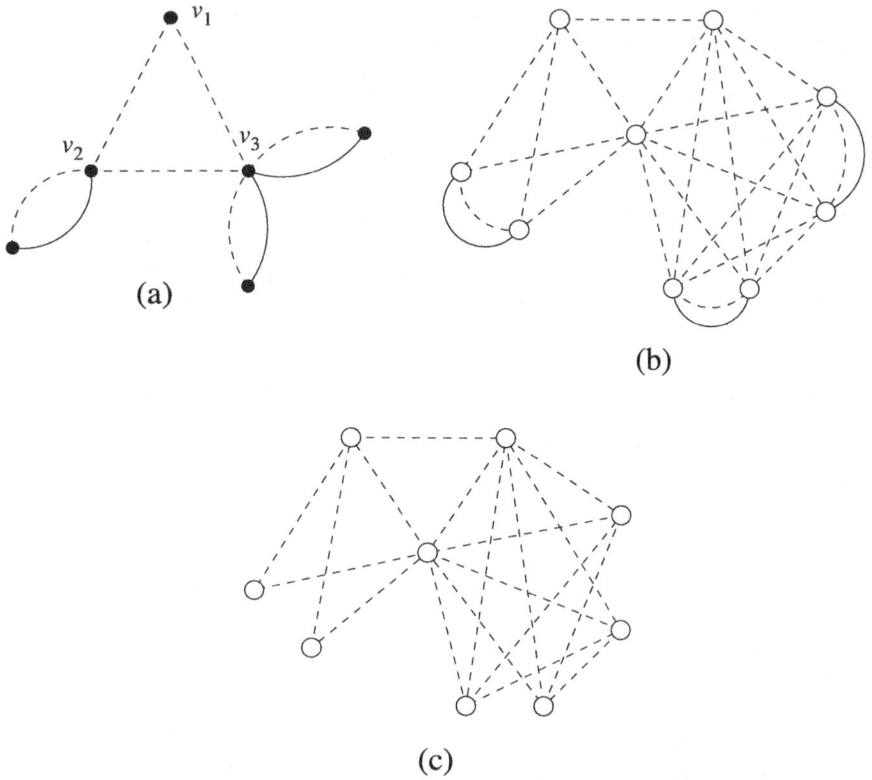

Figure 8. A signed graph $-K_3(0, 1, 2)$ and its reduced line graph, which is a generalised line graph with all negative signs.

4. As a directed graph is a bidirected all-positive graph $(+\Gamma, \eta)$, it has a signed line graph. The positive part $\Lambda^+(+\Gamma, \eta)$ is a directed graph as well; it is precisely the Harary–Norman line digraph of [13].

5.7 *History*

The two definitions of a line graph of a signed graph are from [39], which has been on the verge of being written for more than two decades.

The interpretation of graphs represented by D_∞, including Hoffman's generalised line graphs (Example 5.6.2), as line graphs was first stated in [36, Example 2]. It is implicit in the geometrical representation in [3], in Vijayakumar's geometrical representation by D_n, and also in Godsil and Royle's presentation [9, Section 12.1], based on the seminal paper [3], which talks of cancelling double edges but without recognising the cancellation as due to opposite signs.

References

[1] Robert P. Abelson and Milton J. Rosenberg, Symbolic psycho-logic: a model of attitudinal cognition, *Behavioral Sci.*, **3** (1958) 1–13.

[2] Devadas Acharya B., Spectral criterion for cycle balance in networks, *J. Graph Theory*, **4** (1980) 1–11. MR 81e:05097. Zbl. 445.05066.

[3] Cameron P. J., Goethals J. M., Seidel J. J. and Shult E. E., Line graphs, root systems, and elliptic geometry, *J. Algebra*, **43** (1976) 305–327. MR 56 #182. Zbl. 337.05142. Reprinted in [26, pp. 208–230].

[4] Dorwin Cartwright and Frank Harary, Structural balance: a generalization of Heider's theory, *Psychological Rev.*, **63** (1956) 277–293. Reprinted in: Dorwin Cartwright and Alvin Zander, eds., Group Dynamics: Research and Theory, second ed., pp. 705–726. Harper and Row, New York, 1960. Also reprinted in: Samuel Leinhardt, ed., Social Networks: A Developing Paradigm, pp. 9–25. Academic Press, New York (1977).

[5] Seth Chaiken, A combinatorial proof of the all minors matrix tree theorem, *SIAM J. Algebraic Discrete Methods*, **3** (1982) 319–329. MR 83h:05062. Zbl. 495.05018.

[6] Chawathe P. D. and Vijayakumar G. R., A characterisation of signed graphs represented by root system D_∞, *European J. Combin.*, **11** (1990) 523–533. MR 91k:05071. Zbl. 764.05090.

[7] Jack Edmonds and Ellis L. Johnson, Matching: a well-solved class of integral linear programs. In: Richard Guy *et al.*, eds., Combinatorial Structures and Their Applications (Proc. Calgary Int. Conf., Calgary, 1969), pp. 89–92. Gordon and Breach, New York (1970). MR 42 #2799. Zbl. 258.90032.

[8] A. M. H. Gerards, Graphs and Polyhedra: Binary Spaces and Cutting Planes. CWI Tract, 73. Centrum voor Wiskunde en Informatica, Amsterdam (1990). MR 92f:52027. Zbl. 727.90044.

[9] Chris Godsil and Gordon Royle, Algebraic Graph Theory, Springer-Verlag, New York (2001). MR 2002f:05002. Zbl. 968.05002.

[10] F. Harary, On the notion of balance of a signed graph. *Michigan Math. J.*, **2** (1953–54) 143–146 and addendum preceding p. 1. MR 16, 733h. Zbl. 056.42103.

[11] F. Harary, Structural duality, *Behavioral Sci.*, **2** (1957) 255–265. MR 24B #B851.

[12] F. Harary, On the measurement of structural balance, *Behavioral Sci.*, **4** (1959) 316–323. MR 22 #3696.

[13] Frank Harary and Robert Z. Norman, Some properties of line digraphs, *Rend. Circ. Mat. Palermo* (2) **9** (1960), 161–168. MR 24 #A693. Zbl. 99, 182e (e: 099.18205).

[14] Frank Harary, Robert Z. Norman and Dorwin Cartwright, Structural Models: An Introduction to the Theory of Directed Graphs, Wiley, New York (1965). MR 32 #2345. Zbl. 139.41503.

[15] Alan J. Hoffman, On graphs whose least eigenvalue exceeds $-1-\sqrt{2}$, *Linear Algebra Appl.*, **16** (1977), 153–165. MR 57 #9607. Zbl. 354.05048.

[16] Yaoping Hou, Jiongsheng Li and Yongliang Pan, On the Laplacian eigenvalues of signed graphs, *Linear Multilinear Algebra*, **51** (2003) 21–30. MR 2003j:05084. Zbl. 1020.05044.

[17] Yury J. Ionin and Hadi Kharaghani, Balanced generalized weighing matrices and conference matrices. In: Charles J. Colbourn and Jeffrey H. Dinitz, eds., Handbook of Combinatorial Designs, second ed., Ch. V.6, pp. 306–312. Chapman & Hall/CRC Press, Boca Raton, Fla. (2007).

[18] R. H. Jeurissen, The incidence matrix and labellings of a graph, *J. Combin. Theory Ser. B*, **30** (1981) 290–301. MR 83f:05048. Zbl. 409.05042, 457.05047.

[19] Dénes König, Theorie der endlichen und unendlichen Graphen, Akademische Verlagsgesellschaft, Leipzig, 1936. Repr. Chelsea, New York (1950). MR 12, 195. Zbl. 013.22803. English trans. by Ronald C. Read, Theory of Finite and Infinite Graphs. Birkhäuser, Boston (1990). MR 91f:01026. Zbl. 695.05015.

[20] J. H. van Lint and J. J. Seidel, Equilateral point sets in elliptic geometry, *Proc. Koninkl. Ned. Akad. Wetenschap. Ser. A*, **69** (= *Indag. Math.*, **28**) (1966), 335–348. MR 34 #685. Zbl. 138.41702. Reprinted in [26, pp. 3–16].

[21] James McKee and Chris Smyth, Integer symmetric matrices having all their eigenvalues in the interval $[-2, 2]$, *J. Algebra*, **317** (2007) 260–290. MR 2008j:15038. Zbl. 1140.15007.

[22] Cyriel van Nuffelen, On the rank of the incidence matrix of a graph. Colloque sur la Theorie des Graphes (Bruxelles, 1973). *Cahiers Centre Etudes Rech. Oper.*, **15** (1973) 363–365. MR 50 #162. Zbl. 269.05116.

[23] E. Sampathkumar, Point signed and line signed graphs, *Nat. Acad. Sci. Letters (India)*, **7** (1984) 91–93. Zbl. 552.05051.

[24] J. J. Seidel, Strongly regular graphs with $(-1, 1, 0)$ adjacency matrix having eigenvalue 3. *Linear Algebra Appl.*, **1** (1968) 281–298. MR 38 #3175. Zbl. 159.25403. Reprinted in [26, pp. 26–43].

[25] J. J. Seidel, A survey of two-graphs. In: Colloquio Internazionale sulle Teorie Combinatorie (Rome, 1973), Vol. I, pp. 481–511. Accademia Nazionale dei Lincei, Rome, 1976. MR 58 #27659. Zbl. 352.05016. Reprinted in [26, pp. 146–176].

[26] J. J. Seidel, Geometry and Combinatorics: Selected Works of J. J. Seidel, D. G. Corneil and R. Mathon, eds. Academic Press, Boston (1991). MR 92m:01098. Zbl. 770.05001.

[27] N. M. Singhi and G. R. Vijayakumar, Signed graphs with least eigenvalue < -2, *European J. Combin.*, **13** (1992) 219–220. MR 93e:05069. Zbl. 769.05065.

[28] Tadeusz Sozański, Processus d'équilibration et sous-graphes équilibrés d'un graphe signé complet, *Math. Sci. Humaines*, No. 55 (1976) 25–36, 83. MR 58 #27613.

[29] W. T. Tutte, On chain-groups and the factors of graphs. In: L. Lovász and Vera T. Sós, eds., Algebraic Methods in Graph Theory, (Proc. Colloq., Szeged, 1978), Vol. 2, pp. 793–818. Colloq. Math. Soc. János Bolyai, 25. János Bolyai Math. Soc., Budapest, and North-Holland, Amsterdam, 1981. MR 83b:05104. Zbl. 473.05023.

[30] G. R. Vijayakumar, Signed graphs represented by D_∞. *European J. Combin.*, **8** (1987) 103–112. MR 88b:05111. Zbl. 678.05058.

[31] G. R. Vijayakumar, Signed graphs represented by root system E_8, Combinatorial Math. and Appl. (Proc., Calcutta, 1988), *Sankhya. Ser. A*, **54** (1992) 511–517. MR 94d:05072. Zbl. 882.05118.

[32] G. R. Vijayakumar, Representation of signed graphs by root system E_8, *Graphs Combin.*, **10** (1994) 383–388. MR 96a:05128. Zbl. 821.05040.

[33] G. R. Vijayakumar and N. M. Singhi, Some recent results on signed graphs with least eigenvalues ≥ -2. In: Dijen Ray-Chaudhuri, ed., Coding Theory and Design Theory. Part I: Coding Theory (Proc. Workshop IMA Program Appl. Combin., Minneapolis, 1987–88), pp. 213–218. Springer-Verlag, New York (1990). MR 91e:05069. Zbl. 711.05033.

[34] Thomas Zaslavsky, Characterizations of signed graphs, *J. Graph Theory*, **5** (1981) 401–406. MR 83a:05122. Zbl. 471.05035.

[35] Thomas Zaslavsky, Signed graphs, *Discrete Appl. Math.*, **4** (1982) 47–74. Erratum, *Discrete Appl. Math.*, **5** (1983) 248. MR 84e:05095. Zbl. 503.05060.

[36] Thomas Zaslavsky, Line graphs of switching classes. In: Report of the XVIIIth O.S.U. Denison Maths Conference (Granville, Ohio, 1984), pp. 2–4. Dept. of Mathematics, Ohio State University (1984).

[37] Thomas Zaslavsky, A mathematical bibliography of signed and gain graphs and allied areas, *Electronic J. Combin.*, Dynamic Surveys in Combinatorics (1998), No. DS8 (electronic). MR 2000m:05001a. Zbl. 898.05001. Current version on the Web at www.math.binghamton.edu/zaslav/Bsg/.

[38] Thomas Zaslavsky, Glossary of signed and gain graphs and allied areas, *Electronic J. Combin.*, Dynamic Surveys in Combinatorics (1998), No. DS9 (electronic). MR 2000m:05001b. Zbl. 898.05002. Current version on the Web at www.math.binghamton.edu/zaslav/Bsg/.

[39] Thomas Zaslavsky, Line graphs of signed graphs and digraphs. In preparation.

[40] Thomas Zaslavsky, What is a strongly regular signed graph? In preparation.

Proc. ICDM 2008, RMS-Lecture Notes Series No. 13, 2010, pp. 231–241.

On Cordiality of One Point Unions and One Edge Unions of Some Classes of Graphs

Ujwala Deshmukh[1] and N. B. Limaye[2,*]

[1]Mithibhai College of Arts, and Chauhan Institute of Science, J.V.P.D. Scheme, Vile Parle, Vile Parle(east), Mumbai 400 056, India
[2]Department of Mathematics, I.I.T. Bombay, Powai, Mumbai 400 076, India
e-mail: ujwala_deshmukh@rediffmail.com; nirmala_limaye@yahoo.co.in

Abstract. In this paper we prove that one edge unions of shells of arbitrary sizes, one point unions of complete bipartite graphs of arbitrary sizes as well as one edge unions of complete bipartite graphs of arbitrary sizes are cordial. This proves a conjecture of G. Sethuraman and P. Selvaraju.

Keywords. Graph labeling, cordial labeling, one point union, one edge union.

2000 Subject Classification: 05C78

1. Introduction

For a simple graph G, we denote the vertex set and the edge set of G by $V(G)$ and $E(G)$ respectively. A vertex labeling of G is a map $f : V(G) \longrightarrow \{0, 1\}$. For each edge $e = xy$, we can assign the label $f(e) = |f(x) - f(y)|$. The labeling f is said to be a **cordial labeling** if the number $v_f(0)$ of vertices labeled 0 and the number $v_f(1)$ of vertices labeled 1 differ by at most 1, and the number $e_f(0)$ of edges labeled 0 and the number $e_f(1)$ of edges labeled 1 differ by at most 1. If such a cordial labeling exists, then the graph G is said to be cordial. A binary labeling which is not cordial is refered to as non-cordial labeling.

A shell graph $H(n, n-3)$ is the graph obtained from the cycle C_n of order n by adding $n-3$ concurrent chords, where these chords are concurrent at a vertex on the cycle. This vertex is called the apex of the shell. **One point union** of k shells of various sizes is obtained by taking k shells and identifying the apex. **One edge union** of k shells of various sizes is obtained by taking k shells and identifying an edge adjacent to the apex. The edge obained by this identification is called the **apex-edge.** The edges other than the apex edge are called **non-apex edges**. One point union as well as one edge union of complete bipartite graphs can be simmilarly defined.

Andar *et al.* [1] proved that one point union of shells is cordial. This was a generalisation of a result by Sze-Chin Shee and Yong Song Ho where they proved that one point union of copies of $H(n, n-3)$ is cordial. G. Sethuraman and P. Selvaraju [4]

*This work was done when this author was Emeritus Fellow in the scheme of University Grants Commission, Goverment of India.

proved that one edge union of copies of $H(n, n-3)$ is cordial and they conjectured that one edge union of k copies of shells of arbitrary sizes is cordial. They also proved that one point union of k copies of a complete bipartite graph is cordial and posed a similar conjecture about arbitrary sizes. In this paper we prove their conjectures.

2. One edge union of shell graphs

We first introduce many binary vertex labelings of shell graphs, some of which are cordial. Using these labelings we prove that a *double shell*, that is, a one edge union of two shells, is cordial. For a labeling f of G by $v_f(i), i = 0, 1$, (respectively $e_f(i)$), we mean the number of vertices (respectively edges) with label i. These are called the *vertex numbers* and the *edge numbers*.

Let $V(H(n, n-3)) = \{u, v_1, \ldots, v_{n-1}\}$, where u is the apex and let $E(H(n, n-3)) = \{uv_i | 1 \leq i \leq n-1\} \cup \{v_i v_{i+1} | 1 \leq i \leq n-2\}$. Clearly $|E(H(n, n-3))| = 2n-3$. Consider the binary labelings of S_n defined as follows:

(i) $g_1(u) = 0, g_1(v_i) = 1$, if $i \equiv 1, 2 \bmod 4$ and $g_1(v_i) = 0$, if $i \equiv 0, 3 \bmod 4$. (Cordial for all n.)

(ii) $g_2(u) = 0, g_2(v_i) = 0$, if $i \equiv 1, 2 \bmod 4$ and $g_2(v_i) = 1$, if $i \equiv 0, 3 \bmod 4$. (Cordial iff $n \equiv 1 \bmod 4$.)

(iii) $g_3(u) = 0, g_3(v_i) = 1$, if $i \equiv 1, 2 \bmod 4$ and $g_3(v_i) = 0$, if $i \equiv 0, 3 \bmod 4$, $i \neq n-3, n-1$. Let $g_3(v_{n-3}) = 0$ and $g_3(v_{n-1}) = 1$. (Cordial for all n.)

(iv) $g_4(u) = 0, g_4(v_i) = 1$, if $i \equiv 1, 2 \bmod 4$ and $g_4(v_i) = 0$, if $i \equiv 0, 3 \bmod 4$, $i \neq n-2$. Let $g_4(v_{n-2}) = 0$. (Cordial iff $n \equiv 1, 2, 3 \bmod 4$.)

(v) $g_5(u) = 0, g_5(v_i) = 1$, if $i \equiv 1, 2 \bmod 4$ and $g_5(v_i) = 0$, if $i \equiv 0, 3 \bmod 4$, $i \neq n-1$. Let $g_5(v_{n-1}) = 0$. (Cordial iff $n \equiv 0, 1, 3 \bmod 4$.)

(vi) $g_6(u) = 0, g_6(v_i) = 1$, if $i \equiv 1, 2 \bmod 4$, and $g_6(v_i) = 0$, if $i \equiv 0, 3 \bmod 4$, $i \neq n-2$. Let $g_6(v_{n-2}) = 1$. (Cordial if $n \equiv 1 \bmod 4$.)

Table 1 gives the data of the labelings used in the next theorem:
The one edge union of k shells of sizes $n_1, n_2, \ldots, n_k$ is denoted by $EMS(n_1, n_2, \ldots, n_k)$. A double shell $EMS(n_1, n_2)$ is called **homogeneous** if n_1 and n_2 have the same parity.

Table 1.

Labeling f	n	$v_f(0)$	$v_f(1)$	$e_f(0)$	$e_f(1)$
g_1	$4t+3$ $2t$	$2t+1$ t	$2t+2$ t	$4t+1$ $2t-2$	$4t+2$ $2t-1$
g_3	$4t+1$	$2t+1$	$2t$	$4t-1$	$4t$
	$4t+3$	$2t+1$	$2t+2$	$4t+1$	$4t+2$
g_5	$4t+3$	$2t+2$	$2t+1$	$4t+1$	$4t+2$
g_6	$4t+1$	$2t$	$2t+1$	$4t-1$	$4t$

Table 2.

n_1	h_1	n_2	h_2	$v_f(0)$	$v_f(1)$	$e_f(0)$	$e_f(1)$
$2t_1$	g_1	$2t_2$	g_1	t_1+t_2-1	t_1+t_2-1	n_1+n_2-4	n_1+n_2-3
$4t_1+3$	g_1	$4t_2+3$	g_5	$2t_1+2t_2+2$	$2t_1+2+2t_2$	$4t_1+4t_2+2$	$4t_1+4t_2+3$
$4t_1+1$	g_3	$4t_2+3$	g_1	$2t_1+1+2t_2$	$2t_1+2t_2+1$	$4t_1+4t_2$	$4t_1+4t_2+1$
$4t_1+1$	g_3	$4t_2+1$	g_6	$2t_1+2t_2$	$2t_1+2t_2$	$4t_1+4t_2-2$	$4t_1+4t_2-1$
$2t_1$	g_1	$4t_2+1$	g_3	t_1+2t_2	t_1+2t_2-1	$2t_1+4t_2-3$	$2t_1+4t_2-2$
$2t_1$	g_1	$4t_2+3$	g_3	t_1+2t_2	t_1+2t_2+1	$2t_1+4t_2-1$	$2t_1+4t_2$

Theorem 1. *All double shells $EMS(n_1, n_2)$ are cordial for all $n_1, n_2 \geq 4$.*

Proof. Let the two individual shells involved in the one edge union be H_1 and H_2, and let their sizes be n_1 and n_2 respectively. Clearly, $|V(EMS(n_1, n_2))| = n_1 + n_2 - 2$ and $|E(EMS(n_1, n_2))| = 2(n_1 + n_2) - 7$. Hence the number of vertices is even when this multiple shell is homogeneous. We assign a labeling h_i to H_i, $i = 1, 2$.

Table 2 shows the choices of h_1, h_2 according to the values of n_1, n_2. From the values of $v_f(i), e_f(i), i = 0, 1$, it is clear that the resulting labeling f is cordial.

This completes the proof. ⋆

It follows that single shells as well as double shells are cordial. We use this to prove the cordiality of one edge union of arbitrary number of shell graphs. We note here that for all the homogeneous double shells, the above theorem gives a cordial labeling which assigns the label 1 to one extra edge, that is, $e_f(1) = e_f(0) + 1$. This means that the number of non-apex edges with label 0 is the same as the number of non-apex edges with label 1. The apex edge always gets the label 1. This fact can be used to prove the following result.

Theorem 2. *The one edge union $EMS(n_1, \ldots, n_k)$ of k shells of sizes $n_1, n_2, \ldots, n_k$ is cordial.*

Proof. Consider $EMS(n_1, n_2, \ldots, n_k)$. First, form all possible homogeneous double shells out of the k shells and assign them the labeling given by Theorem 1. Here the apex edge continues to get the label 1. Since in all these double shells the vertex numbers are all equal and the number of non-apex edges with the label 0 is the same as the number of non-apex edges with the label 1, for the final counting of the vertex numbers as well as the edge numbers of the final labeling, we observe that these vertices as well as non-apex edges labeled so far do not give any problem. It is possible that some shells are still to be labeled after forming homogeneous pairs. There are only three possibilities:

(1) Only one shell of an even order is remaining.
(2) Only one shell of an odd order is remaining.
(3) One non-homogeneous double shell is remaining.

For all these, we have a cordial labeling assigning the label 1 to the apex edge. Hence we can complete the labeling of $EMS(n_1, \ldots, n_k)$ which is a cordial labeling. ⋆

Example 1. Suppose we have $EMS(6,14,12,7,9,11)$. There are 47 vertices and 95 edges. We form two homogeneous shells $EMS(6,14)$ and $EMS(7,9)$. In $EMS(6,14)$, assign the labeling g_1 to bothe the shells. These labelings will produce 9 vertices with the label 0 and 9 vertices with the label 1. The apex edge gets the label 1 and the number of non-apex edges with the label 0 is 16, which is the same as the non-apex edges with the label 1. In $EMS(7,9)$, assign the labeling g_1 to the shell $H(7,4)$ and the labeling g_3 to the shell $H(9,6)$. These labelings will produce 7 vertices with the label 0 and 7 vertices with the label 1. The apex edge gets the label 1, and the number of non-apex edges with the label 0 is 11, which is the same as the non-apex edges with the label 1.

Now the shells $H(12,9)$ and $H(11,8)$ remain to be labeled. The last row of the previous table shows that we have to assign the labeling g_1 to the shell $H(12,9)$ and the labeling g_3 to the shell $H(11,8)$. The label 1 of the apex edge remains unchanged. There are additional 10 vertices with the label 0, and 11 vertices with the label 1. Similarly there are 19 non-apex edges with the label 0, and 19 non-apex edges with the label 1.

If f denotes the resulting labeling, then the final vertex numbers and edge numbers are $v_f(0) = 23, v_f(1) = 24, e_f(0) = 47, e_f(1) = 48$. Thus, f is a cordial labeling. ⋆

3. Complete bipartite graphs

For the complete bipartite graph $K_{a,b}$, we take the vertex set and the edge set as follows: $V(K_{a,b}) = \{u_1, u_2, \ldots, u_a\} \cup \{v_1, v_2, \ldots, v_b\}$ and $E(K_{a,b}) = \{u_i v_j | 1 \leq i \leq a, 1 \leq j \leq b.\}$.

We will classify the complete bipartite graphs as follows:

Type I. $K_{2m+1,2n+1}$. We define labeling h, h_1 as follows.

Let $h(u_i) = 0, 1 \leq i \leq m$ and $h(u_i) = 1, m+1 \leq i \leq 2m+1$, while $h(v_j) = 0, 1 \leq j \leq n+1$ and $h(v_j) = 1, n+2 \leq j \leq 2n+1$. Clearly, $v_h(0) = m+n+1 = v_h(1)$ and $e_h(0) = 2mn + m + n$, $e_h(1) = 2mn + m + n + 1$. (cordial labeling)
Let $h_1(u_i) = 0, 1 \leq i \leq m+1$ and $h_1(u_i) = 1, m+2 \leq i \leq 2m+1$, while $h_1(v_j) = 0, 1 \leq j \leq n+1$ and $h_1(v_j) = 1, n+2 \leq j \leq 2n+1$. Clearly, $v_{h_1}(0) = m+n+2, v_{h_1}(1) = m+n$ and $e_{h_1}(0) = 2mn+m+n+1, e_h(1) = 2mn+m+n$. (non-cordial labeling.)

Type II(A). $K_{2m,2n+1}$. We define three labelings h_2, h_3, h_4 as follows.

Let $h_2(u_i) = 0, 1 \leq i \leq m$ and $h_2(u_i) = 1, m+1 \leq i \leq 2m$, while $h_2(v_j) = 0, 1 \leq j \leq n+1$ and $h_2(v_j) = 1, n+2 \leq j \leq 2n+1$. Clearly, $v_{h_2}(0) = m+n+1, v_{h_2}(1) = m+n$ and $e_{h_2}(0) = 2mn + m = e_{h_2}(1)$. (cordial labeling.)

Let $h_3(u_i) = 0, 1 \leq i \leq m-1$ and $h_3(u_i) = 1, m \leq i \leq 2m$, while $h_3(v_j) = 0, 1 \leq j \leq n+1$ and $h_3(v_j) = 1, n+2 \leq j \leq 2n+1$. Clearly, $v_{h_3}(0) = m+n, v_{h_3}(1) = m+n+1$ and $e_{h_3}(0) = 2mn+m-1, e_{h_3}(1) = 2mn+m+1$. (non-cordial labeling.)

Let $h_4(u_i) = 0, 1 \leq i \leq m+1$ and $h_4(u_i) = 1, m+2 \leq i \leq 2m$, while $h_4(v_j) = 0, 1 \leq j \leq n+1$ and $h_4(v_j) = 1, n+2 \leq j \leq 2n+1$. Clearly, $v_{h_4}(0) = m+n+2, v_{h_4}(1) = m+n-1$ and $e_{h_4}(0) = 2mn+m+1, e_{h_4}(1) = 2mn+m-1$. (non-cordial labeling.)

Type II(B). $K_{2m+1,2n}$. We define three labelings $\tilde{h}_2, \tilde{h}_3, \tilde{h}_4$ as follows.

Let $\tilde{h}_2(u_i) = 0, 1 \leq i \leq m+1$ and $\tilde{h}_2(u_i) = 1, m+2 \leq i \leq 2m+1$, while $\tilde{h}_2(v_j) = 0, 1 \leq j \leq n$ and $\tilde{h}_2(v_j) = 1, n+1 \leq j \leq 2n$. Clearly, $v_{\tilde{h}_2}(0) = m+n+1, v_{\tilde{h}_2}(1) = m+n$ and $e_{\tilde{h}_2}(0) = 2mn+m = e_{\tilde{h}_2}(1)$. (cordial labeling.)
Let $\tilde{h}_3(u_i) = 0, 1 \leq i \leq m+1$ and $\tilde{h}_3(u_i) = 1, m+2 \leq i \leq 2m+1$, while $\tilde{h}_3(v_j) = 0, 1 \leq j \leq n-1$ and $\tilde{h}_3(v_j) = 1, n \leq j \leq 2n$. Clearly, $v_{\tilde{h}_3}(0) = m+n, v_{\tilde{h}_3}(1) = m+n+1$ and $e_{\tilde{h}_3}(0) = 2mn+m-1, e_{\tilde{h}_3}(1) = 2mn+m+1$. (non-cordial labeling.)
Let $\tilde{h}_4(u_i) = 0, 1 \leq i \leq m+1$ and $\tilde{h}_4(u_i) = 1, m+2 \leq i \leq 2m+1$, while $\tilde{h}_4(v_j) = 0, 1 \leq j \leq n+1$ and $\tilde{h}_4(v_j) = 1, n+2 \leq j \leq 2n$. Clearly, $v_{\tilde{h}_4}(0) = m+n+2, v_{\tilde{h}_4}(1) = m+n-1$ and $e_{\tilde{h}_4}(0) = 2mn+m+1, e_{\tilde{h}_4}(1) = 2mn+m-1$. (non-cordial labeling.)

We note that the labeling $\tilde{h}_i$ of $K_{2m+1,2n}$ is essentially the same as the labeling h_i of $K_{2n,2m+1}$. We list them separately to be able to use in some one point and one edge unions.

Type III. $K_{2m,2n}$. We define labelings h_5, h_6, h_7 as follows.

Let $h_5(u_i) = 0, 1 \leq i \leq m$ and $h_5(i) = 1, m+1 \leq i \leq 2m$, while $h_5(v_j) = 0, 1 \leq j \leq n$ and $h_5(j) = 1, n+1 \leq j \leq 2n$. Clearly, $v_{h_5}(0) = m+n = v_{h_5}(1)$ and $e_{h_5}(0) = 2mn = e_{h_5}(1)$. (cordial labeling.)
Let $h_6(u_i) = 0, 1 \leq i \leq m$ and $h_6(u_i) = 1, m+1 \leq i \leq 2m$, while $h_6(v_j) = 0, 1 \leq j \leq n+1$ and $h_6(v_j) = 1, n+2 \leq j \leq 2n$. Clearly, $v_{h_6}(0) = m+n+1, v_{h_6}(1) = m+n-1$ and $e_{h_6}(0) = 2mn = e_{h_6}(1)$. (non-cordial labeling.)
Let $h_7(u_i) = 0, 1 \leq i \leq m-1$ and $h_7(i) = 1, m \leq i \leq 2m$, while $h_7(v_j) = 0, 1 \leq j \leq n+1$ and $h_7(v_j) = 1, n+2 \leq j \leq 2n$. Clearly, $v_{h_7}(0) = m+n = v_{h_7}(1)$ and $e_{h_7}(0) = 2mn-2, e_{h_7}(1) = 2mn+2$. (non-cordial labeling).

Table 3 gives the data of these labelings:

Table 3.

Graph	Labeling θ	$v_\theta(0)$	$v_\theta(1)$	$e_\theta(0)$	$e_\theta(1)$
$K_{2m+1,2n+1}$	h	$m+n+1$	$m+n+1$	$2mn+m+n$	$2mn+m+n+1$
	h_1	$m+n+2$	$m+n$	$2mn+m+n+1$	$2mn+m+n$
$K_{2m,2n+1}$	h_2	$m+n+1$	$m+n$	$2mn+m$	$2mn+m$
	h_3	$m+n$	$m+n+1$	$2mn+m-1$	$2mn+m+1$
	h_4	$m+n+2$	$m+n-1$	$2mn+m+1$	$2mn+m-1$
$K_{2m,2n}$	h_5	$m+n$	$m+n$	$2mn$	$2mn$
	h_6	$m+n+1$	$m+n-1$	$2mn$	$2mn$
	h_7	$m+n$	$m+n$	$2mn-2$	$2mn+2$

We have not given the vertex numbers and edge numbers of $\tilde{h}_i, 2 \leq i \leq 4$ in the above table because they are the same as those of $h_i, 2 \leq i \leq 4$.

3.1 *One point union of complete bipartite graphs*

If $G_1, \ldots, G_t$ are complete bipartite graphs involved in one point union we denote their vertices by double sufixes, that is, as $u_{1,1}, u_{2,1}, \ldots$. The point of identification is obtained by identifying the vertices $u_{1,1}, u_{2,1}, \ldots, u_{t,1}$. Note that all the labelings mentioned here assign the label 0 to this vertex.

A double bipartite graph is one point union of two complete bipartite graphs. Such a double bipartite graph is called **homogeneous** if both the bipartite graphs are of the same type.

Theorem 3. *A homogeneous double bipartite graph is cordial.*

Proof. Let G be the one point union of two complete bipartite graphs G_1 and G_2 which are of the same type. We assign labelings to these two graphs and call the combined labeling f. This is possible since all the labelings used here assign the label 0 to the vertices which are identified.

Case 1. $G_i = K_{2m_i+1,2n_i+1}, i = 1, 2$. Assign the labeling h to G_1 and the labeling h_1 to G_2. Clearly, $v_f(0) = m_1 + n_1 + m_2 + n_2 + 2, v_f(1) = m_1 + n_1 + m_2 + n_2 + 1$ and $e_f(0) = 2m_1n_1 + m_1 + n_1 + 2m_2n_2 + m_2 + n_2 + 1 = e_f(1)$.

Case 2(a). $G_i = K_{2m_i,2n_i+1}, i = 1, 2$. Assign the labeling h_3 to G_1 and the labeling h_4 to G_2. Clearly, $v_f(0) = m_1 + n_1 + m_2 + n_2 + 1, v_f(1) = m_1 + n_1 + m_2 + n_2$ and $e_f(0) = 2m_1n_1 + m_1 + 2m_2n_2 + m_2 = e_f(1)$.

Case 2(b). $G_1 = K_{2m_1,2n_1+1}$ and $G_2 = K_{2m_2+1,2n_2}$. We assign the labeling h_3 to G_1 and the labeling $\tilde{h}_4$ to G_2. The vertex numbers and edge numbers remain unchanged.

Case 3. $G_i = K_{2m_i,2n_i}, i = 1, 2$. Assign the labeling h_5 to G_1 and the labeling h_6 to G_2. Clearly, $v_f(0) = m_1 + n_1 + m_2 + n_2, v_f(1) = m_1 + n_1 + m_2 + n_2 - 1$ and $e_f(0) = 2m_1n_1 + 2m_2n_2 = e_f(1)$.

This shows that in all the cases the homogeneous double bipartite graph is cordial. ⋆

Example 2. The following example shows why the labelings $\tilde{h}_i, 2 \leq i \leq 4$ are necessary. Let $G_1 = K4, 5$ and $G_2 = K_{5,4}$. Consider one point union of two copies of $K_{4,5}$. The resulting graph is again bipartite with partite sets of sizes 7 and 10. On the other hand the one point union of G_1 and G_2 has partite sets of sizes 8 and 9. Naturally they will need different labelings. ⋆

One notes that when the bipartite graphs involved in one point union (or in one edge union) are of type II(B), the sizes of partite sets in the resulting graph change. However, the total number of vertices and the total number of edges remain the same. When we find a cordial labeling the vertex numbers and the edge numbers also remain unchanged.

Remark 1. One observes that the cordial labelings obtained in this theorem exhibit one common feature, namely, $v_f(0) = v_f(1) + 1$ and $e_f(0) = e_f(1)$. Moreover, the common apex has the label 0. Thus, the points other than the apex as well as the edges receive the labels 0 and 1 equitably. ⋆

Theorem 4. *One point union of arbitrary number of complete bipartite graphs is cordial.*

Proof. Let G be a one point union of a family of complete bipartite graphs. First form all the possible homogeneous pairs and assign them the labeling described in Theorem 3. Call this labeling of the subgraph g_1. Thus, $v_{g_1}(0) = v_{g_1}(1)+1$, $e_{g_1}(0) = e_{g_1}(1)$. Because of the aforesaid remark, the apex has received the label 0 and all the other vertices as well as edges are labeled equitably. So for the labeling of the remainder of the graph, we can ignore all these vertices as well as edges. The remainder falls into the following cases:

Case 1. Only one complete bipartite graph remains to be labeled. In this case, we can assign the labeling h, h_2 or $\tilde{h}_2$, h_5 if it is of type I, II(A), II(B) or III respectively. This will complete the cordial labeling of G.

Case 2. Suppose two complete graphs F_1, F_2 remain to be labeled.

If $F_1 = K_{2m_1+1,2n_1+1}$ and $F_2 = K_{2m_2,2n_2}$, then assign the labeling h to F_1 and the labeling h_6 to F_2. Call this labeling g_2. Clearly, $v_{g_2}(0) = m_1 + n_1 + m_2 + n_2 + 1 = v_{g_2}(1)+1$, $e_{g_2}(0) = 2n_1m_1+m_1+n_1+1+2m_2n_2 = e_{g_2}(1)-1$. If the resulting labeling of G is f, then $v_f(0) = v_{g_1}(0)+v_{g_2}(0)-1 = v_{g_1}(1)+1+v_{g_2}(1)+1-1 = v_f(1)+1$ and $e_f(0) = e_{g_1}(0) + e_{g_2}(0) = e_{g_1}(1) + e_{g_2}(1) - 1 = e_f(1) - 1$.

If $F_1 = K_{2m_1+1,2n_1+1}$ and $F_2 = K_{2m_2,2n_2+1}$ (respectively $K_{2m_2+1,2n_2}$), assign the labeling h to F_1 and the labeling h_2 (respectivelt $\tilde{h}_2$) to F_2. Call this labeling g_2. Clearly, $v_{g_2}(0) = m_1 + n_1 + m_2 + n_2 + 1 = v_{g_2}(1)$, $e_{g_2}(0) = 2n_1m_1 + m_1 + n_1 + 2m_2n_2 + m_2 = e_{g_2}(1) - 1$. If the resulting labeling of G is f, then $v_f(0) = v_{g_1}(0)+v_{g_2}(0)-1 = v_{g_1}(1)+1+v_{g_2}(1)-1 = v_f(1)$ and $e_f(0) = e_{g_1}(0)+e_{g_2}(0) = e_{g_1}(1) + e_{g_2}(1) - 1 = e_f(1) - 1$.

If $F_1 = K_{2m_1,2n_1}$ and $F_2 = K_{2m_2,2n_2+1}$ (respectively $K_{2m_2+1,2n_2}$) assign the labeling h_5 to F_1 and the labeling h_2 (respectively $\tilde{h}_2$) to F_2. Call this labeling g_2. Clearly, $v_{g_2}(0) = m_1 + n_1 + m_2 + n_2 = v_{g_2}(1)$, $e_{g_2}(0) = 2n_1m_1 + m_2 + 2m_2n_2 = e_{g_2}(1)$. If the resulting labeling of G is f then $v_f(0) = v_{g_1}(0) + v_{g_2}(0) - 1 = v_{g_1}(1) + 1 + v_{g_2}(1) - 1 = v_f(1)$ and $e_f(0) = e_{g_1}(0) + e_{g_2}(0) = e_{g_1}(1) + e_{g_2}(1) = e_f(1)$.

It is now clear that in this case the graph G is cordial.

Case 3. Suppose three complete bipartite graphs F_1, F_2, F_3 of Type I, II(A) (or II(B)), III respectively, remain to be labeled. Assign the labeling h to F_1, the labeling h_2 (respectively $\tilde{h}_2$) to F_2 and the labeling h_6 to F_3. Call the resulting labeling g_2. Then $v_{g_2}(0) = m_1 + n_1 + 1 + m_2 + n_2 + m_3 + n_3 = v_{g_2}(1) - 1$, $e_{g_2}(0) = 2m_1n_1 + m_1 + n_1 + 2m_2n_2 + m_2 + 2m_3n_3 = e_{g_2}(1) - 1$. If the resulting labeling of G is f, then $v_f(0) = v_{g_1}(0) + v_{g_2}(0) - 1 = v_{g_1}(1) + 1 + v_{g_2}(1) - 1 = v_f(1)$ and $e_f(0) = e_{g_1}(0) + e_{g_2}(0) = e_{g_1}(1) + e_{g_2}(1) - 1 = e_f(1) - 1$. So in this case also G is cordial. ⋆

3.2 *One edge union of complete bipartite graphs*

If K_{a_r,b_r}, $1 \le r \le t$, is a family of complete bipartite graphs with partite sets $\{u_{r,1}, \ldots u_{r,a_r}\}$ and $\{v_{r,1}, \ldots, v_{r,b_r}\}$, we identify the edges $u_{r,1}v_{r,b_r}$, $1 \le r \le t$, to obtain the one edge union of these graphs. All the labelings used in this paper give the labels 0 and 1 to $u_{r,1}$, v_{r,b_r} respectively. This means the apex edge gets the label 1 in all the labelings.

Remark 2. In 1987, Cahit [2] proved that an Eulerain graph G whose number of edges is $\equiv 2 \bmod 4$, is not cordial.

Consider a set of complete bipaprtite graphs $G_r = K_{2m_r,2n_r}$, $1 \le r \le 4t+3$, that is, consider $4t+3$ complete bipartite graphs of Type III. Identify an edge to obtain one edge union G of these graphs. The vertices used for identification have degrees $\sum_{r=1}^{4t+3} 2m_r - 4t - 2$ and $\sum_{r=1}^{4t+3} 2n_r - 4t - 2$ respectively. Other vertices have even degree. This means G is Eulerian. The total number of edges is $\sum_{r=1}^{4t+3}(4m_rn_r - 1) + 1 = \sum_{r=1}^{4t+3} 4m_rn_r - 4t - 2$. This number is clearly $\equiv 2 \bmod 4$. Hence G is not cordial.

⋆

Theorem 5. *One edge union of t number of complete bipartite graphs of the same type is cordial if they all belong to Type I or Type II and if they are of Type III with $t \not\equiv 3 \bmod 4$.*

Proof. Let $G_r = K_{a_r,b_r}$, $1 \le r \le t$, be complete bipartite graphs. Let G be the graph obtained by forming the one edge union of these graphs.

Case 1. They are all of Type I. Let $a_r = 2m_r + 1, b_r = 2n_r + 1, 1 \le r \le t$. Let f be the binary labeling of G obtained by assigning the labeling h to all G_r, $1 \le r \le t$. Using induction on r, one can see that $v_f(0) = \sum_{r=1}^{t}(m_r + n_r) + 1 = v_f(1)$ and $e_f(0) = \sum_{r=1}^{t}(2m_rn_r + m_r + n_r)$, $e_f(1) = \sum_{r=1}^{t}(2m_rn_r + m_r + n_r) + 1$. Hence G is cordial.

Case 2. They are all of Type II. Here some of them may be of Type II(A) and some may be of Type II(B). Only the labelings differ for them. But the vertex numbers and the edge numbers remain the same. Let $G_r = K_{2m_r,2n_r+1}$. if it is of type II(A) and $G_r = K_{2m_r+1,2n_r}$ if it is of Type II(B). Let f be the labeling obtained by assigning the labeling h_2 (or $\tilde{h}_2$ in case it is of Type II(B)) to G_r when r is odd and the labeling h_3 (or $\tilde{h}_3$) to G_r when r is even. Then, $|V(G)| = \sum_{r=1}^{t}(2m_r + 2n_r - 1) + 2$ and $|E(G)| = \sum_{r=1}^{t}(4m_rn_r + 2m_r - 1) + 1$.

If $t = 2s$, then $v_f(0) = \sum_{r=1}^{2s}(m_r + n_r) - (s-1) = v_f(1)$ and $e_f(0) = \sum_{r=1}^{t}(2m_rn_r + m_r) - s$, whereas, $e_f(1) = \sum_{r=1}^{t}(2m_rn_r + m_r) - (s-1)$.

If $t = 2s+1$, then $v_f(0) = \sum_{r=1}^{2s}(m_r + n_r) - (s-1) = v_f(1) + 1$ and $e_f(0) = \sum_{r=1}^{t}(2m_rn_r + m_r) - s$, whereas, $e_f(1) = \sum_{r=1}^{t}(2m_rn_r + m_r) - s$, that is, $e_f(0) = e_f(1)$. Hence G is cordial.

Case 3. They are all of Type III but $t \not\equiv 3 \bmod 4$. If t is equal to 1, we have the labeling h_5, which is a cordial labeling. If $t = 2$, assign the labeling h_5 to both G_1 and G_2. Call this labeling g. Clearly, $v_g(0) = v_g(1)$ and $e_g(0) = e_g(1) + 1$, that is, g is a

Table 4.

Types of bipartite graphs	Number of graphs	Number of vertices	Relation between $v_f(0)$ and $v_f(1)$	Relation between $e_f(0)$ and $e_f(1)$
Type I	t	$2T$	$v_f(0) = v_f(1)$	$e_f(1) = e_f(0) + 1$
Type II	$2s$	$2T$	$v_f(0) = v_f(1)$	$e_f(1) = e_f(0) + 1$
	$2s + 1$	$2T + 1$	$v_f(0) = v_f(1) + 1$	$e_f(1) = e_f(0)$
Type III	$4s$	$2T$	$v_f(0) = v_f(1)$	$e_f(1) = e_f(0) + 1$
	$4s + 1$	$2T$	$v_f(0) = v_f(1)$	$e_f(1) = e_f(0)$
	$4s + 2$	$2T$	$v_f(0) = v_f(1)$	$e_f(1) = e_f(0) - 1$
labeling g_3	$4s + 3$	$2T$	$v_f(0) = v_f(1)$	$e_f(1) = e_f(0) - 2$
labeling g_4	$4s + 3$	$2T$	$v_f(0) = v_f(1)$	$e_f(1) = e_f(0) + 2$

cordial labeling. If $t = 4$, assign the labeling h_5 to G_1, G_2, G_3 and the labeling h_7 to G_4. Call this labeling θ. One can check that $v_\theta(0) = \sum_{r=1}^{4} m_r + n_r - 3 = v_\theta(1)$ and $e_\theta(0) = \sum_{r=1}^{4} 2m_r n_r - 2$, $e_\theta(1) = \sum_{r=1}^{4} 2m_r n_r - 1$. Hence θ is a cordial labeling.

Finally, let $t = 4s + k$. Assign the labeling θ to the group $G_{4i+1}, G_{4i+2}, G_{4i+3}, G_{4i+4}, 0 \leq i \leq s - 1$. Call this labeling f. One can check that both $v_f(0)$ and $v_f(1)$ are equal to $(\sum_{j=0}^{s-1}(\sum_{\alpha=1}^{4} m_{4j+\alpha} + n_{4j+\alpha} - 3 - 1)) + 1 = (\sum_{i=1}^{4s}(m_i + n_i)) - 4s + 1$. Whereas, $e_f(0) = (\sum_{j=0}^{s-1}(\sum_{\alpha=1}^{4} 2m_{4j+\alpha}n_{4j+\alpha}) - 2) = \sum_{i=1}^{4s} 2m_i n_i - 2s$ and $e_f(1) = (\sum_{j=0}^{s-1}(\sum_{\alpha=1}^{4} 2m_{4j+\alpha}n_{4j+\alpha}) - 2) + 1 = \sum_{i=1}^{4s} 2m_i n_i - 2s + 1$.

This shows that the resulting labeling is cordial. If $k = 0$, we have obtained the required labeling.

If $k = 1$, one bipartite graph remains to be labeled. Assign the labeling h_5 to this bipartite graph. The above mentioned values of $v_f(0)$ and $v_f(1)$ increase by $m_t + n_t - 1$ and the values of $e_f(0)$ and $e_f(1)$ increase by $2m_t n_t$ and $2m_t n_t - 1$ respectively.

If $k = 2$, two bipartite graphs remains to be labeled. Assign the labeling h_5 to both. The above mentioned values of $v_f(0)$ and $v_f(1)$ increase by $m_{t-1}+n_{t-1}+m_t+n_t-2$ and the values of $e_f(0)$ and $e_f(1)$ increase by $2m_{t-1}n_{t-1} + 2m_t n_t$ and $2m_{t-1}n_{t-1} + 2m_t n_t - 2$ respectively. Since we have one extra edge with label 1 earlier, this still gives a cordial labeling with $e_f(0) = e_f(1) + 1$.

This completes the proof of the Theorem. ⋆

Remark 3. One edge union of t number of complete bipartite graphs is called *homogeneous* if all the complete bipartite graphs involved in it are of the same type. The labelings obtained in Theorem 5 are referred by f_1, f_2, f_3. If we have $4s+3$ graphs of Type III, after labeling $4s+2$ of them as shown in Theorem 5, we can assign the labeling h_5 to the additional graph to get a labeling g_3 such that $v_{g_3}(0) = v_{g_3}(1)$, $e_{g_3}(1) = e_{g_3}(0) - 2$. On the other hand, if we assign the labeling h_7 to the additional graph, then we obtain a labeling g_4 such that $v_{g_4}(0) = v_{g_4}(1)$, $e_{g_4}(1) = e_{g_4}(0) + 2$.

Table 4 gives the data for all the homogeneous one edge unions of complete bipartite graphs.

Andar *et al.* have defined the index $i(G)$ of cordiality of a graph G to be $\min\{|e_f(0) - e_f(1)|$, where the minimum is taken over all the binary labelings of $V(G)$ which satisfy the condition $|v_f(0) - v_f(1)| \leq 1$. Then for a homogeneous one edge union G of $4s + 3$ complete bipartite graphs of Type III, we have $i(G) = 2$.

Remark 4. We alter one of the labelings constructed in Theorem 5. Consider the *Case* 2 in which $2s + 1$ graphs of Type II are considered. If we assign the labeling h_2 to G_r when r is even and the labeling h_3 to G_r when r is odd and call the resulting labeling f_2', then one can check that

$$v_{f_2'}(0) = \left(\sum(m_i + n_i)\right) - s, v_{f_2'}(1) = \left(\sum(m_i + n_i)\right) - (s-1),$$

$$e_{f_2'}(0) = \left(\sum(2m_in_i + m_i)\right) - (s+1), e_{f_2'}(1) = \left(\sum(2m_in_i + m_i)\right) - (s-1).$$

Hence $v_{f_2'}(0) = v_{f_2'}(1) - 1, e_{f_2'}(1) = e_{f_2'}(0) + 2$. Although, this is not a cordial labeling, we will use this in the following Theorem. ⋆

Theorem 6. *Let G be one edge union of complete bipartite graphs. If there are t_1, t_2, t_3 complete bipartite graphs of Type I, II, III, respectively, then G is cordial if the number $t_3 \not\equiv 3(mod\, 4)$.*

Proof. Let G be one edge union of number of complete bipartite graphs. First form homogeneous one edge unions G_1, G_2, G_3 of complete bipartite graphs of Type 1, 2, 3 respectively. Assign them the labeling f_i, $1 \leq i \leq 3$ given by Theorem 5. Call the resulting labeling f. Note that this is possible since each f_i assigns the same labels to the vertices obtained by identifying the two vertices in all the graphs involved in this one edge union. Let t_i, be the number of graphs involved in G_i, $1 \leq i \leq 3$. Clearly, if one of t_1, t_2, t_3 is 0, then $v_f(0) = \sum v_{f_i}(0) - 1, v_f(1) = \sum v_{f_i}(1) - 1, e_f(0) = \sum e_{f_i}(0), e_f(1) = \sum e_{f_i}(1) - 1$, and if none of the t_1, t_2, t_3 is 0, then $v_f(0) = \sum_1^3 v_{f_i}(0) - 2, v_f(1) = \sum_1^3 v_{f_i}(1) - 2$ and $e_f(0) = \sum_1^3 e_{f_i}(0), e_f(1) = \sum_1^3 e_{f_i}(1) - 2$. Table 5 gives the relationship between $v_f(0), v_f(1)$ and $e_f(0), e_f(1)$ which can be varified by the table in Remark 3.

The only cases not covered by this table are the following:

If $t_1 = 0, t_2 = 2s_2 + 1, t_3 = 4s_3 + 2$, then assign the labeling f_2' (defined in Remark 4) to G_2 and the labeling f_3 to G_3. One can check that $v_f(0) = v_f(1) - 1, e_f(0) = e_f(1)$.

If $t_1 = t, t_2 = 2s_2 + 1, t_3 = 4s_3 + 2$, then assign the labeling f_1 to G_1, the labeling f_2' (defined in Remark 4) to G_2 and the labeling f_3 to G_3. One can check that $v_f(0) = v_f(1) - 1, e_f(0) = e_f(1) - 1$.

This completes the proof of the Theorem. ⋆

Finally, let us consider the case when the number t_3 of bipartite graphs of type III is $\equiv 3(\text{mod } 4)$. Here we have to put a restriction on t_2.

Theorem 7. *If G is a one edge union of t_1, t_2, t_3 complete bipartite graphs of Type I, II, III respectively, then G is cordial if t_2 is odd and $t_3 = 4s_3 + 3$.*

Proof. Let G_i, $1 \leq i \leq 3$ be the one edge unions of complete bipartite graphs involved in G. Assign the labelings f_1 to G_1, the labeling f_2' to G_2 and the labeling g_3 to G_3. Call the resulting labeling f.

Table 5.

t_1	t_2	t_3	Relation between $v_f(0)$ and $v_f(1)$	Relation between $e_f(0)$ and $e_f(1)$
t	$2s_2$	0	$v_f(0) = v_f(1)$	$e_f(1) = e_f(0) + 1$
t	$2s_2 + 1$	0	$v_f(0) = v_f(1) + 1$	$e_f(1) = e_f(0)$
t	0	$4s_3$	$v_f(0) = v_f(1)$	$e_f(1) = e_f(0) + 1$
t	0	$4s_3 + 1$	$v_f(0) = v_f(1)$	$e_f(1) = e_f(0)$
t	0	$4s_3 + 2$	$v_f(0) = v_f(1)$	$e_f(1) = e_f(0) - 1$
0	$2s_2$	$4s_3$	$v_f(0) = v_f(1)$	$e_f(1) = e_f(0) + 1$
0	$2s_2$	$4s_3 + 1$	$v_f(0) = v_f(1)$	$e_f(1) = e_f(0)$
0	$2s_2$	$4s_3 + 2$	$v_f(0) = v_f(1)$	$e_f(1) = e_f(0) - 1$
0	$2s_2 + 1$	$4s_3$	$v_f(0) = v_f(1) + 1$	$e_f(1) = e_f(0)$
0	$2s_2 + 1$	$4s_3 + 1$	$v_f(0) = v_f(1) + 1$	$e_f(1) = e_f(0) - 1$
t	$2s_2$	$4s_3$	$v_f(0) = v_f(1)$	$e_f(1) = e_f(0) + 1$
t	$2s_2$	$4s_3 + 1$	$v_f(0) = v_f(1)$	$e_f(1) = e_f(0)$
t	$2s_2$	$4s_3 + 2$	$v_f(0) = v_f(1)$	$e_f(1) = e_f(0) - 1$
t	$2s_2 + 1$	$4s_3$	$v_f(0) = v_f(1) + 1$	$e_f(1) = e_f(0)$
t	$2s_2 + 1$	$4s_3 + 1$	$v_f(0) = v_f(1) + 1$	$e_f(1) = e_f(0) - 1$

If $t_1 = 0$, then $v_f(0) = v_{f_2'}(0) + v_{g_3}(0) - 1 = v_{f_2'}(1) - 1 + v_{g_3}(1) - 1 = v_f(1) - 1, e_f(0) = e_{f_2'}(0) + e_{g_3}(0) = e_{f_2'}(1) - 2 + e_{g_3}(1) + 2 = e_f(1) + 1$.

If $t_1 \neq 0$, then $v_f(0) = v_{f_1}(0) + v_{f_2'}(0) + v_{g_3}(0) - 2 = v_{f_1}(1) + v_{f_2'}(1) - 1 + v_{g_3}(1) - 2 = v_f(1) - 1$ and $e_f(0) = e_{f_1}(0) + e_{f_2'}(0) + e_{g_3}(0) = e_{f_1}(1) - 1 + e_{f_2'}(1) - 2 + e_{g_3}(1) + 2 = e_f(1) + 1$. This completes the proof. ⋆

References

[1] Mahesh Andar, Samina Boxwala and N. B. Limaye, A note on cordial labeling of multiple shells, *Trends Math.*, (2002) 77–80.

[2] I. Cahit, Recent results and open problems on cordial graphs, R. Bodendiek (Ed.) Contemporary Methods in Graph Theory, Wissenschaftsverlag, Mannheim (1990) 209–230.

[3] J. A. Gallian, A dynamic survey of graph labeling, *Electronic J. Comb.*, **15** (2008), DS6, www.combinatorics.org

[4] G. Sethuraman and P. Selvaraju, One edge union of shell graphs and one vertex union of complete bipartite graphs are cordial, *Discrete Mathematics*, **259** (2002) 343–350.

Proc. ICDM 2008, RMS-Lecture Notes Series No. 13, 2010, pp. 243–252.

Some Cycle Related Cordial Graphs in the Context of Vertex Switching

S. K. Vaidya, Sweta Srivastav, V. J. Kaneria and Kailas Kanani

Department of Mathematics, Saurashtra University, Rajkot 360 005
e-mail: samirkvaidya@yahoo.co.in; lali_shweta2003@yahoo.co.in;
kaneria_vinodray_j@yahoo.co.in; kananikkk@yahoo.co.in

Abstract. In the present work we discuss cordial labeling of some cycle related graphs in the context of vertex switching.

Keywords. Cordial graph, cycle, cycle with one chord, cycle with twin chords, vertex switching.

Subject Classification: 05C78

1. Introduction

We begin with simple, finite, connected and undirected graph $G = (V, E)$. Here elements of sets V and E are known as vertices and edges respectively. In the present work C_n denotes cycle with n vertices while $d(v)$ denotes the degree of any vertex $v \in V$. For all other terminology and notations we follow Harary [1]. We will give brief summary of definitions which are useful for the present investigations.

Definition 1.1. *A* chord *of a cycle C_n is an edge joining two non-adjacent vertices.*

Definition 1.2. *Two chords of a cycle are said to be twin chords if they form a triangle with an edge of the cycle C_n. For positive integers n and p with $5 \leq p + 2 \leq n$, $C_{n,p}$ is the graph consisting of a cycle C_n with a pair of twin chords with which the edges of C_n form cycle C_p, C_3 and C_{n+1-p} without chords.*

Definition 1.3. *If the vertices of the graph are assigned values subject to certain conditions then it is known as* graph labeling.

Labeled graph have variety of applications in coding theory, particularly for missile guidance codes, design of good radar type codes and convolution codes with optimal autocorrelation properties. Labeled graph plays vital role in the study of X-ray crystallography, communication network and to determine optimal circuit layouts. A systematic study of various applications of graph labeling is carried out by Bloom and Golomb [4].

For detail survey on graph labeling one can refer Gallian [2]. According to Beineke and Hegde [3] graph labeling serves as a frontier between number theory and structure of graphs.

Definition 1.4. *Let $G = (V, E)$ be a graph. A mapping $f : V(G) \rightarrow \{0,1\}$ is called binary vertex labeling of G and $f(v)$ is called the label of the vertex v of G under f.*

For an edge $e = uv$, the induced edge labeling $f^* : E(G) \rightarrow \{0, 1\}$ is given by $f^*(e) = |f(u) - f(v)|$. Let $v_f(0), v_f(1)$ be the number of vertices of G having labels 0 and 1 respectively under f and let $e_f(0), e_f(1)$ be the number of edges having labels 0 and 1 respectively under f^*.

Definition 1.5. *A binary vertex labeling of a graph G is called a* cordial labeling *if $|v_f(0) - v_f(1)| \leq 1$ and $|e_f(0) - e_f(1)| \leq 1$. A graph G is* cordial *if it admits cordial labeling.*

The concept of cordial labeling was introduced by Cahit [5].

There are three types of problems that can be considered in this area.

1. How cordiality is affected under various graph operations.
2. Construct new families of cordial graphs by finding suitable labelings.
3. Given a graph theoretic property P characterize the class of graphs with property P that are cordial.

Many researchers have studied cordiality of graphs. e.g., Cahit [5] proved that tree is cordial. In the same paper he proved that K_n is cordial if and only if $n \leq 3$. Ho *et al.* [6] proved that unicyclic graph is cordial unless it is C_{4k+2}. Andar *et al.* [7] proved cordiality of multiple shells. Vaidya *et al.* [8–10] have also discussed the cordiality of various graphs. The present work is aimed to discuss first type of problems.

Definition 1.6. *A vertex switching G_v of a graph G is obtained by taking a vertex v of G, removing all edges incident to v and adding edges joining v to every vertex not adjacent to v in G.*

In the present investigations we have tried to connect vertex switching and cordiality of graphs.

2. Main results

Theorem 2.1. *Vertex switching of cycle C_n is cordial.*

Proof. Let $G = C_n$ and $v_1, v_2, \ldots, v_n$ be successive vertices of C_n. G_v denotes the vertex switching of G with respect to the vertex v of G. Here note that in each of the following cases the labeling pattern starts from the switched vertex which is considered as v_1 and label the vertices in clockwise direction. To define binary vertex labeling $f : V(G_v) \rightarrow \{0, 1\}$ following cases to be considered.

Case 1. $n \equiv 0, 1, 2 \pmod 4$

In this case, we define labeling f as:

$$f(v_i) = 0; \quad \text{if } i \equiv 0, 1 \pmod 4$$

$$= 1; \quad \text{if } i \equiv 2, 3 \pmod 4, 1 \leq i \leq n.$$

Table 1.

b	Vertex conditions	Edge conditions
0	$v_f(0) = v_f(1)$	$e_f(0) + 1 = e_f(1)$
1.3	$v_f(0) = v_f(1) + 1$	$e_f(0) = e_f(1) + 1$
2	$v_f(0) = v_f(1)$	$e_f(0) = e_f(1) + 1$

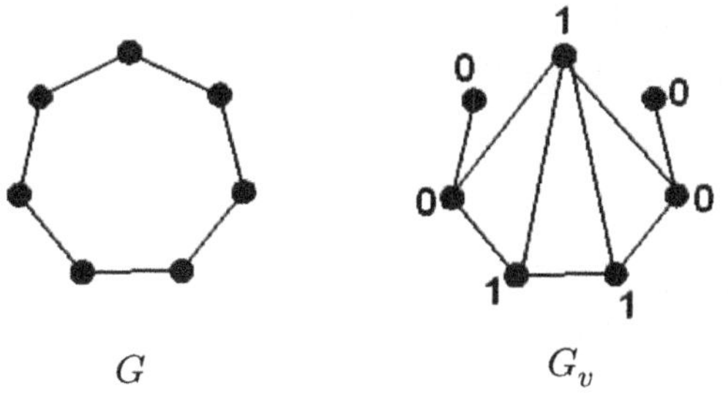

Figure 1.

Case 2. $n \equiv 3 (\text{mod } 4)$
In this case, we define labeling f as:

$$f(v_i) = 0; \quad \text{if } i \equiv 2, 3 (\text{mod } 4)$$
$$= 1; \quad \text{if } i \equiv 0, 1 (\text{mod } 4), \ 1 \leq i \leq n.$$

The labeling pattern defined above covers all the possibility of vertex switching. In each case the graph under consideration satisfies the conditions for cordiality as shown in Table 1 (where $n = 4a + b, a \in N$). i.e. G_v admits cordial labeling.

Illustration 2.2. Consider cycle C_7 (It is the case related to $n \equiv 3 (\text{mod } 4)$). The labeling pattern is shown in Figure 1.

Theorem 2.3. *Vertex switching of cycle C_n with one chord is cordial where chord form a triangle with cyclic edges.*

Proof. Let $G = C_n$ and $v_1, v_2, \ldots, v_n$ be successive vertices of C_n. G_v denotes the vertex switching of G with respect to the vertex v of G. Here note that in each of the following cases the labeling pattern starts from the switched vertex which is considered as v_1 and label the vertex in clockwise direction. To define binary vertex labeling $f : V(G_v) \to \{0, 1\}$ following cases to be considered.

Case A. Vertex switching of a vertex having $d(v) = 2$ and it is adjacent with both end vertices of chord.

Subcase 1. $n \equiv 0 (\text{mod } 4)$
In this case, we define labeling f as:

$$f(v_1) = 0, \ f(v_n) = 1 \text{ and}$$
$$f(v_i) = 0; \quad \text{if } i \equiv 0, 3 (\text{mod } 4)$$
$$= 1; \quad \text{if } i \equiv 1, 2 (\text{mod } 4), \ 2 \leq i \leq n - 1.$$

Subcase 2. $n \equiv 1 (\text{mod } 4)$
In this case, we define labeling f as:

$$\begin{aligned} f(v_i) &= 0; \quad \text{if } i \equiv 2, 3 (\text{mod } 4) \\ &= 1; \quad \text{if } i \equiv 0, 1 (\text{mod } 4), 1 \leq i \leq n. \end{aligned}$$

Subcase 3. $n \equiv 2 (\text{mod } 4)$
In this case, we define labeling f as:

$$\begin{aligned} &f(v_1) = 0, \ f(v_n) = 1 \text{ and} \\ &f(v_i) = 0; \quad \text{if } i \equiv 2, 3 (\text{mod } 4) \\ &\qquad = 1; \quad \text{if } i \equiv 0, 1 (\text{mod } 4), 2 \leq i \leq n - 1. \end{aligned}$$

Subcase 4. $n \equiv 3 (\text{mod } 4)$
In this case, we define labeling f as:

$$\begin{aligned} &f(v_1) = 0, \ f(v_{n-1}) = 1, \ f(v_n) = 0 \text{ and} \\ &f(v_i) = 0; \quad \text{if } i \equiv 2, 3 (\text{mod } 4) \\ &\qquad = 1; \quad \text{if } i \equiv 0, 1 (\text{mod } 4), 2 \leq i \leq n - 2. \end{aligned}$$

Case B. Vertex switching of a vertex having $d(v) = 3$.

Subcase 1. $n \equiv 0 (\text{mod } 4)$
In this case, we define labeling f as:

$$\begin{aligned} f(v_i) &= 0; \quad \text{if } i \equiv 1, 2 (\text{mod } 4) \\ &= 1; \quad \text{if } i \equiv 0, 3 (\text{mod } 4), 1 \leq i \leq n. \end{aligned}$$

Subcase 2. $n \equiv 1 (\text{mod } 4)$
In this case, we define labeling f as:

Subcase 2(a). For extreme left end vertex of the chord:

$$\begin{aligned} &f(v_n) = 0 \text{ and} \\ &f(v_i) = 0; \quad \text{if } i \equiv 2, 3 (\text{mod } 4) \\ &\qquad = 1; \quad \text{if } i \equiv 0, 1 (\text{mod } 4), 1 \leq i \leq n - 1. \end{aligned}$$

Subcase 2(b). For extreme right end vertex of the chord:

$$\begin{aligned} f(v_i) &= 0; \quad \text{if } i \equiv 0, 1 (\text{mod } 4) \\ &= 1; \quad \text{if } i \equiv 2, 3 (\text{mod } 4), 1 \leq i \leq n. \end{aligned}$$

Subcase 3. $n \equiv 2 (\text{mod } 4)$
In this case, we define labeling f as:

$$\begin{aligned} &f(v_1) = 1, \ f(v_n) = 0 \text{ and} \\ &f(v_i) = 0; \quad \text{if } i \equiv 0, 1 (\text{mod } 4) \\ &\qquad = 1; \quad \text{if } i \equiv 2, 3 (\text{mod } 4), 2 \leq i \leq n - 1. \end{aligned}$$

Subcase 4. $n \equiv 3 (\text{mod } 4)$

In this case, we define labeling f as:

$$f(v_i) = 0; \quad \text{if } i \equiv 1, 2 (\text{mod } 4)$$
$$= 1; \quad \text{if } i \equiv 0, 3 (\text{mod } 4), 1 \leq i \leq n.$$

Case C. Vertex switching of a remaining vertex which are having $d(v) = 2$.

Subcase 1. $n \equiv 0 (\text{mod } 4)$

In this case, we define labeling f as follows:

First label both end vertices of chord by 0. For remaining vertices define labeling f as

$$f(v_1) = 1$$
$$f(v_i) = 0; \quad \text{if } i \equiv 0, 1 (\text{mod } 4)$$
$$= 1; \quad \text{if } i \equiv 2, 3 (\text{mod } 4), 2 \leq i \leq n.$$

Subcase 2. $n \equiv 1, 2, 3 (\text{mod } 4)$

In this case, we define labeling f as:

$$f(v_i) = 0; \quad \text{if } i \equiv 0, 1 (\text{mod } 4)$$
$$= 1; \quad \text{if } i \equiv 2, 3 (\text{mod } 4), 1 \leq i \leq n.$$

The labeling pattern defined above covers all the possibility of vertex switching. In each case A, B and C the graph under consideration satisfies the conditions for cordiality as shown in following Table 2, Table 3 and Table 4 respectively (where $n = 4a + b, a \in N$). i.e. In each case G_v admits cordial labeling.

Remark 2.4. It is also possible to develop the pattern for any position of chord and cordiality can be discussed accordingly.

Table 2.

b	Vertex conditions	Edge conditions
0.2	$v_f(0) = v_f(1)$	$e_f(0) = e_f(1)$
1	$v_f(0) + 1 = v_f(1)$	$e_f(0) = e_f(1)$
3	$v_f(0) = v_f(1) + 1$	$e_f(0) = e_f(1)$

Table 3.

b	Vertex conditions	Edge conditions
0.2	$v_f(0) = v_f(1)$	$e_f(0) = e_f(1)$
1.3	$v_f(0) + 1 = v_f(1)$	$e_f(0) = e_f(1)$

Table 4.

b	Vertex conditions	Edge conditions
0.2	$v_f(0) = v_f(1)$	$e_f(0) = e_f(1)$
1	$v_f(0) = v_f(1) + 1$	$e_f(0) = e_f(1)$
3	$v_f(0) + 1 = v_f(1)$	$e_f(0) = e_f(1)$

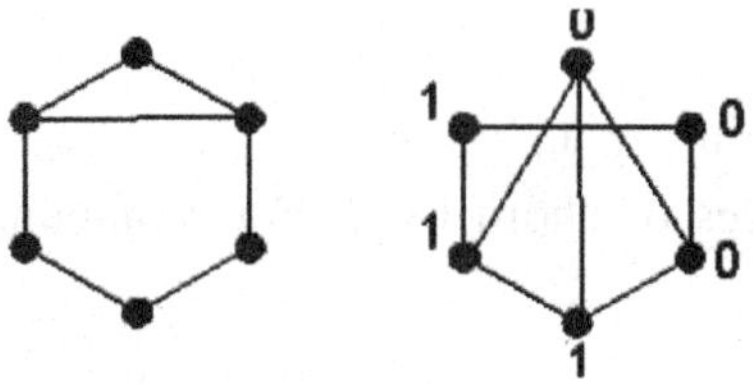

G G_v

Figure 2.

Illustration 2.5. Consider cycle C_6 with one chord (It is the case related to *Case A*, $n \equiv 2 (\text{mod } 4)$). The labeling pattern is as shown in Figure 2.

Theorem 2.6. *Vertex switching of cycle C_n with twin chords is cordial where chords form two triangles and cycle C_{n-2}.*

Proof. Let $G = C_n$ be the cycle with twin chords, where chords form two triangles and one cycle C_{n-2}. Here number of vertices $p = n$ and number of edges $q = n + 2$. Let $v_1, v_2, \ldots, v_n$ be successive vertices of G. Let $e_1 = v_n v_2$ and $e_2 = v_n v_3$ be the chords of cycle C_n. G_v will denote the vertex switching of G with respect to the vertex v of G. Here note that in each of the following cases the labeling pattern starts from the switched vertex which is considered as v_1 and label the vertex in clockwise direction. To define binary vertex labeling $f : V(G_v) \to \{0, 1\}$ following cases to be considered.

Case A. Vertex switching of a vertex having $d(v) = 2$ and it is adjacent with both end vertices of chord.

Subcase 1. $n \equiv 0 (\text{mod } 4)$

In this case, we define labeling f as:

$$
\begin{aligned}
f(v_1) &= 0,\ f(v_n) = 1 \text{ and} \\
f(v_i) &= 0; \quad \text{if } i \equiv 0, 3 (\text{mod } 4) \\
&= 1; \quad \text{if } i \equiv 1, 2 (\text{mod } 4), 2 \le i \le n - 1.
\end{aligned}
$$

Subcase 2. $n \equiv 1 (\text{mod } 4)$

In this case, we define labeling f as:

$$
\begin{aligned}
f(v_1) &= 0 \text{ and} \\
f(v_i) &= 0; \quad \text{if } i \equiv 2, 3 (\text{mod } 4) \\
&= 1; \quad \text{if } i \equiv 0, 1 (\text{mod } 4), 2 \le i \le n.
\end{aligned}
$$

Subcase 3. $n \equiv 2 (\text{mod } 4)$
In this case, we define labeling f as:

$$\begin{aligned} f(v_1) &= 0,\ f(v_2) = 1 \text{ and} \\ f(v_i) &= 0; \quad \text{if } i \equiv 1, 2 (\text{mod } 4) \\ &= 1; \quad \text{if } i \equiv 0, 3 (\text{mod } 4), 3 \leq i \leq n. \end{aligned}$$

Subcase 4. $n \equiv 3 (\text{mod } 4)$
In this case, we define labeling f as:

$$\begin{aligned} f(v_i) &= 0; \quad \text{if } i \equiv 1, 2 (\text{mod } 4) \\ &= 1; \quad \text{if } i \equiv 0, 3 (\text{mod } 4), 1 \leq i \leq n. \end{aligned}$$

Case B. Vertex switching of a vertex having $d(v) = 3$.

Subcase 1. $n \equiv 0 (\text{mod } 4)$
In this case, we define labeling f as:

$$\begin{aligned} f(v_i) &= 0; \quad \text{if } i \equiv 0, 1 (\text{mod } 4) \\ &= 1; \quad \text{if } i \equiv 2, 3 (\text{mod } 4), 1 \leq i \leq n. \end{aligned}$$

Subcase 2. $n \equiv 1, 2 (\text{mod } 4)$
In this case, we define labeling f as:

$$\begin{aligned} f(v_1) &= 0,\ f(v_2) = 1 \text{ and} \\ f(v_i) &= 0; \quad \text{if } i \equiv 1, 2 (\text{mod } 4) \\ &= 1; \quad \text{if } i \equiv 0, 3 (\text{mod } 4), 3 \leq i \leq n. \end{aligned}$$

Subcase 3. $n \equiv 3 (\text{mod } 4)$
In this case, we define labeling f as:

$$\begin{aligned} f(v_1) &= 0 \text{ and} \\ f(v_i) &= 0; \quad \text{if } i \equiv 0, 3 (\text{mod } 4) \\ &= 1; \quad \text{if } i \equiv 1, 2 (\text{mod } 4), 2 \leq i \leq n. \end{aligned}$$

Case C. Vertex switching of a vertex which is having $d(v) = 4$.

Subcase 1. $n \equiv 0, 1, 3 (\text{mod } 4)$
In this case, we define labeling f as:

$$\begin{aligned} f(v_i) &= 0; \quad \text{if } i \equiv 1, 2 (\text{mod } 4) \\ &= 1; \quad \text{if } i \equiv 0, 3 (\text{mod } 4), 1 \leq i \leq n. \end{aligned}$$

Subcase 2. $n \equiv 2 (\text{mod } 4)$
In this case, we define labeling f as:

$$\begin{aligned} f(v_i) &= 0; \quad \text{if } i \equiv 0, 1 (\text{mod } 4) \\ &= 1; \quad \text{if } i \equiv 2, 3 (\text{mod } 4), 1 \leq i \leq n. \end{aligned}$$

Case D. Vertex switching of remaining vertices which are having $d(v) = 2$.

Subcase 1. $n \equiv 0 (\text{mod } 4)$
In this case, we define labeling f as follows:
First label both end vertices of chord by 0. For remaining vertices define labeling f as

$$
\begin{aligned}
f(v_1) &= 1\\
f(v_i) &= 0; \quad \text{if } i \equiv 0, 1 (\text{mod } 4)\\
&= 1; \quad \text{if } i \equiv 2, 3 (\text{mod } 4), 2 \leq i \leq n.
\end{aligned}
$$

Subcase 2. $n \equiv 1, 2, 3 (\text{mod } 4)$
In this case, we define labeling f as:

$$
\begin{aligned}
f(v_i) &= 0; \quad \text{if } i \equiv 0, 1 (\text{mod } 4)\\
&= 1; \quad \text{if } i \equiv 2, 3 (\text{mod } 4), 1 \leq i \leq n.
\end{aligned}
$$

The labeling pattern defined above covers all the possibility of vertex switching. In each case A, B, C and D the graph under consideration satisfies the conditions for cordiality as shown in following Table 5, Table 6, Table 7 and Table 8 respectively (where $n = 4a + b, a \in N$). i.e. In each case G_v admits cordial labeling.

Illustration 2.7. Consider cycle C_6 with twin chord (It is the case related to *Case A*, $n \equiv 2 (\text{mod } 4)$). The labeling pattern is shown in Figure 3.

Table 5.

b	Vertex conditions	Edge conditions
0	$v_f(0) = v_f(1)$	$e_f(0) + 1 = e_f(1)$
1.3	$v_f(0) = v_f(1) + 1$	$e_f(0) + 1 = e_f(1)$
2	$v_f(0) = v_f(1)$	$e_f(0) = e_f(1)$

Table 6.

b	Vertex conditions	Edge conditions
0.2	$v_f(0) = v_f(1)$	$e_f(0) + 1 = e_f(1)$ or $e_f(0) = e_f(1) + 1$
1	$v_f(0) + 1 = v_f(1)$	$e_f(0) + 1 = e_f(1)$ or $e_f(0) = e_f(1) + 1$
3	$v_f(0) = v_f(1) + 1$	$e_f(0) + 1 = e_f(1)$ or $e_f(0) = e_f(1) + 1$

Table 7.

b	Vertex conditions	Edge conditions
0	$v_f(0) = v_f(1)$	$e_f(0) = e_f(1) + 1$
1	$v_f(0) = v_f(1) + 1$	$e_f(0) + 1 = e_f(1)$
2	$v_f(0) = v_f(1)$	$e_f(0) = e_f(1) + 1$
3	$v_f(0) = v_f(1) + 1$	$e_f(0) = e_f(1) + 1$

Table 8.

b	Vertex conditions	Edge conditions
0.2	$v_f(0) = v_f(1)$	$e_f(0) + 1 = e_f(1)$ or $e_f(0) = e_f(1) + 1$
1	$v_f(0) = v_f(1) + 1$	$e_f(0) + 1 = e_f(1)$ or $e_f(0) = e_f(1) + 1$
3	$v_f(0) + 1 = v_f(1)$	$e_f(0) + 1 = e_f(1)$ or $e_f(0) = e_f(1) + 1$

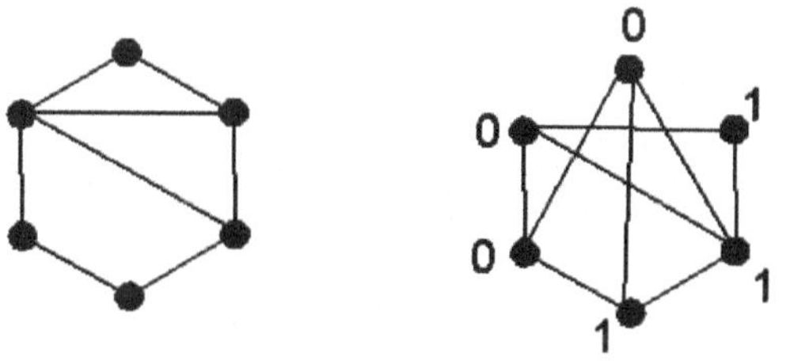

G G_v

Figure 3.

3. Concluding remarks

Labeled graph is the topic of current interest for many researchers as it has several diversified applications. We discuss here cordial labeling in the context of graph operation namely vertex switching. This approach is novel and contributes three new graphs to the theory of cordial graphs. As we referred earlier cycle C_n where $n \equiv 2 \pmod 4$ is not cordial while vertex switching of all cycles is cordial which is the salient feature of present investigations. The derived labeling pattern is demonstrated by means of sufficient illustrations which provide better understanding of the derived results. The results reported here are new and will add new dimension to the theory of cordial graphs.

Acknowledgement

The authors are thankful to the anonymous referee for their useful suggestions and kind comments.

References

[1] F. Harary, Graph theory, Addison Wesley, Reading, Massachusetts (1972).

[2] J. A. Gallian, A dynamic survey of graph labeling, *The Electronics Journal of Combinatorics*, **14** (2007) $\sharp DS6$, 1–180.

[3] L. W. Beineke and S. M. Hegde, Strongly Multiplicative graphs, Discuss, *Math. Graph Theory*, **21** (2001) 63–75.

[4] G. S. Bloom and S. W. Golomb, Application of numbered undirected graphs, *Proceedings of IEEE*, **165(4)** (1977) 562–570.

[5] I. Cahit, Cordial Graphs: A weaker version of graceful and harmonious Graphs, *Ars Combinatoria*, **23** (1987) 201–207.

[6] Y. S. Ho, S. M. Lee and S. C. Shee, Cordial labeling of unicyclic graphs and generalized Petersen graphs, *Congress. Numer.*, **68** (1989) 109–122.

[7] M. Andar, S. Boxwala and N. B. Limaye, A Note on cordial labeling of multiple shells, *Trends Math.*, (2002) 77–80.

[8] S. K. Vaidya, G. V. Ghodasara, Sweta Srivastav and V. J. Kaneria, Cordial labeling for two cycle related graphs, *The Mathematics Student*, **76** (2007) 237–246.

[9] S. K. Vaidya, G. V. Ghodasara, Sweta Srivastav and V. J. Kaneria, Some new cordial graphs, *International Journal of Scientific Computing*, **2(1)** (2008) 81–92.

[10] S. K. Vaidya, Sweta Srivastav, G. V. Ghodasara and V. J. Kaneria, Cordial labeling for cycle with one chord and its related graphs, *Indian J. of Mathematics and Mathematical Sciences*, **4(2)** (2008) 145–156.

List of Referees

S.D. Agashe
J.L.R. Alfonsin
S. Arumugam
R. Balakrishnan
R.B. Bapat
N.S. Bhave
G.S. Bloom
J.A. Bondy
G. Chartrand
T. Chelvam
S.A. Choudam
B.K. Dass
M. Deza
Hai Q. Dinh
Alan Frieze
Yuan Gao
Ferenc Gecseg
J.R. Griggs
Purnima Gupta
S.T. Hedetniemi
S.M. Hegde
V.A. Hiremath
V. Jeyakumar
M.I. Jinnah
G.O.H. Katona
A.K. Lal
R. Laskar
M. Lasvergnas
N.B. Limaye
D. Michalak
J. Nešetřil
Pratima Panigrahi
Alexandre Pinlou
Venkatesh Raman
Andre Raspaud
A. Rosa
Rahul Roy
D. Rautenbach
S.S. Sane
R.S. Sankaranarayana
S.K. Simic
T. Singh
M. Sonntag
N. Sridharan
C.R. Subramanian
V. Swaminathan
S. Thomasse
Z. Tuza
G.R. Vijayakumar
T. Zaslavsky
Ping Zhang

Made in United States
North Haven, CT
05 November 2026

21764890R00143